WILLIAM L. QUIRIN ADELPHI UNIVERSITY

Probability and Statistics

Harper & Row, Publishers
New York, Hagerstown, San Francisco, London

Sponsoring Editor: Charlie Dresser
Project Editor: Lois Lombardo
Designer: Emily Harste
Production Supervisor: Marion Palen
Compositor: Syntax International Pte. Ltd.
Printer and Binder: The Maple Press Company
Art Studio: J & R Technical Services Inc.
Cover Photo: Fundamental Photographs from the Granger Collection

**Probability
and Statistics**

Library of Congress Cataloging in Publication Data

Quirin, William L
    Probability and statistics.

    Includes index.
    1. Probabilities. 2. Mathematical statistics.
I. Title.
QA273.Q54   519.2   77-15594
ISBN 0-06-045293-5

To my wife Diane

$P(\text{I will love you forever}) = 1$

Contents

Preface

This book is designed for a reasonably paced two-semester or two-quarter course in probability and statistics given at the junior-senior level to students who have had the standard calculus sequence, including infinite series and multiple integration. The manuscript itself and the lecture notes from which the manuscript grew were classroom tested several times. The author found that the first seven chapters, which contain all the usual topics of discrete probability theory, including Markov chains, can be covered in one semester. The remaining four chapters, which cover continuous probability theory and the usual topics in statistical inference, form a logical unit for a second semester.

The author feels that the text provides an excellent balance between theory and applications. Most of the theorems stated in the text are proved, with all the steps spelled out in detail. Each topic discussed is illustrated with several thoroughly worked out examples, many of which are taken from "real world" situations.

A reasonable number of applied and theoretical exercises are included after (almost) every section, and an additional set of exercises are included at the end of each chapter.

Several examples in the text make use of computer simulation. Students interested in computer programing should be encouraged to run these experiments themselves and to write the programs suggested in several exercises throughout the text.

The tables that appear in Appendices A, B, and C are computer produced by the author. The tables that appear in Appendices D and E are reproduced with the permission of the *Biometrika* trustees, and the table that appears in Appendix F is reproduced with the permission of Oliver & Boyd, Ltd., Edinburgh, Scotland.

The author would like to thank his colleagues Ida Sussman and Fred Pohle for reading the original manuscript, suggesting improvements, and providing examples which now appear in the text, and former students whose suggestions and comments helped shape the text in its present form.

William L. Quirin

Probability
and Statistics

Introduction

1.1 Probability and Statistics

The related fields of probability and statistics have grown in prominence in recent years, mostly because of their many and varied applications to the physical, social, and biological sciences. A wide spectrum of disciplines, such as medicine, politics, business, linguistics, and criminology, to name just a few, all make use of probabilistic reasoning and statistical inference.

We are all familiar with the use of the word "probability." We have all used the word, or one of its many cognates, on many occasions. When we purchase a lottery ticket, we speak of the *likelihood* that our ticket will be a winner. Weather forecasters tell us that the "*chance* of rain tommorrow is 20 percent." The odds board at a racetrack is filled with numbers that reflect the probability of the different horses winning the race.

On the other hand, most people are probably not familiar with the meaning of the word "statistics" when it is used in reference to the discipline of mathematical statistics. The word "statistics" usually conjures up images of large tables of data, of averages, medians, and modes, and of bar graphs and line graphs. We shall see, however, that statistics far transcends such simple arithmetic study of tables and that its real power lies in the interpretation of the data and in the inferences that can be validly drawn from the data.

Perhaps the most widely known example of the use of probability and statistics occurs on television on election night when political candidates are projected winners on the basis of only a small percentage of the returns. We shall discuss how this is done later in the book.

Both probability and statistics deal with what are called random or nondeterministic phenomena or experiments. In their science courses students become familiar with deterministic experiments. Given a well-defined set of conditions under which such experiments are to be performed, scientific laws and formulas can then be applied to predict the exact outcomes of the experiments. For random or nondeterministic experiments, no such prediction is possible, even if the conditions under which the experiments are to be performed are specifically defined.

We are all familiar with the randomness of games of chance. Although we know that a coin must land either heads or tails, there is no way to predict with complete confidence the outcome of a single toss of the coin. Similarly, although we know that a machine produces 5 percent defective items, there is no way to predict whether or not the next item it produces will be defective.

In our study of probability and statistics, we shall be interested in the collection of all possible outcomes of a given random experiment. We shall call this set either the *population* or the *sample space* of outcomes of the experiment. Subcollections of this collection will be referred to as either *samples* taken from the population or as *events* in the sample space.

Probability and statistics approach the study of random experiments from rather inverse viewpoints. Probability, on the one hand, assumes that the structure of the population is known and attempts to determine the likelihood of particular samples or events occurring. For example, the sample space of outcomes for the roll of two dice is well known. A typical probability concern is, for example, whether the sum of two dice turns out to be, say, seven. Statistics, on the other hand, starts with the fact that a particular sample has occurred and attempts from the information in that sample to make inferences about the unknown structure of the population from which the sample was taken. Political straw polls are a familiar example of this process. In an attempt to make an educated guess concerning the structure of the population, that is, the percentage of voters favoring a particular issue or candidate, the opinions of a small sample (the straw poll) are solicited. The structure of the sample is then attributed to the population as a whole. Since samples must be chosen at random, to insure the integrity and lack of bias of the experiment, it can be seen that probability will play an important role in statistical inference.

The study of probability had its origins in the sixteenth and seventeenth centuries, and it was the interest in games of chance that sparked that study. Fermat and Pascal, because of their acquaintance with the gambler and amateur mathematician Chevalier de Méré, became interested in problems like the following:

1. How many tosses of a single die are required so that there is a better than even chance of at least one six occurring?
2. Suppose that two people are playing a game in which the first to win four matches is the winner. If player A has already won three matches and if player B has won two matches, and the players agree to end the game, then how should the stakes be divided?

The ideas of Fermat and Pascal on these and similar problems were among the first contributions to the study of probability. They were soon followed by contributions from such notable mathematicians as Huygens, Newton, de Moivre, Laplace, and Gauss.

The modern theory of probability is due to the Russian mathematician Kolmogorov, who presented an axiomatic approach to the study of probability in 1933. Although his approach, which uses measure theory, is beyond the scope of this text, we shall present a set-theoretic approach which is quite similar.

Since our populations will be the *sets* of all possible outcomes of our random experiments and since it will be necessary to *count* the number of different possible outcomes of these experiments, the remainder of this chapter consists of a review of the basics of set theory and a study of the various techniques of counting.

1.2 Set Theory

We assume the reader to be familiar with the basic concepts of set theory, which we now briefly review.*

* For more details and examples, refer to P. R. Halmos, *Naive Set Theory* (Princeton, N.J.; Van Nostrand, 1960).

By a *set*, we mean a collection of objects called the *elements* of the set. We write that the element x belongs to the set S as $x \in S$. Otherwise, we write $x \notin S$. For example, the collection of eligible voters in New York City is a set. Any person who lives within the confines of the city and is eligible to vote is an element of our set.

The totality of all elements which might conceivably belong to any of the sets under discussion is called the *universal set*, denoted U. In our example, U might be the set of all people living in the United States. A set A is called a *subset* of a set B, written $A \subset B$, if every element of A is also an element of B. That is, $x \in A$ implies that $x \in B$. For example, the set of all female eligible voters in New York City is a subset of the set of all eligible voters in New York City. The empty set $\varnothing$, the set which contains no elements, is a subset of every set.

A set which can be placed in a one-to-one correspondence with a subset $\{1, 2, 3, \ldots, n\}$ of positive integers is called a *finite* set. All other sets are called *infinite* sets. The set of all eligible voters in New York City is in fact a finite set, whereas the set of all the real numbers on the number line is an infinite set. Any infinite set which can be placed in a one-to-one correspondence with the set $\{1, 2, 3, \ldots\}$ of natural numbers is called a *countable* set and is said to be *countably infinite*. All other infinite sets are called *uncountable* sets and are said to be *uncountably infinite*. The rational numbers are countable, whereas the irrational numbers are uncountable.

Given a set A, we can form another set, called the *complement* of A and denoted $C(A)$, as follows:

$$C(A) = \{x \in U : x \notin A\}$$

That is, $C(A)$ contains precisely those elements in the universal set which are not in A. The complement of the set

$$A = \text{eligible voters in Queens County}$$

in the universal set of all eligible voters in New York City is the set of all eligible voters living in all New York City's other counties.

If B is a subset of A, then the *relative complement* of B in A, denoted $A - B$, is defined as follows:

$$A - B = \{x \in A : x \notin B\}$$

Given two sets A and B, we define the *union* $A \cup B$ and the *intersection* $A \cap B$ of A and B as follows:

$$A \cup B = \{x \in U : x \in A \quad \text{and/or} \quad x \in B\}$$

and

$$A \cap B = \{x \in U : x \in A \quad \text{and} \quad x \in B\}$$

That is, $A \cap B$ contains only those elements which are in both A and B, while $A \cup B$ contains any elements which is in either A or B or in both.

For example, if

A = all eligible male voters in New York City

and

B = all eligible voters in Queens County

then $A \cap B$ is the set of all eligible male voters in Queens County, while $A \cup B$ is the set of those eligible voters in New York City who are male or who live in Queens County or both.

The operations of union and intersection both satisfy commutative and associative laws, and they are related by the distributive laws

D1: $A \cap (B \cup C) = (A \cap B) \cup (A \cap C)$

and

D2: $A \cup (B \cap C) = (A \cup B) \cap (A \cup C)$

Even more important are De Morgan's laws, which relate complementation to union and intersection,

DM1: $C(A \cup B) = C(A) \cap C(B)$

and

DM2: $C(A \cap B) = C(A) \cup C(B)$

It is easier to remember De Morgan's laws when we realize that they are merely expressing obvious linguistic identities. Since "either" is the word that best signifies union, while "and" or "both" can be used to signify intersection we can rewrite De Morgan's laws as

DM1: not either = neither

and

DM2: not both = not one or the other

Using the A and B from the preceding example, we find that $C(A \cup B)$ is the set of all eligible female voters who live in some county other than Queens, while $C(A \cap B)$ is the set of those eligible voters in New York City who either are female or live in some county other than Queens.

Two sets A and B are called *exclusive* or *disjoint* if $A \cap B = \varnothing$. Note that any set and its complement are exclusive.

The definitions of union and intersection can be extended to any collection of sets, whether that collection be finite, countable, or uncountable. To belong to the union of a collection of sets, an element must belong to at least one of the sets. To belong to the intersection, an element must belong to every one of the sets. For example, if $A_1, A_2, \ldots$ is a countable collection of sets, then we can define

$$\bigcup_{k=1}^{\infty} A_k = \{x \in U : x \in A_k \qquad \text{for at least one } k\}$$

and

$$\bigcap_{k=1}^{\infty} A_k = \{x \in U : x \in A_k \quad \text{for every } k\}$$

De Morgan's laws extend to finite, countable, and uncountable unions and intersections. The complement of a union is always the intersection of the complements of the sets, and the complement of an intersection is always the union of the complements of the sets. For example, should the collection be countable, we would have

$$\text{DM1: } C\left(\bigcup_{k=1}^{\infty} A_k\right) = \bigcap_{k=1}^{\infty} C(A_k)$$

and

$$\text{DM2: } C\left(\bigcap_{k=1}^{\infty} A_k\right) = \bigcup_{k=1}^{\infty} C(A_k)$$

A collection of sets is said to be *mutually exclusive* if $A \cap B = \varnothing$ for any two sets A and B in the collection.

The *Cartesian product* of the sets $A_1, A_2, \ldots, A_n$ is the set, denoted $A_1 \times A_2 \times \cdots \times A_n$, which is defined by

$$\{(a_1, a_2, \ldots, a_n) : a_k \in A_k \quad \text{for all } k\}$$

This is simply the set of all n-tuples whose kth coordinate must always be an element from the kth set A_k.

Sets and their complements, unions, and intersections can be visualized by means of *Venn diagrams*, in which the set is represented by the interior of a circle which is contained within a rectangle that represents the universal set. The complement of the set is then represented by the exterior of the circle within the rectangle. Figure 1.1 depicts the sets A and B together with their union and intersection. The union is shaded top right to bottom left, while the intersection is shaded top left to bottom right.

Figure 1.1

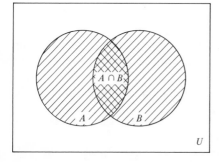

We shall use the notation $n(A)$ to denote the number of elements in the set A. Clearly, if $A \cap B = \emptyset$, then $n(A \cup B) = n(A) + n(B)$. More generally, we have

Proposition 1.1

$$n(A \cup B) = n(A) + n(B) - n(A \cap B)$$

Proof: From Figure 1.2, we observe that the following are true:

(a) $A \cap B$ and $A \cap C(B)$ are disjoint sets whose union is A;
(b) $A \cap B$ and $C(A) \cap B$ are disjoint sets whose union is B;
(c) $A \cap B$, $A \cap C(B)$, and $C(A) \cap B$ are mutually exclusive sets whose union is $A \cup B$.

We conclude immediately from (a) and (b) that

$$n(A) = n(A \cap B) + n(A \cap C(B))$$

and

$$n(B) = n(A \cap B) + n(C(A) \cap B)$$

Adding both sides of these two equations and then subtracting $n(A \cap B)$ from both sides, we obtain

$$n(A) + n(B) - n(A \cap B) = n(A \cap B) + n(A \cap C(B)) + n(C(A) \cap B)$$

but (c) implies that

$$n(A \cup B) = n(A \cap B) + n(A \cap C(B)) + n(C(A) \cap B)$$

and so the desired result follows.

Note that the term subtracted in this formula merely accounts for the fact that those elements common to A and B were counted twice in the sum $n(A) + n(B)$.

Figure 1.2

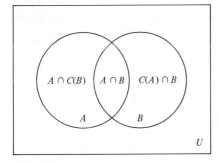

Example 1 A certain college is offering two courses in health education this semseter. One of the courses has a registration of 35, while the other course has 30. There are 5 students who signed up for both courses. We would like to determine how many different students are taking health education courses this semester. If we let

A = the set of students taking course 1

and

B = the set of students taking course 2

then we have $n(A) = 35$, $n(B) = 30$, and $n(A \cap B) = 5$. Substituting these values into our formula, we find

$$n(A \cup B) = 35 + 30 - 5 = 60$$

Since $A \cup B$ is the set of those students taking either course 1 or course 2 or both, we have solved the problem.

Example 2 The registrar made a mistake and scheduled two courses for the same room at the same time. If the roster for one class contained 28 names, while the roster for the other class contained 24 names, and if, in fact, 40 different students came to the classroom at the assigned time, then how many students signed up for both courses? If we let

A = students who signed up for course 1

and

B = students who signed up for course 2

then we have $n(A) = 28$ and $n(B) = 24$. We are also told that $n(A \cup B) = 40$. Substituting these three values into our formula, we get

$$40 = 28 + 24 - n(A \cap B)$$

Solving for $n(A \cap B)$, we find that there were 12 students who signed up for both courses.

Many times, it is extremely difficult, if not altogether impossible, to count the number of elements in the intersection of two sets directly, as we have just done. An alternate method is to count the number of elements *not* in the desired intersection, which can be done by means of De Morgan's laws and Proposition 1.1, as the following example demonstrates.

Example 3 In a group of 100 math majors, 42 are taking a course in probability, 35 are taking a course in algebra, while 32 are taking neither course. We wish to determine the number of students taking both courses. We let

P = the set of students taking probability

and

A = the set of students taking algebra

We are given that $n(P) = 42$ and $n(A) = 35$, and we wish to find $n(P \cap A)$. Since $C(P \cap A) = C(P) \cup C(A)$, we have

$$n(C(P \cap A)) = n(C(P) \cup C(A))$$
$$= n(C(P)) + n(C(A)) - n(C(P) \cap C(A))$$

but $C(P) \cap C(A)$ is the set of students taking neither course, and we are given that $n(C(P) \cap C(A)) = 32$. Hence,

$$n(C(P \cap A)) = 58 + 65 - 32 = 91$$

Consequently,

$$n(P \cap A) = 100 - 91 = 9$$

Another approach to this problem would be to recognize that $C(P) \cap C(A) = C(P \cup A)$, and, since $n(C(P) \cap C(A)) = 32$, $n(C(P \cup A)) = 32$; hence, $n(P \cup A) = 68$. Then, since

$$n(P \cup A) = n(P) + n(A) - n(P \cap A)$$

we have

$$n(P \cap A) = n(P) + n(A) - n(P \cup A)$$
$$= 42 + 35 - 68 = 9.$$

EXERCISES

1. If A is the set of all thoroughbred race horses and B is the set of all grey horses, describe $A \cup B$, $A \cap B$, $C(A)$, $C(B)$, $C(A \cup B)$, $C(A \cap B)$, $C(A) \cap B$, and $A \cap C(B)$.
2. If $n(A) = 40$, $n(B) = 20$, and $n(A \cap B) = 10$, calculate $n(A \cup B)$, $n(C(A) \cap C(B))$, and $n(C(A) \cup C(B))$. Assume $n(U) = 60$.
3. If 40 of the 90 Democrats in a political club are women, and there are a total of 70 women in the club, how many members of the club are either women or Democrats?
4. In a group of 75 teenage girls, 25 have blond hair, 18 have blue eyes, while 35 have neither of these traits. How many in the group have both blond hair and blue eyes?
5. There are 65 books on a bookshelf which have either a dedication or a foreword. Of these, 47 have a dedication and 58 have a foreword. How many have both?

1.3 Principles of Counting

We come now to the two basic principles of counting, which we shall use on numerous occasions throughout this text. Very often, the two are used side by side in the solution of a counting problem.

Proposition 1.2 The First Principle of Counting

Suppose that a certain job can be done by completing a sequence of m steps. If

(a) the first step can be completed in any of N_1 different ways;
(b) upon completion of the first step, and regardless of how the first step

was completed, the second step can then be completed in any of N_2 different ways;

(c) inductively, upon completion in succession of the first k steps, and regardless of how these k steps were completed, the $(k + 1)$st step can then be completed in any of N_{k+1} different ways,

then the job itself can be completed in any of $N_1 \cdot N_2 \cdots \cdot N_m$ different ways.

Proof: The proof is a simple application of the principle of mathematical induction and is left as an exercise.

To facilitate visualization of the principles of counting, we introduce the concept of the *tree diagram*. The branches of the tree are broken into segments, each of which indicates the outcome of one of the steps in the process outlined above. The first segment of any branch indicates the outcome of the first step, the second segment the subsequent outcome of the second step, and so forth. Emanating from the end of any segment but the last is a number of segments equal to the number of ways the next step might be done. The total number of distinct branches in the tree represents the total number of distinct ways that the job can be completed.

For example, if a job can be done by a sequence of two steps of which the first can be done in any of $N_1 = 3$ different ways and the second can be done in any of $N_2 = 2$ different ways, then the tree diagram in Figure 1.3 represents the $N_1 \cdot N_2 = 6$ different ways the job can be completed.

Figure 1.3

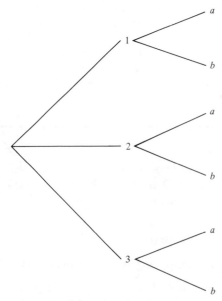

Example 1 If we first roll a green die and next roll a red die, then there are a total of $6 \cdot 6 = 36$ possible outcomes. $N_1 = 6$ since the green die can show any of six numbers, and then $N_2 = 6$ since the red die can also show any of the six numbers.

Example 2 If by a two-letter "word" we mean *any* combination of two letters, then there are a total of $26 \cdot 26 = 676$ different two-letter "words" that can be formed with the letters of the alphabet. The first letter can be chosen in any of $N_1 = 26$ different ways, and the choice of the first letter, by our definition, places no restriction on the choice of the second letter, which consequently can also be chosen in any of $N_2 = 26$ different ways. If we make the restriction that no letter can be used twice, then, having chosen the first letter, we are restricted to only $N_2 = 25$ different ways to choose the second letter. Hence, there are $26 \cdot 25 = 650$ different two-letter words with no letter repeated.

Example 3 If a state's license plate consists of two distinct letters followed by four digits the first of which cannot be a zero, then the state could print as many as $26 \cdot 25 \cdot 9 \cdot 10 \cdot 10 \cdot 10 = 5{,}850{,}000$ different license plates.

Proposition 1.3 The Second Principle of Counting

If a job can be done by any one of m different methods and if method k $(1 \leq k \leq m)$ can be completed in any of N_k different ways, then the job itself can be completed in any of $N_1 + N_2 + \cdots + N_m$ different ways.

Proof: Exercise

We emphasize the difference between the two principles of counting. The first principle (multiplication) applies when the job is to be done by a pre-scribed *sequence* of steps, while the second principle (addition) applies when there is a *choice* of ways to complete the job. In the first case, we must do this, *and then* do that, *and then*, etc., while in the second case, we must either do this, *or* do that, *or*, etc. The "and" indicates multiplication, and the "or" indicates addition.

Example 4 We wish to calculate the number of distinct "words" containing no more than three letters. In forming such a word, we have a *choice* of using either one, two, *or* three letters. This indicates that the second principle of counting will be used. Now, $N_1 = 26$, $N_2 = (26)^2$, and $N_3 = (26)^3$, so the total number of such words is $N_1 + N_2 + N_3 = 26 + 676 + 17576 = 18278$. Notice that the first principle was used to calculate N_2 and N_3.

Example 5 Suppose that we wish to form two-letter "words" in which the first letter must be a vowel and the second letter a consonant which must occur later in the alphabet than the vowel which it is to follow. Clearly, the vowel can be chosen in any of five different ways, but can the first principle be applied? No, because the number of different ways that the consonant can be chosen, after the vowel has been chosen, depends on which vowel was chosen. If the vowel chosen was A, then $N_2 = 21$, but if the vowel chosen was U, then $N_2 = 5$. Therefore, we cannot say that there are a certain number N_2 of ways to complete the second step (choose the consonant) regardless of how the first step (choose the vowel) has been completed. Hence, the first principle does not apply. However, the second principle does apply, because, in forming such words, we have a *choice* of which vowel to use. We have $m = 5$ choices, and the values $N_1 = 21$, $N_2 = 18$, $N_3 = 15$, $N_4 = 10$, and $N_5 = 5$ represent the number of consonants following each of the five vowels in the alphabet. As a result, we can form a total of $21 + 18 + 15 + 10 + 5 = 69$ different two-letter "words" satisfying the given condition. Figure 1.4 is the tree diagram for this process.

Figure 1.4

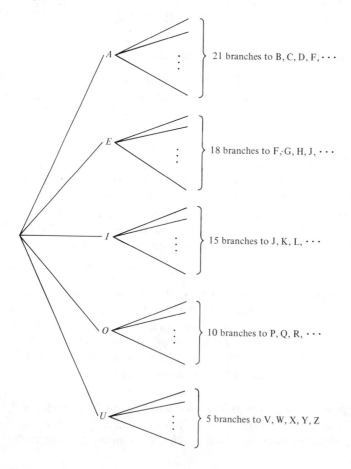

21 branches to B, C, D, F, $\cdots$

18 branches to F, G, H, J, $\cdots$

15 branches to J, K, L, $\cdots$

10 branches to P, Q, R, $\cdots$

5 branches to V, W, X, Y, Z

We include the following example as another use of a tree diagram in a situation where the principles of counting cannot be used.

Example 6   Margaret and Bobby play a tennis match in which the first to win two consecutive games or a total of three games wins the match. How many different game-by-game outcomes are possible? We construct the tree diagram in Figure 1.5, which indicates the progression of the match. Each segment of the tree represents the outcome of a different game in the match. The ten different branches of the tree represent the ten possible game-by-game outcomes of the match, which are

MM	BB
BMM	MBB
MBMM	BMBB
MBMBM	BMBMB
BMBMM	MBMBB

EXERCISES

1. Prove Proposition 1.2.
2. Prove Proposition 1.3.
3. How many four-letter "words" have repeated letters?
4. How many five-letter "words" can be formed such that vowels and consonants alternate?
5. How many different ways can a student answer a ten question true-false test?
6. Voters must choose from among three candidates for president, two candidates for governor, and four candidates for mayor, and must vote Yes or No on a bond issue. How many different ways can voters mark their ballots?
7. A man must choose his outfit for the day from among 4 pairs of trousers, 6 shirts, and 12 pairs of socks. How many different outfits can he possibly put together?
8. A horseplayer likes three horses in the first race and four horses in the second race. How many daily double tickets must he buy to cover all possibilities?

Figure 1.5

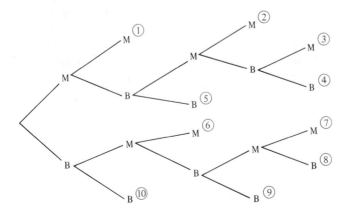

1.4 Permutations

In our study of probability and statistics, we shall frequently be interested in taking a sample of some size from our population. At times, we shall be interested in the order in which we choose our sample elements. Most of the time, however, order will not matter, and we will be interested only in which elements constitute our sample. In this section, we shall discuss ordered samples. Unordered samples will be discussed in Section 1.5.

DEFINITION

Suppose that S is a set consisting of n elements and that k is a positive integer not exceeding n. An ordered k-tuple $(x_1, x_2, \ldots, x_k)$, in which the coordinates x_i are distinct elements of S, is called a *permutation* (or *arrangement*) of k elements from the n-element set S. The total number of such permutations is denoted P_k^n.

We note that some people restrict the use of the word permutation to the case $n = k$. A permutation would then be an arrangement (ordering) of the entire set S.

Example 1 If $S = \{1, 2, 3\}$, then $(1, 2)$ and $(2, 1)$ are different permutations of two elements taken from S. There are a total of P_2^3 such permutations. Different permutations of the entire set S are $(1, 2, 3)$ and $(3, 2, 1)$. There are a total of P_3^3 such permutations.

We now derive two formulas for calculating P_k^n.

Proposition 1.4

$$P_k^n = n(n - 1)(n - 2) \cdots (n - k + 1)$$

Proof: The proof is a simple application of the first principle of counting. The k-tuple (arrangement) is formed in a sequence of k steps, one step for each position. The first position can be filled with any of the n elements of S, and so $N_1 = n$. Having determined the first position, there are $n - 1$ elements of S remaining as candidates for the second position. Hence, $N_2 = n - 1$. Similarly, $N_3 = n - 2$, and so forth. When we reach the kth position, $k - 1$ positions have already been filled, leaving $n - (k - 1)$ elements of S as candidates for the kth position. Hence, $N_k = n - k + 1$. The result now follows upon invoking the first principle of counting.

Corollary 1.5

$$P_k^n = \frac{n!}{(n - k)!}$$

Proof: This formula can be obtained by both multiplying and dividing the preceding formula by $(n - k)!$.

We point out that the first formula obtained above is both easier to remember and easier to use than the second formula, and we note that it always contains exactly k terms.

Corollary 1.6

$P_n^n = n!$

Proof: Substitute $k = n$ into Corollary 1.5. (*Note:* $0! = 1$)

Example 2 (a) $P_2^5 = 5 \cdot 4 = 20$. Since $k = 2$, there are two terms, starting with $n = 5$. Using Corollary 1.5 instead, we would also find that $P_2^5 = 5!/(5 - 2)! = 5!/3! = 120/6 = 20$. (b) $P_5^{10} = 10 \ 9 \cdot 8 \cdot 7 \cdot 6 = 30,240$, there being $k = 5$ terms starting with $n = 10$. Alternatively, $P_5^{10} = 10!/(10 - 5)! = 30,240$.

Example 3 If five pieces of paper are inscribed with the letters A, B, C, D, and E, then a total of $P_5^5 = 5! = 120$ five-letter "words" may be formed. How many of these have the B occurring immediately after the A? To answer this question, we might imagine that the A and B papers have been taped together (side by side). The problem then reduces to one of calculating the number of ways the four remaining pieces of paper can be arranged in different orders. This number is just $P_4^4 = 4! = 24$.

Example 4 Given the six digits 2, 3, 5, 6, 7, and 9 and assuming that repetitions are not permitted, then a total $P_3^6 = 6 \cdot 5 \cdot 4 = 120$ different three-digit numbers can be formed. How many of these numbers are less than 400? Recognizing that the first digit in all such numbers is restricted, we divide the problem of finding all such numbers into two steps:

Step 1: Choose the first digit.
Step 2: Choose the remaining two digits.

Since the first digit must be a 2 or a 3, we have $N_1 = 2$, and then, since the last two digits are arbitrary from the remaining five digits, we have $N_2 = P_2^5 = 5 \cdot 4 = 20$. Thus, there are a total of $2 \cdot 20 = 40$ ways to form a number less than 400 from the given digits. Likewise, if we wish to determine how many of our three-digit numbers are odd, we recognize that it is now the last digit which is restricted and proceed according to the following two steps:

Step 1: Choose the last digit.
Step 2: Choose the remaining two digits.

Since the last digit must be 3 or 5 or 7 or 9, we have $N_1 = 4$. As above, the other two digits are arbitrary, so $N_2 = 20$. Consequently, there are $4 \cdot 20 = 80$ ways to form an odd number from the given six digits. To make things more interesting (and more difficult), suppose that we wish to determine how many of these three-digit numbers are *both* odd *and* less than 400. In this case, both the first and the third digits are restricted, and we must proceed according to the following *three* steps:

Step 1: Choose the first digit.
Step 2: Choose the third digit.
Step 3: Choose the remaining (second) digit.

Unfortunately, the first principle of counting cannot be applied here, because the choice of a first digit has an effect on the choice of a third digit. If we choose 3 to be the first digit, then we cannot choose 3 to be the third digit, leaving us with only three possibilities for the third digit. On the other hand, had we chosen 2 to be the first digit, we would have had (all) four choices for the third digit. The solution to our problem lies in the tree diagram in Figure 1.6, which combines both the first and second principles of counting. Thus, there are $1 \cdot 4 \cdot 4 + 1 \cdot 3 \cdot 4 = 16 + 12 = 28$ different ways to form three-digit numbers which are odd and less than 400.

Figure 1.6

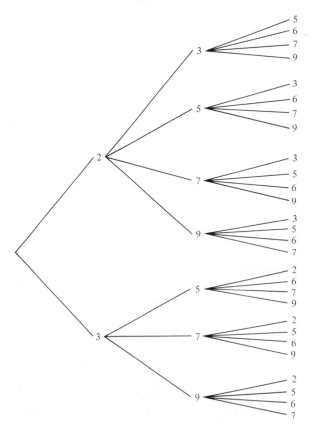

Example 5   We wish to determine the number of ways that seven people might seat themselves in a row of seven chairs. If we imagine that the people are numbered 1, 2, 3, 4, 5, 6, and 7, then any arrangement of the seven people in the seven chairs determines a permutation of the set $S = \{1, 2, 3, 4, 5, 6, 7\}$. Hence, the answer to our question is $P_7^7 = 7! = 5040$. To make the problem more difficult, suppose that there are eight chairs and only seven people. Obviously, one of the chairs must be left empty, and the seven people must arrange themselves in the other seven chairs. This suggests that we follow the following sequence of steps:

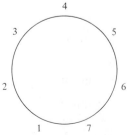

Step 1: Determine which chair is to be empty.
Step 2: Determine how the seven people will arange themselves in the other seven chairs.

Clearly, $N_1 = 8$ and $N_2 = 5040$ by the same argument as above; so the seven people may seat themselves in any of $8 \cdot 5{,}040 = 40{,}320$ different ways in the eight chairs.

Example 6   Suppose that we wish to determine the number of different ways seven people could be seated in seven chairs arranged around a circular table. Here the word "different" can be ambiguous. Are we to consider the two arrangements in Figure 1.7 to be different? In both arrangements, each person has the same two people sitting to his (or her) left or right. Although the two seatings look different, they are basically the same arrangement of the seven people, and we shall agree to consider them to be the same. As a result, where person 1, say, is sitting is of no consequence. What matters is where the other six people are sitting in relationship to person 1. This is essentially the same as seating these six people in a row of six chairs. Reasoning as in Example 5, we find there are $P_6^6 = 6! = 720$ such ways.

EXERCISES

1. Calculate P_3^7, P_5^8, P_7^{11}, and P_{10}^{20}.
2. Write a computer program that will calculate P_k^n. As n and k get large, you will have to consider the problem of computer overflow.

Figure 1.7

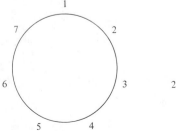

3. In a field of eight horses, how many different "triples" are possible? (A triple is the first three finishers in order.)
4. In how many ways can the winner and first, second, third, and fourth runnerups be chosen in the Miss America contest? (Assume there are 50 contestants.)
5. In how many ways can a club consisting of 30 people choose a president, vice-president, secretary, and treasurer?
6. In how many ways can an artist display four of his eight paintings on the wall in his studio?
7. A traveling salesman has a choice of visiting five of the 20 cities in his territory this week. In how many different ways can he arrange his traveling schedule for the week?
8. Referring back to Example 4 in the text, how many of these three-digit numbers are even? How many are even and larger than 600? In how many of these numbers are the three digits in ascending order from left to right?

1.5 Combinations

In the preceding section, we discussed samples in which the order that the elements were chosen was of some importance. In this section, we will discuss samples in which the major consideration is *which* elements make up the sample, and is not which order they were chosen.

DEFINITION

Suppose that S is a set containing n elements and k is a nonnegative integer not exceeding n. A k-element subset of S is called a *combination* of k of the n elements of S, and the number of such combinations is denoted by either $C_k{}^n$ or the more familiar $\binom{n}{k}$.

Note that there is no reference to the *order* of the k elements in a combination. Whereas $(1, 2)$ and $(2, 1)$ were regarded as different permutation of two elements from the set $S = \{1, 2, 3\}$, $\{1, 2\}$ and $\{2, 1\}$ are regarded as the same combination (two-element subset).

We now discuss some computational formulas for $C_k{}^n$.

Proposition 1.7

$$C_k{}^n = \frac{n!}{k!(n-k)!} \qquad \text{for any } 1 \leq k \leq n$$

Proof: There are as many ways of forming a permutation of k of the n elements of the set S as there are ways of completing the following two steps:

Step 1: Choose a k-element subset of S.
Step 2: Arrange this k-element subset into some specific order.

By the first principle of counting, we then have

$$P_k{}^n = C_k{}^n \cdot P_k{}^k = C_k{}^n \cdot k!$$

since the first step can be done in any of C_k^n ways, and the second step can be done in any of $P_k^k = k!$ ways. As a result,

$$C_k^n = \frac{P_k^n}{k!} = \frac{n!}{k!(n-k)!}$$

Notice that this formula also works in the case $k = 0$, since $\varnothing$ is the only 0-element subset of S.

Cancellation in this formula yields the more useful formula

Corollary 1.8

$$C_k^n = \frac{n(n-1)(n-2)\cdots(n-k+1)}{k!}$$

This expression for C_k^n has k descending terms in both its numerator and in its denominator, the numerator starting with n and the denominator starting with k.

It is also simple to verify that

Corollary 1.9

$$C_k^n = C_{n-k}^n \qquad \text{for any } 0 \le k \le n$$

This identity is stating the obvious fact that there are as many ways to choose k of the n elements of S as there are ways *not* to choose the other $n - k$ elements of S. That is, choosing k of the n elements of S for inclusion in a sample is the same as choosing $n - k$ elements of S for exclusion from the sample.

If we were to use Corollary 1.8 to evaluate C_k^n, both numerator and denominator would contain k terms. In the equivalent form C_{n-k}^n, both would contain $n - k$ terms. Hence, it is wiser to choose the smaller of k and $n - k$ and evaluate the corresponding expression. In other words, instead of evaluating

$$C_7^{10} = \frac{10 \cdot 9 \cdot 8 \cdot 7 \cdot 6 \cdot 5 \cdot 4}{7 \cdot 6 \cdot 5 \cdot 4 \cdot 3 \cdot 2 \cdot 1}$$

it is quicker to evaluate the equivalent

$$C_3^{10} = \frac{10 \cdot 9 \cdot 8}{3 \cdot 2 \cdot 1} = 120$$

Corollary 1.10

$$C_k^n = C_k^{n-1} + C_{k-1}^{n-1} \qquad \text{if } n \ge 2 \text{ and } 1 \le k \le n - 1$$

Proof: Exercise

This formula gives rise to what is known as *Pascal's Triangle*, which is illustrated in Figure 1.8. Notice that each interior element of the triangle is obtained by adding the two elements diagonally above it, which is exactly what Corollary 1.10 is saying.

It is because we can write the binomial expansion as

$$(a + b)^n = \sum_{k=0}^{n} C_k^{\,n} \cdot a^k \cdot b^{n-k}$$

that the terms $C_k^{\,n}$ are called the *binomial coefficients.*

Example 1 Students must answer eight of the ten questions on an examination. They therefore have $C_8^{\,10} = C_2^{\,10} = (10 \cdot 9)/(2 \cdot 1) = 45$ choices of different sets of eight questions to answer. If the students are required to answer the first three questions, then they must choose five of the last seven questions, giving them $C_5^{\,7} = C_2^{\,7} = (7 \cdot 6)/(2 \cdot 1) = 21$ choices of different sets of eight questions to answer. Suppose, finally, that the students are required to answer at least four of the first five questions. How many choices do they now have? The students have basically two strategies as to how they will answer eight questions:

Strategy 1: Answer four of the first five questions, then four of the last five questions.

Strategy 2: Answer all of the first five questions, then only three of the last five questions.

Since the students have a choice of strategies available, the second principle of counting will be used. Should they employ the first strategy, they would have a choice of $C_4^{\,5} \cdot C_4^{\,5} = 25$ ways to answer eight questions. On the other

Figure 1.8

1

1 1

1 2 1

1 3 3 1

1 4 6 4 1

1 5 10 10 5 1

1 6 15 20 15 6 1

1 7 21 35 35 21 7 1

⋮

hand, should they employ strategy 2, they would have a choice of $C_5{}^5 \cdot C_3{}^5 = 10$ ways to answer eight questions. By the second principle of counting then, the students have a choice of $25 + 10 = 35$ different ways to answer eight of the ten questions.

Example 2 How many different poker hands might you be dealt? Since a poker hand is a five-element subset of the fifty-two–element (card) deck, there are

$$C_5{}^{52} = \frac{52 \cdot 51 \cdot 50 \cdot 49 \cdot 48}{5 \cdot 4 \cdot 3 \cdot 2 \cdot 1} = 2,598,960$$

different poker hands possible.

Example 3 How many poker hands are flushes? (A flush is a hand in which all five cards are of the same suit.) To form a flush, we may proceed by the following two steps:

Step 1: Choose a suit.
Step 2: Choose five cards of this suit.

Since there are four suits, $N_1 = 4$, and since there are 13 cards of each suit, $N_2 = C_5{}^{13} = 1287$. Consequently, there are $4 \cdot 1287 = 5148$ different flushes possible.

Example 4 How many poker hands are full houses? (A full house is a hand consisting of three cards of one type and two cards of another type.) To form a full house, we may proceed in the following manner:

Step 1: Choose a first type.
Step 2: Choose 3 cards of this type.
Step 3: Choose a second type.
Step 4: Choose 2 cards of this type.

Since there are 13 types and four cards of each type, we have $N_1 = C_1{}^{13} = 13$, $N_2 = C_3{}^4 = 4$, $N_3 = C_1{}^{12} = 12$ (the second type is chosen from the remaining 12 types), and $N_4 = C_2{}^4 = 6$. Therefore, there are $13 \cdot 4 \cdot 12 \cdot 6 = 3744$ different full houses possible.

Example 5 How many poker hands are two pairs? (A two pair hand consists of two cards each of two different types and a fifth card of still another type.) To form such a hand, proceed as above, using the following five steps:

Step 1: Choose a first type.
Step 2: Choose two cards of this type.
Step 3: Choose a second type.

Step 4: Choose two cards of this type. $N_4 = C_2^4 = 6$
Step 5: Choose a fifth card of a third type. 44

As above, $N_1 = 13$, $N_2 = C_2^4 = 6$, $N_3 = 12$, and $N_4 = C_2^4 = 6$. Since there are $52 - 4 - 4 = 44$ cards of a type different from the first four cards chosen, $N_5 = 44$. Thus, it would seem that there are $13 \cdot 6 \cdot 12 \cdot 6 \cdot 44 = 247{,}104$ different hands of two pairs possible. However, this number is incorrect, there are actually only half as many such hands. Where did we go wrong? Figure 1.9 shows a segment of the tree diagram corresponding to the procedure outlined above. The two branches that are fully labeled both correspond to the same hand—the king of hearts, king of diamonds, queen of spades, queen of clubs, and the ace of spades. Since this hand corresponds to two different branches of the tree, it has been counted twice. As a matter of fact, every possible two-pair hand is counted twice by the procedure described above. To correct this error, we must realize that the order in which the two

Figure 1.9

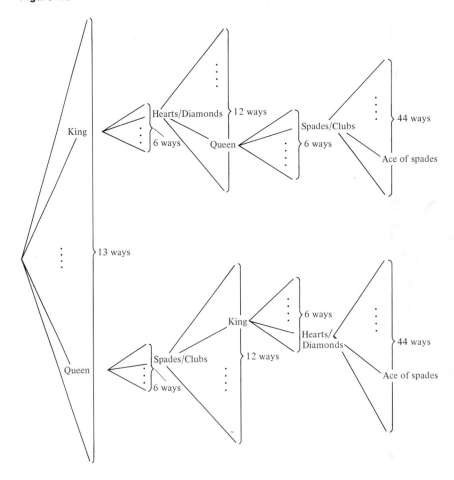

types are chosen is irrelevant. It would be correct to use the following sequence of four steps:

Step 1: Choose two types. $\left(\begin{smallmatrix} 13 \\ 2 \end{smallmatrix}\right)$
Step 2: Choose two cards of one type. $\left(\begin{smallmatrix} 4 \\ 2 \end{smallmatrix}\right) = 6$
Step 3: Choose two cards of the other type. $\left(\begin{smallmatrix} 4 \\ 2 \end{smallmatrix}\right) = 6$
Step 4: Choose a fifth card of a third type. 44

Here, $N_1 = C_2^{13}$, $N_2 = C_2^4$, $N_3 = C_2^4$, and $N_4 = 44$. Hence, there are a total of $78 \cdot 6 \cdot 6 \cdot 44 = 123{,}552$ different two-pair hands possible. Notice that $N_1 = C_2^{13} = (13 \cdot 12)/2$ here is exactly half the product $(N_1 = 13)$ $(N_3 = 12)$ from the first method, and the answer here is exactly half the first answer.

The lesson to be learned from Example 5 is that care must be taken that the sequence of steps we construct in such a case counts exactly once for each possible way to complete the job.

EXERCISES

1. Prove Corollary 1.9.
2. Prove Corollary 1.10.
3. How many poker hands are
   (a) four of a kind;
   (b) straights;
   (c) two of a kind?
4. Prove that a set consisting of n elements has 2^n subsets.
5. Write a computer program to calculate C_k^n. Use the formula in Corollary 1.8. To avoid difficulties with possible computer overflow, have the computer perform k divisions, rather than k multiplications divided by another k multiplications.
6. Calculate C_5^9, C_7^{20}, and C_{10}^{20}.
7. In how many ways can you answer a 20 question true-false test if you must answer exactly ten questions with "true".
8. A political organization has 60 men and 40 women members. How many four-person delegations to the national convention are possible? How many if the delegation must contain two men and two women? How many if the delegation cannot be constituted entirely of members of the same sex?
9. In how many ways can the five finalists be chosen from among the 50 contestants in the Miss America contest?
10. The NIT basketball tournament is considering 20 teams, including 3 teams from the New York area. In how many ways can the twelve-team field be selected? How many ways if at least one New York area team must be in the field?

1.6 Urns and Balls

Many counting problems can be rephrased in terms of urns and balls so that the problem becomes a variation of one of the following two general types:

(a) In how many ways can a given number of balls be drawn from an urn in a prescribed manner?

(b) In how many ways can a given number of balls be placed in a given number of urns in a prescribed manner?

We have already considered problems of type (a) in Sections 1.4 and 1.5. In the poker examples of the preceding section, you might imagine that each of the 52 cards in the deck is represented by a different ball in the urn. Our task then is to draw five balls from the urn (choose five cards from the deck) in such a manner that the resulting sample (hand) is of a certain type—a flush, a full house, and so forth.

We shall return to problems of type (a) in the next section. In this section, we shall consider two special problems of type (b). We start by describing the situation to which the first of these problems applies.

Suppose that we have k numbered urns and n distinguishable balls. Suppose further that we have k nonnegative integers $n_1, n_2, \ldots, n_k$ whose sum is n. We wish to know in how many ways we can place the n balls into the k urns so that n_i balls are in the ith urn, for $1 \leq i \leq k$. We shall use the notation

$$\binom{n}{n_1, n_2, \ldots, n_k}$$

which is meant to be suggestive of our alternative notation for combinations, to denote the answer to this question. The following result is a formula that can be used to evaluate this term.

Proposition 1.11

$$\binom{n}{n_1, n_2, \ldots, n_k} = \frac{n!}{n_1! n_2! \cdots n_k!}$$

Proof: The proof is an application of the first principle of counting. We first choose n_1 balls to be placed in urn 1, then n_2 balls to be placed in urn 2, and so forth. Now there are

$$\binom{n}{n_1}$$ different ways

we could choose n_1 balls to be placed in urn 1. Having done this, there are

$$\binom{n - n_1}{n_2}$$ different ways

we could choose n_2 of the remaining $n - n_1$ balls to be placed in urn 2. Inductively, having filled the first j urns as prescribed, there are

$$\binom{n - (n_1 + n_2 + \cdots + n_j)}{n_{j+1}}$$ different ways

that we can choose n_{j+1} of the remaining $n - (n_1 + n_2 + \cdots + n_j)$ balls to be placed in urn $j + 1$. This process continues until, finally, we find that

there are

$$\binom{n - (n_1 + n_2 + \cdots + n_{k-1})}{n_k} \text{ different ways}$$

of choosing n_k of the remaining $n - (n_1 + n_2 + \cdots + n_{k-1}) = n_k$ (why?) balls to be placed in urn k. Consequently, this last step can be done in just one way (why is this logical?). By the first principle of counting, there are

$$\binom{n}{n_1}\binom{n - n_1}{n_2}\binom{n - (n_1 + n_2)}{n_3} \cdots \binom{n_k}{n_k}$$

different ways to place the n balls into the k urns as prescribed. However,

$$\binom{n}{n_1}\binom{n - n_1}{n_2} = \frac{n!}{n_1!(n - n_1)!} \cdot \frac{(n - n_1)!}{n_2!(n - n_1 - n_2)!}$$

$$= \frac{n!}{n_1!n_2!(n - n_1 - n_2)!}$$

and similar cancellations throughout the expression reduce it to the desired form.

Two special cases of this formula are worth noting:

Corollary 1.12

(a) $\underbrace{\binom{n}{1, 1, \ldots, 1}}_{n \text{ times}} = n!$

(b) $\binom{n}{n_1, n_2} = \binom{n}{n_1} = \binom{n}{n_2}$

Proof: Exercise: If we were to place one ball into each of n urns, as in (a) above, we would actually be forming a permutation of the n balls. Therefore, the result of (a) should not come as a surprise. On the other hand, if $n = n_1 + n_2$ and if we choose n_1 balls to place in urn 1 and n_2 balls to place in urn 2, we are essentially just choosing n_1 of the n balls—those to be placed in urn 1—and so (b) is reasonable.

Example 1 We wish to determine the number of different 11-letter "words" that may be formed from the letters MISSISSIPPI. We imagine 11 balls numbered $1, 2, 3, \ldots, 11$, representing the 11 positions in the "word", and 4 urns, one for each of the letters M, I, S, and P. One ball must be placed in the M urn, since there is one M in the word MISSISSIPPI, four balls must be placed in both the I and S urns, and two in the P urn. Therefore, $n = 11$, $k = 4$, and $n_1 = 1$, $n_2 = n_3 = 4$, and $n_4 = 2$. Each such distribution of the 11 balls into the 4 urns determines a different rearrangement of the letters of MISSISSIPPI and, therefore, a different 11-letter "word". For example, if the balls are placed in the urns as in Figure 1.10, we have formed the "word" PIISISMPSSI. By Proposition 1.11, the total number of words are

$$\binom{11}{1, 4, 4, 2} = \frac{11!}{1!4!4!2!} = 34,650$$

Example 2 Suppose that we wish to determine the number of different arrangements of 12 books on a shelf. The 12 books consist of three identical copies of each of four different titles. We imagine 12 balls, corresponding to the 12 positions on the shelf, which must be placed into four urns, one corresponding to each of the four different titles. Three balls must be placed into each urn, since there are three copies of each title. The total number of ways that this may be done equals

$$\binom{12}{3, 3, 3, 3} = \frac{12!}{3!3!3!3!} = 369,600$$

Example 3 How many different ways are there to deal 13-card bridge hands to each of four players? Imagine 52 balls, one corresponding to each card, which must be placed into one of four urns, corresponding to the four players, with 13 balls going into each urn. The total number of ways to distribute the 52 balls into the four urns in this manner equals

$$\binom{52}{13, 13, 13, 13} = \frac{52!}{(13!)^4}$$

The problem we shall now discuss is also of type (b) and is more general than the preceding problem. It is known as the *classical occupancy problem* and can be stated as follows: "In how many ways can n indistinguishable balls be placed into k different urns?" Since the balls are indistinguishable, we do not care *which* balls go into which urns, nor are we prescribing *how many* balls must go into each urn.

To solve the classical occupancy problem, we first imagine that each of the n balls is marked with the letter B. We then imagine that the k urns are

Figure 1.10

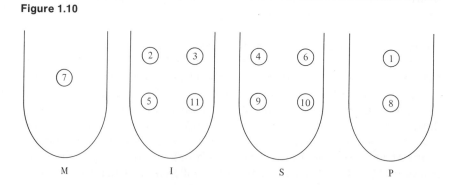

M   I   S   P

placed in a row, thereby creating $k - 1$ "walls" between the contents of the k urns. We further imagine that we have $k - 1$ balls marked W, representing these "walls." Then, any arrangement of these $n + k - 1$ balls into a $n + k - 1$ letter "word" corresponds to a way in which the original n balls can be placed into the k urns. For example, if $n = 5$ and $k = 4$, we would have a total of $5 + 4 - 1 = 8$ balls, five marked B and three marked W. The word BBWWBWBB corresponds to placing five balls into four urns as indicated in Figure 1.11. Our problem has now been reduced to finding the number of $n + k - 1$ letter "words" that can be formed from n B's and $k - 1$ W's. Using the technique of the preceding three examples, we imagine $n + k - 1$ balls numbered $1, 2, 3, \ldots, n + k - 1$, representing the $n + k - 1$ positions in the word, to be placed into either of two urns (a B urn and a W urn) with n balls going into the first urn and $k - 1$ balls going into the second urn. This can be done in any of

$$\binom{n+k-1}{n, k-1} = \binom{n+k-1}{n} = \binom{n+k-1}{k-1} \text{ different ways}$$

Therefore, there are

$n = 5 + 4 - 1 =$

$$\binom{n+k-1}{n} = \binom{n+k-1}{k-1} \text{ solutions}$$

to the classical occupancy problem.

Example 4 In how many ways may three people divide a \$50 reward? Let us imagine that each of the 50 dollars is represented by a ball to be placed in one of three urns, one for each of the three people. Then, according to the result just derived, there are

$$\binom{50+3-1}{50, 2} = \binom{52}{2} = 1326 \text{ different ways}$$

to place the 50 balls into the three urns, and therefore 1326 different ways to divide the \$50.

Figure 1.11

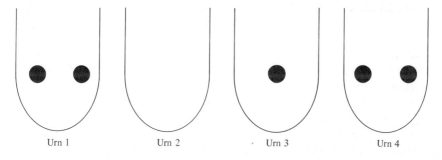

Urn 1   Urn 2   Urn 3   Urn 4

EXERCISES

1. Fill in the details in the proof of Proposition 1.11.
2. In how many ways can 12 people be placed on three teams so that each team has four players?
3. In a series of 11 games, how many ways are there to win six, lose two, and tie three?
4. How many six-digit numbers can be formed from the digits 1, 1, 1, 2, 3, 3? How many of these begin with a two? If we list these numbers in order of increasing magnitude, then how far down the list is the number 321,311?
5. How many arrangements of 15 indistinguishable balls into eight distinguishable urns leave no empty urns?
6. In how many different ways can 20 graduate students attend four seminars which are given at the same time in four different rooms?
7. In how many ways can 100 students register for four sections of the same course? How many if there must be 25 students in each section?

1.7 Methods of Sampling

We conclude this chapter with a brief discussion of several different methods of sampling, phrased in terms of drawing balls from an urn. Suppose that we have an urn that contains n numbered balls. When we say that a sample of k balls can be drawn from the urn, there is some vagueness regarding the manner in which the balls are to be drawn. The sample can be taken from the urn in any of the following ways:

1. The k balls can be drawn as a combination, that is, "simultaneously," if that is physically possible. There are a total of C_k^n different samples that might be drawn in this manner. In such samples, order is usually not considered to be important.
2. The k balls can be drawn one at a time, with each ball being discarded (not placed back in the urn) before the next ball is drawn. Sampling of this type is called *sampling without replacement*. If we wish to consider the order in which such a sample is drawn, then there are a total of P_k^n different possible samples. If we do *not* wish to consider order, then there are only $P_k^n/k! = C_k^n$ different samples. In other words, if order is not considered, then the samples produced by sampling without replacement are precisely the combinations.
3. The k balls can be drawn one at a time with each ball being placed back into the urn before the next ball is drawn. Sampling of this type is called *sampling with replacement*. The order in which such a sample is drawn is usually considered to be of importance. By the first principle of counting, there are a total of n^k such ordered samples, since each ball is drawn from the full urn.
4. Suppose that the n balls in the urn are of various colors, there being n_1 balls of color 1, n_2 balls of color 2, and so forth, and finally n_s balls of color s. Suppose also that our sample of k balls must contain k_1 balls of color 1, k_2 balls of color 2, and so forth, and, finally, k_s balls of color s.

Such a sample is called a *stratified sample*. By the first principle of counting, there are a total of.

$$\binom{n_1}{k_1} \cdot \binom{n_2}{k_2} \cdot \ldots \cdot \binom{n_s}{k_s}$$

different such samples. Many surveys and opinion polls are taken by means of stratified samples, the population being stratified along guidelines that might be geographic, ethnic, religious, economic, and so forth, and the sample stratified in the same pattern as the population.

Example 1 How many seven-digit telephone numbers are possible? Imagine ten balls numbered 1, 2, 3, 4, 5, 6, 7, 8, 9, 0 in an urn. To form a telephone number, we must choose an ordered sample of seven with replacement. According to the formula above, this can be done in a total of 10^7 different ways. (Notice that the terminology "sampling with replacement" was superfluous here and that this problem and almost all other problems concerning this type of sampling can be done using just the first principle of counting.)

Example 2 How many different bridge hands are there that contain exactly three clubs, four spades, four hearts, and two diamonds? Our population here, the 52 cards in the deck, is naturally stratified into the clubs, spades, hearts, and diamonds, and the problem demands that the sample of 13 (the bridge hand) also be stratified. So there are a total of

$$\binom{13}{3} \cdot \binom{13}{4} \cdot \binom{13}{4} \cdot \binom{13}{2} = 11,404,407,300$$

such bridge hands.

Example 3 In a university which has 100 full professors, 80 associate professors, and 120 assistant professors, there are a total of

$$\binom{100}{3} \cdot \binom{80}{2} \cdot \binom{120}{4}$$

different possible committees consisting of three full, two associate, and four assistant professors. The population in this example is stratified by rank, and the problem demands that the sample also be stratified by rank.

EXERCISES

1. Five people enter an elevator in a twenty-story building. In how many ways can they designate five floors? In how many ways five different floors?
2. Two cards are drawn, one after the other, from a deck of 52 cards. How many ways can one draw
   (a) a spade, then a club;
   (b) a spade, then a red card;

(c) a diamond, then another diamond;
(d) a diamond, then a picture card?
Repeat the problem, assuming this time that the first card is placed back into the deck before the second card is drawn.
3. Suppose that there are 30 Democratic governors and 20 Republican governors. In how many ways can three Democrats and two Republicans be elected to an executive committee?
4. How many poker hands contain three picture cards and two sevens?
5. In how many different ways can a doctor choose three of his 18 patients who are suffering from high blood pressure, four of his 24 patients who are suffering from diabetes, and five of his 12 patients who are suffering from angina pectoris for a medical experiment?

ADDITIONAL PROBLEMS

1. In a fraternity consisting of 80 members, 55 are underclassmen and 40 own cars. If ten seniors in the fraternity do not own cars, how many underclassmen do own cars?
2. At a certain high school, 100 seniors are studying mathematics. Of these, 87 are taking twelfth year mathematics, 42 are taking advanced placement mathematics, and 23 are taking a computer course. If 31 are taking both twelfth year math and advanced placement, 15 are taking both twelfth year math and the computer course, and nine are taking both the computer course and advanced placement, how many are taking all three courses?
3. A barrel of 1000 screws contains 20 defectives. How many different batches of ten screws containing no defectives are possible? How many containing at least one defective?
4. How many different teams can a little-league coach field from a roster of 20? How many different batting orders can he use?
5. How many ways can seven balls be placed into ten urns so that each ball is in a different urn?
6. In one morning, nine different people in a town call a local doctor. If there are four doctors in town, how many different selections of doctors could the nine people have made?
7. From a group of five married couples, three people are to be chosen for a psychological experiment. How many ways can the three people be selected? How many if a husband and his wife cannot both be selected?
8. A delegation of four students is selected from a college to attend a National Student Association meeting. If there are 12 eligible students, then how many ways can the delegation be chosen if
(a) two particular students refuse to attend the meeting together; or
(b) two students are married and will only attend if they both can go?

12

Sample Spaces and Probability Functions

2.1 Sample Spaces

The theory of probability, like the theory of Euclidean geometry, is a body of axioms and definitions together with the theorems which can be deduced from them. The undefined objects of this theory, similar to the points and lines of geometry, are the occurrences of random phenomena. The outcomes of random experiments have been studied, particularly by mathematicians and gamblers, for many centuries. The approach of the mathematician, after studying many particular examples of random phenomena, would be to try to create a general mathematical model which would encorporate the basic characteristics of the particular examples. Attempts at creating such models date back at least to Laplace, but it was not until 1933 that a mathematically rigorous set of axioms was developed by Kolmogorov. We now begin our study of a set of axioms quite similar to those of Kolmogorov.

DEFINITION

The *sample space* of a random experiment is the set S of all possible outcomes of the experiment.

We point out that the sample space must be unambiguous and non-repetitive. There must be a clearly defined one-to-one correspondence between the outcomes of the experiment and the elements of the sample space.

Example 1　When a coin is tossed, there are only two possible outcomes, heads or tails. The sample space of this experiment is written $\{H, T\}$.

Example 2　If a coin is tossed twice, or if two coins are tossed, then there are four possible outcomes, which are denoted HH, HT, TH, and TT. The outcome HT, for example, represents the ordered pair (H, T), which means that a head occurred on the first toss and a tail occurred on the second toss. The set $S = \{0, 1, 2\}$ can also be a sample space for this experiment, in which the elements of S indicate the different possible number of heads that might occur in two tosses of a coin. However, the set $S = \{HH, HT, TH, TT, 0, 1, 2\}$ is not a potential sample space because it is ambiguous to which element of the sample space a given outcome of the experiment corresponds.

Example 3　If a single die is rolled, the set of six possible outcomes is represented by the sample space $S = \{1, 2, 3, 4, 5, 6\}$.

Example 4　Suppose that a green die and a red die are rolled in that order. The sample space

$$S_1 = \{(a, b) : a, b = 1, 2, 3, 4, 5, 6\}$$

contains all 36 possible outcomes, with the first coordinate denoting the outcome of the green die and the second coordinate the outcome of the red die. However, S_1 is not the only sample space that could be used with the experiment of rolling two dice. If we are interested only in the number of 6's that occur when two dice are rolled, then the set

$$S_2 = \{0, 1, 2\}$$

could be used as a sample space. If we are interested only in the sum of the two faces, then the set

$$S_3 = \{2, 3, 4, 5, 6, 7, 8, 9, 10, 11, 12\}$$

could be used as a sample space.

Example 5   Suppose that we toss a coin until it comes up heads and record the number of tosses required before that first head occurred. If we denote by ∞ the outcome that a head never occurs, which would seem physically impossible although it is logically conceivable, then the set

$$S = \{\infty, 1, 2, 3, \ldots\}$$

is an example of a sample space which is countable but not finite.

Example 6   Suppose that we shoot an arrow which hits at a circular target, or, more generally, suppose that we choose a point in a circle. Then the set consisting of all the points in the circle (on the target) forms an uncountably infinite sample space.

Example 7   Suppose that we interview k people, asking each their birthday. The set

$$S = \{(a_1, a_2, \ldots, a_k) : 1 \le a_i \le 365, 1 \le i \le k\}$$

in which a_i represents the birthday of the ith person interviewed, is a sample space for this experiment. To simplify calculations, we agree to ignore people born on February 29. Since the birthdays of the k people are in no way related, we see by the first principle of counting that $n(S) = (365)^k$.

DEFINITION

A subset E of the sample space S is called an *event*. If E is of the form $\{x\}$, where $x \in S$, then E is called a *simple event*. The event $\emptyset$ is called the *impossible event* while the event S is called the *sure event*. An event E is said to *occur* if the experiment results in an outcome $x \in E$.

If E is an event, then the set $C(E)$ represents the event that "E does not occur". If A and B are events, then the set $A \cap B$ represents the event that "both A and B occur," while the set $A \cup B$ represents the event that "either A occurs or B occurs or possibly both occur."

DEFINITION

Events A and B are called *exclusive* if they cannot occur simultaneously; that is, $A \cap B = \varnothing$. The events $A_1, A_2, A_3, \ldots, A_n$ are called *mutually exclusive* if every pair A_i and A_j is exclusive.

Example 8   In Example 2 above, the event that both coins turn up the same corresponds to the subset $E = \{HH, TT\}$ of S.

Example 9   In Example 4 above, the event that the green die comes up a three corresponds to the subset

$$A = \{(3, 1), (3, 2), (3, 3), (3, 4), (3, 5), (3, 6)\}$$

of S_1. The event that the sum of the two dice is 11 corresponds to the subset

$$B = \{(5, 6), (6, 5)\}$$

The events A and B are exclusive since $A \cap B = \varnothing$.

Due to a technicality in mathematical logic which we shall not attempt to explain, if the sample space S is uncountably infinite, then *not every* subset of S can be an event. In this case, then, we must restrict the term "event" to a certain collection of subsets of S, which we shall call the set of events, and which we shall denote ξ. Not just any collection of subsets of S will suffice, however. If it makes sense to talk about whether an event E occurs, then it also makes sense to talk about whether the event E does not occur. The nonoccurrence of E is equivalent to the occurrence of $C(E)$. Therefore, if E is an event, then $C(E)$ should also be an event. Likewise, if A and B are events, and it makes sense to talk about whether or not A occurs or B occurs, then it should also make sense to talk about whether A and B occur simultaneously, or about whether either of A or B occur. That is, whenever A and B are events, we should expect $A \cap B$ and $A \cup B$ to also be events.

As it turns out, the conditions listed above are not quite sufficient. In order that our theory of probability be successful, we must require that the set of events satisfy the following stronger set of conditions:

1. $S \in \xi$.
2. If $E \in \xi$, then $C(E) \in \xi$.
3. If $E_1, E_2, \ldots$ is either a finite or countable collection of events, then both $\bigcap_{i=1}^{\infty} E_i$ and $\bigcup_{i=1}^{\infty} E_i$ must be events.

Of course, if S itself is either finite or countable, and we let ξ be the set of all subsets of S, then the three conditions are trivially satisfied.

EXERCISES

Determine a sample space for each of the following situations:
1. A die is rolled, and then a coin is tossed.
2. Three dice are rolled.
3. A card is drawn from a conventional fifty-two–card deck.

4. A poker hand is dealt from the fifty-two–card deck.
5. A two-digit number is formed from the digits 0, 1, 2, 3, 4, 5, 6, 7, 8, 9.
6. A voter must choose between three presidential, two gubernatorial, and four mayoral candidates.
7. A child has a penny, nickel, dime, and quarter in his pocket. He reaches into his pocket and takes out two coins.
8. An interviewer asks passersby two questions: whether or not they are married; whether or not they own a car.

2.2 Probability Functions

Given a random experiment and a sample space of possible outcomes of the experiment, it is only natural to ask "will this particular event occur?" It is here that we frequently use the terms "chance," "likelihood," and "probability." We are looking for a numerical value that will be a measure of the frequency of occurrence of the event.

Before the time of Kolmogorov, there were two popular approaches to assigning a numerical value that would measure the "probability" of an event occurring. Both approaches failed because they could not give rise to a rigorous mathematical model with a wide variety of applications.

The first approach is due to Laplace, who considered only random experiments with a finite number of outcomes which he postulated to be equally likely. He then defined the probability of an event E to be $n(E)/n(S)$, where $n(E)$ is the number of simple events in the sample space corresponding to the occurrence of E, while $n(S)$ is the total number of simple events (elements) in the sample space S. While Laplace's theory is fine as far as it goes, it does not take into account either random experiments which have an infinite number of possible outcomes or random experiments with a finite number of outcomes which are *not* equally likely.

The second approach, the so-called relative frequency approach, is more empirical in nature. It requires that the random experiment be observed over a long series of trials, with the outcome of each trial being recorded. If an event E occurs k times in n trials, then the relative frequency f_n for the event is simply k/n for the n trials. If the values f_n approach some limit (become stable) as n gets larger and larger, then this limiting value is defined to be the probability of the event E. There is no guarantee, of course, that such a limit will exist. Even if it did exist, it might be difficult in practice to find it since n must remain finite. Needless to say, it would be difficult to adapt this approach to a theoretical study of probabilities.

To illustrate how the relative frequency approach works and to illustrate some of the problems or questions that might arise using this approach, we have used the technique of computer simulation. The experiment of rolling one die, rolling two dice, and tossing a fair coin have each been simulated a large number of times. The results are contained in the following examples.

Example 1 A single fair die has been rolled 100,000 times, with the face that occurred on top being recorded each roll. The results of the experiment, in terms of both the frequency and relative frequency of each of the six faces at various stages of the experiment, can be found in Table 2.1. Notice that after even as many as 10,000 rolls, we cannot be sure that the die is fair—it conceivably could be biased so as to favor face 1 at the expense of face 3. However, after 100,000 rolls, the six relative frequencies are within 0.004 of each other and seem to be leveling off at approximately one-sixth, or 0.1666, each. Therefore, on the basis of 100,000 rolls, we would seem to be able to postulate the obvious—that each of the six faces is equally likely to occur over the long run. Of course, it is rather tedious to have to roll the die 100,000 times to reach this conclusion, and it is only natural to wonder at this stage if the same conclusion (or its opposite, that the die is biased) could be reached on the basis of a much smaller sample. This is a typical problem in the field of statistical inference, and it shall be studied later in the book.

Example 2 Two fair dice have been rolled a total of 100,000 times, with the sum of the two top faces being recorded each roll. The frequencies and relative frequencies of the different possible sums, at various stages of the experiment, appear in Table 2.2. Unlike the preceding example, however, the relative

---

Table 2.1
Simulation of 100,000 Rolls of a Fair Die

Number of rolls	Frequency (Relative frequency)					
	Face 1	Face 2	Face 3	Face 4	Face 5	Face 6
1,000	159 (15.9%)	178 (17.8%)	153 (15.3%)	180 (18.0%)	165 (16.5%)	165 (16.5%)
2,000	335 (16.7%)	342 (17.1%)	320 (16.0%)	352 (17.6%)	316 (15.8%)	335 (16.7%)
3,000	510 (17.0%)	493 (16.4%)	474 (15.8%)	536 (17.9%)	474 (15.8%)	513 (17.1%)
4,000	693 (17.3%)	644 (16.1%)	635 (15.9%)	693 (17.3%)	646 (16.1%)	689 (17.2%)
5,000	854 (17.1%)	808 (16.2%)	806 (16.1%)	871 (17.4%)	817 (16.3%)	844 (16.9%)
10,000	1,768 (17.7%)	1,625 (16.3%)	1,596 (16.0%)	1,694 (16.9%)	1,646 (16.5%)	1,671 (16.7%)
25,000	4,235 (16.9%)	4,147 (16.6%)	4,060 (16.2%)	4,222 (16.9%)	4,152 (16.6%)	4,184 (16.7%)
50,000	8,501 (17.0%)	8,359 (16.7%)	8,257 (16.5%)	8,414 (16.8%)	8,240 (16.5%)	8,229 (16.5%)
100,000	16,857 (16.9%)	16,819 (16.8%)	16,523 (16.5%)	16,663 (16.7%)	16,463 (16.5%)	16,675 (16.7%)

Table 2.2
Simulation of 100,000 Rolls of Two Fair Dice

Number of rolls	Frequency (Relative frequency)					
	Sum 2	Sum 3	Sum 4	Sum 5	Sum 6	Sum 7
1,000	33 (3.3%)	52 (5.2%)	87 (8.7%)	121 (12.1%)	115 (11.5%)	160 (16.0%)
2,000	66 (3.3%)	110 (5.5%)	174 (8.7%)	215 (10.8%)	238 (11.9%)	331 (16.6%)
5,000	162 (3.2%)	280 (5.6%)	438 (8.8%)	560 (11.2%)	652 (13.0%)	842 (16.8%)
10,000	279 (2.8%)	548 (5.5%)	805 (8.1%)	1,119 (11.2%)	1,391 (13.9%)	1,704 (17.0%)
25,000	724 (2.9%)	1,375 (5.5%)	2,059 (8.2%)	2,780 (11.1%)	3,610 (14.4%)	4,143 (16.6%)
50,000	1,412 (2.8%)	2,824 (5.6%)	4,087 (8.2%)	5,520 (11.0%)	7,130 (14.3%)	8,286 (16.6%)
100,000	2,782 (2.8%)	5,710 (5.7%)	8,291 (8.3%)	11,008 (11.0%)	14,104 (14.1%)	16,580 (16.6%)

Number of rolls	Frequency (Relative frequency)				
	Sum 8	Sum 9	Sum 10	Sum 11	Sum 12
1,000	152 (15.2%)	125 (12.5%)	81 (8.1%)	47 (4.7%)	27 (2.7%)
2,000	293 (14.7%)	241 (12.1%)	176 (8.8%)	101 (5.6%)	55 (2.8%)
5,000	659 (13.2%)	561 (11.2%)	447 (8.9%)	263 (5.3%)	136 (2.7%)
10,000	1,423 (14.2%)	1,121 (11.2%)	823 (8.2%)	523 (5.2%)	264 (2.6%)
25,000	3,582 (14.3%)	2,748 (11.0%)	1,970 (7.9%)	1,346 (5.4%)	663 (2.7%)
50,000	7,083 (14.2%)	5,589 (11.2%)	4,017 (8.0%)	2,704 (5.4%)	1,348 (2.7%)
100,000	14,064 (14.1%)	11,128 (11.1%)	8,173 (8.2%)	5,450 (5.5%)	2,710 (2.7%)

frequencies do not all appear to be approaching the same limit. Do we then postulate, for example, that the probability that sum 5 occurs is equal to 0.110, which is the relative frequency after 100,000 rolls? Is it not conceivable that the relative frequency after 200,000 rolls could be 0.108 or 0.111? If we wish to determine the *exact* probabilities, this is obviously not the proper method. It would be better if we had some other way to determine what these probabilities *should* be and used the results of such an experiment to either verify or refute the validity of these probabilities.

Example 3 In this experiment, the process of tossing a coin until it comes up heads is repeated 10,000 times, and the toss on which the first head occurs is recorded. Table 2.3 contains the results at various stages of the experiment. We point out that in the 10,000 trials, it never took more than 12 tosses before the first head occurred. The results of the experiment seem to indicate that the probability that the first head occurs on the kth toss is approximately twice the probability that the first head occurs on the $(k + 1)$st toss. Therefore, the experiment would seem to lead us to the conclusion that the probability that the first head occurs on the kth toss should be $(\frac{1}{2})^k$. We shall see later on that this is correct.

The problem that faces us, then, is to assign numerical values to the events of a sample space in a way that is both intuitively and theoretically pleasing.

Table 2.3
10,000 Tosses of a Coin Until a Head Occurs

	Frequency								
	(Relative frequency)								
Number of trials	*Number of rolls until a head occurs*								
	1	*2*	*3*	*4*	*5*	*6*	*7*	*8*	*9+*
100	39	32	11	9	6	3	0	0	0
	(39.0%)	(32.0%)	(11.0%)	(9.0%)	(6.0%)	(3.0%)	(0%)	(0%)	(0%)
200	97	48	23	16	9	5	1	1	0
	(48.5%)	(24.0%)	(11.5%)	(8.0%)	(4.5%)	(2.5%)	(0.5%)	(0.5%)	(0%)
500	246	120	61	35	19	12	5	1	1
	(49.2%)	(24.0%)	(12.2%)	(7.0%)	(3.8%)	(2.4%)	(1.0%)	(0.2%)	(0.2%)
1,000	503	245	118	69	33	18	8	4	2
	(50.3%)	(24.5%)	(11.8%)	(6.9%)	(3.3%)	(1.8%)	(0.8%)	(0.4%)	(0.2%)
2,000	990	498	235	133	71	43	17	9	4
	(49.5%)	(24.9%)	(11.8%)	(6.7%)	(3.5%)	(2.1%)	(0.8%)	(0.5%)	(0.1%)
5,000	2,533	1,200	637	300	155	92	45	25	13
	(50.7%)	(24.0%)	(12.7%)	(6.0%)	(3.1%)	(1.8%)	(0.9%)	(0.5%)	(0.3%)
10,000	4,953	2,463	1,329	627	318	158	83	40	29
	(49.5%)	(24.6%)	(13.3%)	(6.3%)	(3.2%)	(1.6%)	(0.8%)	(0.4%)	(0.3%)

The results that follow from our definition should contradict neither logic nor experience.

With this in mind, we now state the definition of a probability function. We shall use the symbol $\mathbb{R}$ to denote the set of real numbers.

DEFINITION

Let S be a sample space and ξ a collection of events. A function $P : \xi \to \mathbb{R}$ is called a *probability function*, and $P(E)$ is used to denote the *probability of the event E*, if the following axioms hold:

1. $0 \leq P(E) \leq 1$ for all $E \in \xi$
2. $P(S) = 1$
3. If $E_1, E_2, \ldots$ is a finite or countable sequence of mutually exclusive events, then

$$P(E_1 \cup E_2 \cup \cdots) = P(E_1) + P(E_2) + \cdots$$

We postpone giving any examples until after we have seen how probability functions are constructed. The remainder of this section will be devoted to proving several basic results which follow almost immediately from the definition.

Proposition 2.1

$P(\varnothing) = 0$

Proof: Since $E \cap \varnothing = \varnothing$ for any event E, we have by (3) that

$P(E) = P(E \cup \varnothing) = P(E) + P(\varnothing)$

Canceling $P(E)$ from both sides of this equation yields $P(\varnothing) = 0$.

Proposition 2.2

$P(C(E)) = 1 - P(E)$

Proof: Since $S = E \cup C(E)$ and $E \cap C(E) = \varnothing$, we have by (2) and (3) that

$1 = P(S) = P(E \cup C(E)) = P(E) + P(C(E))$

Subtracting $P(E)$ from both sides of this equation yields the desired result.

Proposition 2.3

If $A \subseteq B$, then $P(A) \leq P(B)$.

Proof: Since $A \subseteq B$, we can write $B = A \cup (B - A)$, where $B - A$ is the relative complement of A in B. However, $B - A \subseteq C(A)$, so that

$A \cap (B - A) = \emptyset$. Hence, by (3), we have

$$P(B) = P(A \cup (B - A)) = P(A) + P(B - A)$$

Now, by (1), $P(B - A)$ is nonnegative, and since $P(A)$ plus this term add up to $P(B)$, $P(A)$ cannot exceed $P(B)$. Therefore, $P(A) \le P(B)$.

Proposition 2.4

$$P(A - B) = P(A) - P(A \cap B)$$

Proof: Since $A = (A - B) \cup (A \cap B)$ and $(A - B) \cap (A \cap B) = \emptyset$ (see Figure 2.1), we have by (3) that

$$P(A) = P((A - B) \cup (A \cap B)) = P(A - B) + P(A \cap B)$$

from which the desired result follows.

Proposition 2.5

$$P(A \cup B) = P(A) + P(B) - P(A \cap B)$$

Proof: Since $A \cup B = (A - B) \cup B$ and $(A - B) \cap B = \emptyset$ (see Figure 2.1 again), we have by (3) that

$$P(A \cup B) = P((A - B) \cup B) = P(A - B) + P(B)$$

The desired result then follows from Proposition 2.4.

Note the similarity between Proposition 2.5 and Proposition 1.1.

EXERCISES

1. Suppose that $A \subseteq B$ and $P(B) = 0$. Prove that $P(A) = 0$.
2. Suppose that $P(A) = 0$. Prove that $P(A \cap B) = 0$ and $P(A \cup B) = P(B)$.
3. Simulate on a computer the experiment of interviewing k people and recording their birthdays. How often do two or more people in the group have the same birthday? How often does at least one person in the group have the same birthday as you?

Figure 2.1

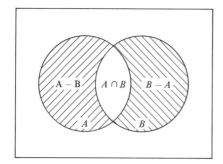

2.3 Finite Probability Spaces

A sample space S together with a probability function P is called a *probability space*. In this section, we consider only those probability spaces for which the sample space S is finite.

Suppose then that $S = \{x_1, x_2, \ldots, x_n\}$. We choose for our collection of events ξ the set of *all* subsets of S, a choice which clearly satisfies our definition for a class of events. How then are we to assign probabilities to the events in ξ? We start by assigning a probability p_k to each simple event $\{x_k\}$ in ξ. These probabilities are usually assigned on the basis of symmetries observed in the experiment or as the result of knowledge gained from repeated observation of the experiment. To be consistent with the first two conditions of our definition of a probability function, these probabilities must be assigned so that each $p_k \geq 0$, and, in addition, that $\sum_{i=1}^{n} p_i = 1$.

Since any event $E \in \xi$ is a union of simple events, we can define the probability of E to be the sum of the probabilities of the simple events belonging to E. That is,

$$P(E) = \sum_{p_k \in E} p_k$$

Example 1 Suppose that an urn contains ten red, six blue, five green, and three yellow balls. If our experiment consists of drawing one ball from the urn, then the obvious choice of a sample space is the set

$$S = \{R, B, G, Y\}$$

Since there are 24 balls in the urn, it would seem logical to assign probabilities as follows:

$p_1 = P(R) = \frac{10}{24} = \frac{5}{12}$

$p_2 = P(B) = \frac{6}{24} = \frac{1}{4}$

$p_3 = P(G) = \frac{5}{24}$

$p_4 = P(Y) = \frac{3}{24} = \frac{1}{8}$

If E is the event "red or blue," then $E = \{R, B\}$, and so $P(E) = P(R) + P(B) = \frac{10}{24} + \frac{6}{24} = \frac{16}{24} = \frac{2}{3}$. If F is the event "not yellow," then $F = \{R, B, G\}$, and so $P(F) = P(R) + P(B) + P(G) = \frac{10}{24} + \frac{6}{24} + \frac{5}{24} = \frac{21}{24} = \frac{7}{8}$.

It turns out that the process described above does, in fact, define a probability function on ξ. Since each simple probability is nonnegative, the probability of an event must always be nonnegative. Also, since the sum of *all* the simple probabilities is 1, we conclude that $P(S) = 1$ and that the probability of an event $E \subseteq S$ cannot exceed 1. Therefore, $0 \leq P(E) \leq 1$. Finally, if

$$A = \{x_{i_1}, x_{i_2}, \ldots, x_{i_k}\}$$

and

$$B = \{x_{j_1}, x_{j_2}, \ldots, x_{j_m}\}$$

and $A \cap B = \varnothing$, then

$$A \cup B = \{x_{i_1}, x_{i_2}, \ldots, x_{i_k}, x_{j_1}, x_{j_2}, \ldots, x_{j_m}\}$$

and, hence,

$$P(A \cup B) = \underbrace{p_{i_1} + p_{i_2} + \cdots + p_{i_k}}_{P(A)} + \underbrace{p_{j_1} + p_{j_2} + \cdots + p_{j_m}}_{P(B)}$$

so that $P(A \cup B) = P(A) + P(B)$. Condition (3) of the definition then follows by mathematical induction.

If the physical characteristics of the experiment suggest that the various outcomes in S all be assigned the same probability, then the probability space is called an *equiprobable* or *uniform* space. In a uniform sample space containing n elements, each simple event is assigned probability $1/n$. If E is an event in such a uniform space and $n(E) = k$, then

$$P(E) = \sum_{x_k \in E} p_k = k\left(\frac{1}{n}\right) = \frac{k}{n}$$

Thus, our axiomatic approach agrees with the "equally likely" approach of Laplace in the case of uniform spaces. Recall that Laplace assumed that all probability spaces were uniform.

We shall use the term "at random" in connection with uniform spaces. An outcome of an experiment is said to occur *at random* if all outcomes of the experiment are equally likely to occur.

What exactly do we mean when we say that the probability of an event is, say, $\frac{1}{3}$? Simply, that if we were to carry out the experiment right now, there would be one chance in three that the given event would occur. Or, if the experiment were performed many times, the given event would occur in approximately one-third of those trials.

Example 2 If one coin is tossed and we use the sample space $S = \{H, T\}$, then we can assign the probabilities $p_1 = P(H) = p$ and $p_2 = P(T) = 1 - p$ for any value $0 \le p \le 1$. Of course, we normally would choose $p = \frac{1}{2}$ unless we suspect that the coin is biased. Suppose, in fact, that we had observed the coin being tossed many times and noticed that tails seemed to come up about twice as often as heads. How would we assign the probabilities p_1 and p_2 now? Under the assumption that $P(T) = 2P(H)$ and $P(H) = p$, we have $P(T) = 2p$. Since $1 = P(H) + P(T) = p + 2p = 3p$, we find that $3p = 1$ or $p = \frac{1}{3}$. Therefore, $P(H) = \frac{1}{3}$ and $P(T) = \frac{2}{3}$.

Example 3 If two fair coins are tossed and we use the sample space $S = \{HH, HT, TH, TT\}$, then we can assume that we have a uniform space and assign the probabilities

$$P(HH) = P(HT) = P(TH) = P(TT) = \tfrac{1}{4}$$

That is, we define $p_i = \frac{1}{4}$, $i = 1, 2, 3, 4$. If E is the event "at least one head," then $E = \{HH, HT, TH\}$ and $P(E) = \frac{1}{4} + \frac{1}{4} + \frac{1}{4} = \frac{3}{4}$. The complementary event "no heads," therefore, has probability $1 - \frac{3}{4} = \frac{1}{4}$. Note that $P(E) + P(C(E)) = \frac{3}{4} + \frac{1}{4} = 1$. Now suppose that the first coin is biased so that it will never come up tails. How then should we assign the probabilities? Since the events TH and TT can never occur, we should let $P(TH) = P(TT) = 0$. If the second coin is unbiased, then the events HH and HT would seem to be equally likely, so we should set $P(HH) = P(HT) = \frac{1}{2}$. Under this assignment of probabilities, we find that $P(E) = \frac{1}{2} + \frac{1}{2} + 0 = 1$, although $E \neq S$, and $P(C(E)) = 0$, although $C(E) \neq \emptyset$. Thus, it is possible for events other than the trivial events $\emptyset$ and S to have probabilities 0 and 1.

Example 4 Suppose that two dice are rolled and we use the sample space $S = \{(a, b): a, b = 1, 2, 3, 4, 5, 6\}$. If we have no reason to suspect that either of the dice is loaded, then we can assume that S is a uniform space. Since $n(S) = 36$, we must assign the probability $p = \frac{1}{36}$ to each simple event. If E is the event "sum of the two faces is seven," then

$$E = \{(1, 6), (2, 5), (3, 4), (4, 3), (5, 2), (6, 1)\}$$

Since $n(E) = 6$, we find that $P(E) = \frac{6}{36} = \frac{1}{6}$

Example 5 A machine has produced 12 items, of which four are defective. We have purchased two of these items and wish to determine the probabilities of the following events:

E_1: Both items purchased are defective.
E_2: Neither item purchased is defective.
E_3: At least one of the items purchased is defective.

A logical sample space for this experiment would be the set S of all possible ways we might have chosen two of the 12 items. Since there are $C_2{}^{12} = 66$ different two-element samples possible, $n(S) = 66$. Since there is nothing to indicate that any one particular sample is more likely to occur than any of the others, we may assume that S is a uniform space. Now $n(E_1) = C_2{}^4 = 6$, the number of ways to choose two of the four defectives available, while $n(E_2) = C_2{}^8 = 28$, the number of ways to choose two of the eight non-defectives. Hence,

$$P(E_1) = \frac{n(E_1)}{n(S)} = \frac{6}{66} = \frac{1}{11}$$

and

$$P(E_2) = \frac{n(E_2)}{n(S)} = \frac{28}{66} = \frac{14}{33}$$

Finally, since $E_3 = C(E_2)$, we have

$$P(E_3) = 1 - P(E_2) = 1 - \frac{14}{33} = \frac{19}{33}$$

Example 6 Suppose that we have recorded the birthdays of k people (see Example 7 in Section 2.1), and we wish to determine the probability that at least two of these people have the same birthday. We have already seen that

$$S = \{(a_1, a_2, \ldots, a_k): 1 \le a_i \le 365, \quad i = 1, 2, \ldots, k\}$$

is a logical sample space for this experiment if we are willing to ignore people who were born on February 29. Since there is no reason to believe that any particular arrangement of the k birthdays should occur more or less frequently than any other, we may assume that S is uniform. (Again, we must ignore distortions caused by February 29.) Instead of calculating $P(E)$, where E is the event "at least two of the people sampled have the same birthday," we will calculate instead the probability $P(C(E))$, where $C(E)$ is the event "none of the people sampled have the same birthday." (It is always easier to calculate the probability of "none" rather than the probability of "at least one.") Since S is uniform, we have

$$P(C(E)) = \frac{n(C(E))}{n(S)}$$

so we must determine the number of ways we can record k birthdays with no repetitions. This is just an arrangement of k of the 365 days 1, 2, 3, ..., 365, so that

$$n(C(E)) = P_k^{365} = 365 \cdot 364 \cdots\cdots (365 - k + 1)$$

and hence,

$$P(C(E)) = \frac{365 \cdot 364 \cdots\cdots (365 - k + 1)}{(365)^k}$$

since, as we have seen in Section 2.1, $n(S) = (365)^k$. By Proposition 2.2, we then have that

$$P(E) = 1 - \frac{365 \cdot 364 \cdots\cdots (365 - k + 1)}{(365)^k}$$

Table 2.4 (opposite) lists the values of $P(E)$ for various values of k. Notice that in a group of only 23 people, there is a better than 50% chance that at least two of the people will have the same birthday.

Example 7 The preceding example is just one application of the following more general problem: Suppose that k balls are placed at random into n urns; what is the probability that each ball is in a different urn? In the preceding example, there were 365 urns, one for each day of the year, and k balls, one for each person interviewed. Each ball was placed into the urn corresponding to the person's birthday. In this more general situation, we choose as sample space the set

$$S = \{(a_1, a_2, \ldots, a_k): 1 \le a_i \le n, \quad i = 1, 2, \ldots, k\}$$

where a_i corresponds to the number of the urn into which the ith ball is placed. Since each ball may be placed into any of the n urns, we have $n(S) = n^k$. If E is the event "each ball in a different urn," then by the first principle of counting, we have

$$n(E) = n \cdot (n - 1) \cdot (n - 2) \cdots (n - k + 1)$$

Since the sample space again may be assumed to be uniform, we have

$$P(E) = \frac{n(E)}{n(S)} = \frac{n \cdot (n - 1) \cdots (n - k + 1)}{n^k}$$

$$= \frac{n!}{n^k(n - k)!}$$

Example 8 Suppose that an urn contains n balls, one of which is colored red while all the others are white. A sample of k balls is drawn at random from the urn, and we wish to find the probability that the red ball is included in the sample. The sample space for the experiment consists of all C_k^n possible samples of k balls that can be drawn from the urn, all of which we assume

Table 2.4
Probabilities That Two or More Among n People Will Have the Same Birthday

n	Probability	n	Probability
2	0.0027	21	0.4437
3	0.0082	22	0.4757
4	0.0164	23	0.5073
5	0.0271	24	0.5383
6	0.0405	25	0.5687
7	0.0562	26	0.5982
8	0.0743	27	0.6269
9	0.0946	28	0.6545
10	0.1170	29	0.6810
11	0.1411	30	0.7063
12	0.1670	40	0.8912
13	0.1944	50	0.9704
14	0.2231	60	0.9941
15	0.2592	70	0.9992
16	0.2836	80	0.9999
17	0.3150	90	1.0000
18	0.3469	100	1.0000
19	0.3791		
20	0.4114		

to be equally likely. If E is the event "red ball is included in the sample," then we need only count $n(E)$. By the first principle of counting, we find that

$$n(E) = C_1{}^1 \cdot C_{k-1}^{n-1}$$

since we must choose the red ball and $k - 1$ of the $n - 1$ white balls. Hence,

$$P(E) = \frac{C_1{}^1 \cdot C_{k-1}^{n-1}}{C_k^n} = \frac{\dfrac{(n-1)!}{(k-1)!(n-1-(k-1))!}}{\dfrac{n!}{k!(n-k)!}} = \frac{k}{n}$$

So $P(E) = k/n$, which is a logical answer since we must choose k balls, each of which seemingly has one chance in n of being the red ball.

Example 9 The preceding example is the model for a class of problems which can be called "inclusion of a specified item" problems. The following problem is a concrete application of the model. Suppose that a club has 20 members, only one of whom is female. A committee of five is to be chosen at random, and we wish to find the probability that the female member is included on the committee. Here, we are choosing $k = 5$ of the $n = 20$ balls in the urn, there being one ball for each member of the club. The red ball is the one representing the female member of the club. Hence, if E is the event "female member chosen on the committee," then

$$P(E) = \frac{k}{n} = \frac{5}{20} = \frac{1}{4}$$

Example 10 An integer is chosen at random from the first 200 positive integers. We wish to calculate the probability that the integer is divisible by either 6 or 8. Let A be the event "divisible by 6" and B the event "divisible by 8." We want to calculate the probability $P(A \cup B)$, and from Proposition 2.5, we know that $P(A \cup B) = P(A) + P(B) - P(A \cap B)$. Our sample space is $S = \{1, 2, 3, \ldots, 200\}$ and is assumed to be uniform. Hence, we must calculate $n(A)$, $n(B)$, and $n(A \cap B)$. But since $\frac{200}{6} = 33\frac{1}{3}$, $n(A) = 33$ (there being 33 integers less than 200 divisible by 6). Similarly, $n(B) = 25$ since $\frac{200}{8} = 25$. Thus, $P(A) = \frac{33}{200}$ and $P(B) = \frac{25}{200}$. Finally, since an integer is divisible by both 6 and by 8 if and only if it is divisible by 24, and since $\frac{200}{24} = 8\frac{1}{3}$, we have $n(A \cap B) = 8$, and, consequently, $P(A \cap B) = \frac{8}{200}$. Therefore,

$$P(A \cup B) = \tfrac{33}{200} + \tfrac{25}{200} - \tfrac{8}{200} = \tfrac{50}{200}$$

so that $P(A \cup B) = \frac{1}{4}$.

Example 11 Two balls are selected at random from an urn containing ten balls numbered $1, 2, \ldots, 10$. We wish to calculate the probability of the event E "sum of the two balls is odd." We shall consider three different ways in which the sample could have been drawn and calculate $P(E)$ in each case.

(a) Suppose first that the two balls are drawn together. Then there are $C_2{}^{10} = 45$ possible combinations of two balls, which we may take to be our sample space. In this case, E can be considered as equivalent to the event "one ball is odd, the other ball is even," and hence

$$n(E) = C_1{}^5 \cdot C_1{}^5 = 25$$

since there are five ways to choose one of the five odd balls and five ways to choose one of the five even balls. Therefore,

$$P(E) = \frac{n(E)}{n(S)} = \frac{25}{45} = \frac{5}{9}$$

(b) If the two balls are drawn without replacement, then the sample space would consist of the $10 \cdot 9 = 90$ possible ordered samples of two balls. The event E can be considered as a union $A \cup B$, in which

A: first ball is odd, second ball is even
B: first ball is even, second ball is odd

Now $n(A) = n(B) = 5 \cdot 5 = 25$, and since $A \cap B = \varnothing$, we have $n(A \cup B) = 50$, and hence

$$P(E) = P(A \cup B) = \frac{n(A \cup B)}{n(S)} = \frac{50}{90} = \frac{5}{9}$$

As we pointed out in Section 1.7, methods (a) and (b) produce essentially the same samples, as long as we do not consider order. Consequently, we should expect the answers to (a) and (b) to be identical.

(c) If the two balls are drawn with replacement, then, using the same S, A, and B as in (b), we find that $n(S) = 10 \cdot 10 = 100$ in this case, while both $n(A)$ and $n(B)$ remain 25. Therefore, $n(A \cup B) = 50$ and

$$P(E) = P(A \cup B) = \frac{n(A \cup B)}{n(S)} = \frac{50}{100} = \frac{1}{2}$$

this time.

EXERCISES

1. Determine the probabilities that the sum of two dice will take on each of the values 2, 3, 4, ..., 12. Compare your answers with the relative frequencies found in Table 2.2.
2. Three horses are entered in a race. Horse A is twice as likely to win as horse B, who is twice as likely to win as horse C. What is the probability that horse A wins the race?
3. What is the probability that a poker hand dealt to you will be
    (a) a full house;
    (b) a flush;
    (c) a straight;
    (d) two pairs;
    (e) three of a kind?

4. What is the probability that in a group of k people there will be at least one whose birthday is the same as yours? Write a computer program to calculate these probabilities for various values of k. How large must k be to insure that there is a 50% or better chance that there will be at least one person whose birthday is the same as yours?

5. From a term project of 24 problems, a professor decides to choose eight to be graded. If you have answer all but one of the problems correctly, what is the probability that you will receive a grade of 100% for the project?

6. Two students and two professors are arranged in a row for a panel discussion. What is the probability that the students and professors alternate in the row?

7. A boy has four coins in his pocket, a penny, a nickel, a dime, and a quarter. He takes two coins out, one after the other. What is the probability that he has more than 11¢ in his hand?

8. A ball is drawn from an urn containing 30 balls numbered 1, 2, ... , 30. The number is recorded and the ball is replaced. A second ball is drawn and its number recorded. What is the probability that the two balls did *not* have the same number?

9. Six married couples are standing in a room. If two people are chosen at random, find the probability that
   (a) they are married to each other;
   (b) one is male and the other is female.

10. The key to a locked door is one of 12 keys in a cabinet. If a person selects two keys at random from the cabinet and takes them to the door, what is the probability that he can open the door?

2.4 Countable Sample Spaces

We have already seen in Example 5 of Section 2.1 that sample spaces need not be finite, that they might, in fact, be countably infinite. If we toss a coin until a head appears, it is possible that the first head might appear on any particular toss, or, in fact, not at all. Thus, we were lead to the sample space $S = \{\infty, 1, 2, 3, \ldots\}$ where the simple event k occurs if the first head occurs on the kth toss, while the simple event ∞ represents the event that a head never occurs.

In the case of countable sample spaces, just as in the case of finite sample spaces, we can choose the set of *all* subsets of S as our set of events. We run into no set-theoretic or logical difficulties if we do so. In order to define a probability function on our countable sample space, we proceed as we did in the finite case.

Suppose that $S = \{x_1, x_2, x_3, \ldots\}$ is a countably infinite sample space. To each simple event $\{x_i\}$ in the sample space, we assign a nonnegative probability p_i with the restriction that $\sum_{i=1}^{\infty} p_i = 1$. If $E \subseteq S$ is an event, we define

$$P(E) = \sum_{x_i \in E} p_i$$

As in the finite case, the axioms for a probability function are satisfied.

Perhaps the expression $\sum_{i=1}^{\infty} p_i$ needs some clarification. It is an infinite sum which, by definition, is equal to the limit of a sequence of partial sums $s_n = \sum_{i=1}^{n} p_i$. In the case of our definition, we require that $\text{limit}_{n \to \infty} s_n = 1$.

One particular infinite sum that we shall use frequently is the geometric sum $\sum_{k=0}^{\infty} r^k$, where $0 < r < 1$. If we take the nth partial sum

$$s_n = 1 + r + r^2 + r^3 + \cdots + r^n$$

and multiply it by r, we get

$$rs_n = r + r^2 + r^3 + \cdots + r^{n+1}$$

Subtracting this term from s_n, we get

$$s_n - rs_n = 1 - r^{n+1}$$

Solving for s_n, we find

$$s_n = \frac{1 - r^{n+1}}{1 - r}$$

Since $0 < r < 1$, $\text{limit}_{n \to \infty} r^{n+1} = 0$, and so $\text{limit}_{n \to \infty} s_n = 1/(1 - r)$. Thus,

$$\sum_{k=0}^{\infty} r^k = \frac{1}{1 - r}$$

Subtracting the first term of the sum from the sum, we have

$$\sum_{k=1}^{\infty} r^k = \frac{1}{1 - r} - 1 = \frac{1 - (1 - r)}{1 - r} = \frac{r}{1 - r}$$

Therefore,

$$\sum_{k=1}^{\infty} r^k = \frac{r}{1 - r}$$

Differentiating the series above term by term with respect to r gives

$$\frac{d(1 + r + r^2 + \cdots)}{dr} = \frac{d\left(\frac{1}{1 - r}\right)}{dr}$$

or

$$1 + 2r + 3r^2 + \cdots = \frac{1}{(1 - r)^2}$$

We then multiply both sides of the equation by r to get

$$r + 2r^2 + 3r^3 + \cdots = \frac{r}{(1 - r)^2}$$

That is,

$$\sum_{k=1}^{\infty} kr^k = \frac{r}{(1 - r)^2}$$

Example 1 We now attempt to assign probabilities to the simple events in the sample space $S = \{\infty, 1, 2, 3, \ldots\}$. We let $p_0 = P(\{\infty\})$ and $p_k = P(\{k\})$, for $k = 1, 2, 3, \ldots$. Now the event $\{1\}$ means that the first head occurs on the first toss, so we should have $p_1 = \frac{1}{2}$. The event $\{2\}$ means that the first head occurs on the second toss, and, therefore, a tail occurs on the first toss. In the sample space $\{HH, HT, TH, TT\}$, the simple event TH has probability $\frac{1}{4}$, so it would seem logical to assign $p_2 = \frac{1}{4}$. Likewise, the event $\{3\}$ corresponds to the simple event TTH in the sample space

$$\{HHH, HHT, HTH, HTT, THH, THT, TTH, TTT\}$$

for three tosses of a coin, so we assign $p_3 = (\frac{1}{2})^3 = \frac{1}{8}$. The pattern is now obvious, so we assign $p_k = (\frac{1}{2})^k$ for any k. Finally, the logical value to assign p_0 is $\lim_{n \to \infty} (\frac{1}{2})^n = 0$, which seems appropriate since the event ∞ seems so unlikely. Since

$$\sum_{i=0}^{\infty} p_i = \sum_{i=0}^{\infty} \left(\frac{1}{2}\right)^i = \frac{\frac{1}{2}}{1 - \frac{1}{2}} = 1$$

we have a valid assignment of probabilities. Suppose now that we wish to calculate the probability of the event E "first head occurs on an even-numbered toss." This event corresponds to the subset $\{2, 4, 6, 8, \ldots\}$ of S, so by definition,

$$P(E) = \left(\frac{1}{2}\right)^2 + \left(\frac{1}{2}\right)^4 + \left(\frac{1}{2}\right)^6 + \cdots$$

that is

$$P(E) = \sum_{k=1}^{\infty} \left(\frac{1}{4}\right)^k = \frac{\frac{1}{4}}{1 - \frac{1}{4}} = \frac{1}{3}$$

As so often happens in probability, this is not the answer that intuition might suggest—$P(E) = \frac{1}{2}$. Can you see the intuitive argument, and why it fails?

Example 2 Suppose that a fair die is rolled until it shows a 6, and we wish to calculate the probability of the event E "more than six tosses are required before the first 6 occurs." We set up a sample space $S = \{\infty, 1, 2, 3, \ldots\}$, as in Example 1, and assign probabilities as follows: $p_1 = \frac{1}{6}$, $p_2 = \frac{5}{6} \cdot \frac{1}{6}$, $p_3 = \frac{5}{6} \cdot \frac{5}{6} \cdot \frac{1}{6}$, and so forth, and, in general, $p_n = (\frac{5}{6})^{n-1} \cdot \frac{1}{6}$, and finally, $p_0 = 0$. This assignment of probabilities could be justified as in Example 1, although it would be more tedious, or they could be justified as follows: For the event $\{n\}$ to occur, we must get anything but a 6 on the first $n - 1$ rolls and then a 6 on the nth roll. On any of the first $n - 1$ rolls, the probability of *not* getting a 6 is $\frac{5}{6}$, while the probability of getting a 6 on the nth roll is $\frac{1}{6}$. We have defined p_n to be the product of these n probabilities. (This will be justified in Chapter 3.) Now that we have assigned what seems to be a logical set of probabilities, we must check that their sum is 1. We have

$$\sum_{k=0}^{\infty} p_k = \sum_{k=1}^{\infty} \left(\frac{5}{6}\right)^{k-1} \cdot \frac{1}{6} \qquad (p_0 = 0)$$

$$= \frac{1}{6} \sum_{k=1}^{\infty} \left(\frac{5}{6}\right)^{k-1}$$

$$= \frac{1}{6} \sum_{k=0}^{\infty} \left(\frac{5}{6}\right)^{k}$$

$$= \frac{1}{6} \left(\frac{1}{1 - \frac{5}{6}}\right) = 1$$

The event E corresponds to the subset $\{7, 8, 9, \ldots\}$ of S, and so

$$P(E) = \sum_{k=7}^{\infty} p_k = \sum_{k=7}^{\infty} \left(\frac{5}{6}\right)^{k-1} \cdot \frac{1}{6}$$

$$= \frac{1}{6} \sum_{k=6}^{\infty} \left(\frac{5}{6}\right)^{k}$$

$$= \frac{1}{6} \cdot \left(\frac{5}{6}\right)^{6} \cdot \sum_{k=0}^{\infty} \left(\frac{5}{6}\right)^{k}$$

$$= \frac{1}{6} \cdot \left(\frac{5}{6}\right)^{6} \cdot \frac{1}{1 - \frac{5}{6}}$$

$$= \left(\frac{5}{6}\right)^{6}$$

so that $P(E) = (\frac{5}{6})^6$. In general, if F is the event "first head occurs after the mth toss," then $P(F) = (\frac{5}{6})^m$.

Example 3 Two people of equal ability play a sequence of games until one of them wins two games in succession. We wish to calculate the probability of the event E "an odd number of games are required." If we denote the two players by A and B, then the set of all possible outcomes of their match can be represented

AA	BB
BAA	ABB
ABAA	BABB
BABAA	ABABB
$\vdots$	$\vdots$

We assign probability $(\frac{1}{2})^n$ to both sequences of n games, $n = 2, 3, 4, \ldots$, since each is a simple event in the uniform space of 2^n outcomes of n games. Note that psychological factors are ignored and each player is assumed to have probability $\frac{1}{2}$ of winning each game. We choose as sample space the set $S = \{\infty, 2, 3, 4, \ldots\}$, where the simple event $\{k\}$ indicates that k games were required, while the simple event $\{\infty\}$ represents the logically conceivable event that neither player ever wins two consecutive games. The occurrence of the simple event $\{k\}$ means that one of the two sequences of length k occurred, so we define

$$p_k = P(\{k\}) = \left(\frac{1}{2}\right)^k + \left(\frac{1}{2}\right)^k = \left(\frac{1}{2}\right)^{k-1}$$

As usual, we let $p_0 = P(\{\infty\}) = 0$. Therefore,

$$p_0 + \sum_{k=2}^{\infty} p_k = \sum_{k=2}^{\infty} \left(\frac{1}{2}\right)^{k-1} = \sum_{k=1}^{\infty} \left(\frac{1}{2}\right)^k = \frac{\frac{1}{2}}{1 - \frac{1}{2}} = 1$$

so we have an acceptable assignment of probabilities. Since the event E corresponds to the subset $\{3, 5, 7, \ldots\}$ of S, we have

$$P(E) = p_3 + p_5 + p_7 + \cdots$$
$$= \left(\frac{1}{2}\right)^2 + \left(\frac{1}{2}\right)^4 + \left(\frac{1}{2}\right)^6 + \cdots$$
$$= \sum_{k=1}^{\infty} \left(\frac{1}{4}\right)^k$$
$$= \frac{\frac{1}{4}}{1 - \frac{1}{4}} = \frac{1}{3}$$

EXERCISES

1. Two people toss a coin until one of them gets a head. What is the probability that the player making the first toss wins the game?
2. Cards are drawn with replacement from a full deck of 52 cards until a heart is drawn. What is the probability that the first heart is drawn after the fourth card?
3. Two players draw cards with replacement from a full deck until one of them draws an ace. What is the probability that the first player to draw wins?
4. Two fair dice are rolled until sum 12 occurs. What is the probability that at least ten rolls will be required?
5. A roulette wheel (containing 36 numbers 1, 2, 3, ..., 36 and two house numbers 0 and 00) is spun until the number 7 comes up. What is the probability that the 7 will come up in the first ten spins?

2.5 Uncountable Sample Spaces

Many random experiments give rise to a sample space which is neither finite nor countable but rather is an uncountable infinite set. In such a case, it is desirable to characterize the sample space geometrically. The outcomes of the experiment can usually be taken to occur uniformly throughout this geometric configuration, and, as a result, probabilities of events can be defined as the proportion of the configuration corresponding to the event. This proportion is expressed in terms of a geometrical measurement such as length, area, or volume, depending on the dimension of the sample space.

In the case of uncountable sample spaces, care must be taken in choosing a set of events. We shall say no more than that it is always possible to choose such a set satisfying the axioms of Section 2.1.

In the uncountable case, it is not necessary to assign values p_i as the probabilities of the simple events of the sample space. For one reason, we would have no method, such as summing an infinite series, to add all of these probabilities to verify that $P(S) = 1$ or to add some of these probabilities to calculate the probability of an event. Another reason, as we shall see in an example later, is that the values of these probabilities would always be zero anyhow.

So our approach to calculating probabilities in the uncountable case will be completely different from the familiar methods of the finite and countable cases.

Example 1 Suppose that an arrow is shot at a circular target, and we wish to calculate the probability that it lands closer to the center of the target than to the circumference of the target. We assume the archer is unskilled and is no more likely to hit any one point on the target than any other. If we assume that the target has radius r, then our sample space, which consists of all the points on the target, can be thought of as a circle of radius r. The area of the sample space is therefore πr^2. The event E "closer to the center than to the circumference" is represented by the circle concentric with the sample circle and having radius half as large, as depicted in Figure 2.2. Hence,

Figure 2.2

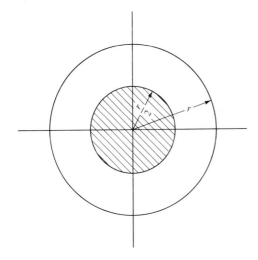

the area of E is $\pi(r/2)^2 = \pi r^2/4$. The proportion of the area of the sample space represented by E is then $(\pi r^2/4)/\pi r^2 = \frac{1}{4}$. We, therefore, say that $P(E) = \frac{1}{4}$.

Example 2 The 7:30 train to New York departs at times uniformly distributed between 7:25 and 7:40. Mr. Lindlay lives five minutes from the railroad station and leaves home at 7:30. It should seem obvious that the probability that he catches the train is $\frac{1}{3}$. As a sample space, we have the interval [7:25, 7:40] of possible departure times, which by assumption is uniform. The event E "Mr. Lindlay catches the train" corresponds to the subinterval [7:35, 7:40] whose length is five-fifteenths, or one-third, the length of the sample interval. Hence, $P(E) = \frac{1}{3}$.

Example 3 On the real line $\mathbb{R}$, two points a and b are chosen at random such that $0 \le a \le 3$ and $-2 \le b \le 0$. We wish to find the probability that the distance between a and b is greater than 3. This problem is more difficult than the preceding two because the sample space is not readily apparent. Although the problem seems to be phrased in terms of one dimension, it is actually easier to solve in terms of two dimensions. For a sample space, we choose the rectangle

$$S = \{(x, y): 0 \le x \le 3 \quad \text{and} \quad -2 \le y \le 0\}$$

Choosing the points a and b as specified is equivalent to choosing a point (a, b) in S. If d represents the difference between a and b, then the event $E = \{d > 3\}$ consists of those points in the rectangle below the line $x - y = 3$, as depicted in Figure 2.3. Since the area of the sample space is $3 \cdot 2 = 6$, while the area corresponding to the event E is $\frac{1}{2} \cdot 2 \cdot 2 = 2$, the proportion

Figure 2.3

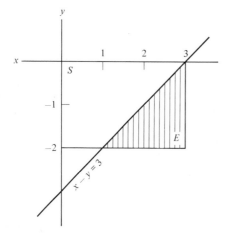

of the area of the sample space corresponding to E is $\frac{2}{6} = \frac{1}{3}$, and hence $P(E) = \frac{1}{3}$.

Example 4 A stick of length L is broken at a random point. This is equivalent to choosing a point at random in the sample interval $[0, L]$. If $0 \leq a < b \leq L$ then the break occurs between a and b with probability $(b - a)/L$, the proportion of the sample interval occurring between a and b. What is the probability that the break will occur at some specific point c? Since $\{c\} \subseteq (c - 1/n, c + 1/n)$ for any value n and presuming that the interval $(c - 1/n, c + 1/n)$ is an event, we find by Proposition 2.3 that

$$P(\{c\}) \leq P\left(c - \frac{1}{n}, c + \frac{1}{n}\right) = \frac{2/n}{L} = \frac{2}{nL}$$

Thus, $P(\{c\}) \leq 2/nL$ for *any* value of n, and since $\lim_{n \to \infty} (2/nL) = 0$, we must have $P(\{c\}) = 0$. Thus, although it is possible for the break to occur at this point, the probability of that happening is 0.

EXERCISES

1. A point is picked at random inside a square. What is the probability that it lies inside the inscribed circle which is concentric with the square?
2. A point is picked at random inside a circle of radius 1. What is the probability that the point does not lie within a square of diagonal 2 with the same center as the circle?
3. What is the probability that a point chosen at random in a right circular cone is in the upper half of the cone?
4. If the coefficient a in the quadratic $x^2 + ax + 4$ is chosen at random in the interval $[0, 6]$, what is the probability that the quadratic has real roots?
5. Three points are selected at random on the circumference of a circle. What is the probability that the points lie on a semicircle?

ADDITIONAL PROBLEMS

1. In a row of ten trees, four adjacent trees are diseased. Should one suspect that the disease is spreading from tree to tree?
2. Suppose that you hold the following poker hand: the Jack of clubs, King of spades, and the seven, ten, and King of hearts. Which of the following three strategies would be the wisest choice:
   (a) hold the two kings and try for three (or more) kings;
   (b) hold the three hearts and try for a flush;
   (c) discard the seven of hearts and one of the kings and try for a straight?
3. A high school senior applies for admission to college A and college B. If his probability of being admitted to A is 0.7, his probability of being rejected at B is 0.5, and the probability of at least one of his applications being rejected is 0.6, what is the probability that he will be admitted to at least one of the colleges?
4. Find the probability that in a bridge hand each player has exactly one ace.
5. People passing a street corner are questioned with regard to their marital status and ownership of a car. If it is assumed the probability that a person is neither married nor owns a car is $\frac{1}{4}$, and the probability that a person either owns a car or

is married, but not both, is $\frac{1}{3}$, what is the probability that the next person questioned is married and owns a car? Is it possible to determine the probability that this person is married?

6. A waiting room has six seats arranged in a row. Suppose that three people enter the room and choose their seats at random. What is the probability that they sit with no empty seats between them? That there is exactly one empty seat between any two of them?

7. One card is drawn from each of two ordinary decks of 52 cards. What is the probability that at least one of the two cards is the ace of spades?

8. A three-digit number is formed from the digits 1, 2, 3, . . . , 9, 0. What is the probability that there is a nine in either the first or last position of the number but not both?

9. A game warden inspects a catch of ten fish by examining two which he selects at random. What is the probability that a fisherman will be arrested with a catch of ten that includes three undersized fish?

10. Ten men drive their cars to the city each day and park in one of three lots. If they park in lots chosen at random, what is the probability that there will be four cars in lot A, four cars in lot B, and two cars in lot C?

11. A political meeting consists of six people from each of 52 countries. Five people are selected at random to serve on a committee. What is the probability that they are from five different countries?

12. A magazine prints the photographs of four movie stars and, in scrambled order, a baby picture of each. What is the probability that a reader guessing at random matches at least two of the pictures?

13. A group of $n > 2$ people sit at round table. What is the probability that a certain two people will sit next to each other?

14. The letters of the word MISSISSIPPI are arranged in some order. What is the probability that the four S's are consecutive?

Conditional Probability

3.1 Definition and the Multiplication Theorem

If you were to make a bet on the outcome of the World Series, on whom you bet and how much you bet could very well be different if you were to place your bet *after* the first game of the series had been played, rather than before the series began. Knowledge of what happened in the first game would almost certainly affect your appraisal of who was likely to win the series.

Similarly, in an arbitrary probability space, the fact that a certain event E has already occurred will most likely affect the chances of another event A occurring. For example, if we were to roll two fair dice, we would figure that we had one chance in nine of rolling sum 9 because exactly 4 of the 36 possible outcomes have sum 9. Now suppose that we had seen one of the two dice. If we had seen a 1 or a 2, then we would be certain that the sum of the two dice could not possibly be 9. However, suppose that we had seen a 6. How should we revise our estimate of the likelihood of getting sum 9? If one of the dice was a 6, then only the following 11 elements of the original sample space remain relevant:

$$\begin{Bmatrix} (1,6) & (2,6) & (3,6) & (4,6) & (5,6) & (6,6) \\ & (6,1) & (6,2) & (6,3) & (6,4) & (6,5) \end{Bmatrix}$$

We know that the other 25 elements did *not* occur. Of the 11 elements that might have occurred, only two, $(6,3)$ and $(3,6)$, have sum 9. Therefore, we can now estimate our chances of obtaining sum 9 as 2 in 11, which is somewhat higher than 1 in 9.

In order to further motivate a formal definition for this type of situation, we suppose that we have a uniform space S and proceed as we did in the example above. Knowledge that an event E has occurred restricts all future considerations to a new "conditional sample space," the set E itself. The elements of $C(E)$ have not occurred, and so are irrelevant. If we wish to find the probability that an event A will occur, we would have to determine the proportion of the elements in E which are favorable to A occurring. Since the sample space is uniform, this proportion equals $n(E \cap A)/n(E)$.

If the original sample space were not uniform, we would not be able to use the above proportion. However, assuming that $P(E) \neq 0$, we can assign revised probabilities to the elements of the conditional sample space E. We do this in such a manner that the relative magnitudes of the probabilities of the simple events in E are not altered. If $x \in E$, we let

$$P'(\{x\}) = \frac{P(\{x\})}{P(E)}$$

Now P' is a probability function on E because

$$\sum_{x \in E} P'(\{x\}) = \sum_{x \in E} \frac{P(\{x\})}{P(E)}$$

$$= \frac{1}{P(E)} \sum_{x \in E} P(\{x\})$$

$$= \frac{1}{P(E)} \cdot P(E) = 1$$

As a result of our definition of P', we have

$$P'(A) = \sum_{x \in A \cap E} P'(\{x\})$$

$$= \sum_{x \in A \cap E} \frac{P(\{x\})}{P(E)}$$

$$= \frac{1}{P(E)} \sum_{x \in A \cap E} P(\{x\})$$

$$= \frac{P(A \cap E)}{P(E)}$$

We are now ready to state our definition of conditional probability.

DEFINITION

Let S be a probability space with probability function P, and let E be an event such that $P(E) \neq 0$. The probability that an event A occurs given that an event E has occurred is called the *conditional probability of A given E*, denoted $P(A|E)$, and defined as $P(A \cap E)/P(E)$.

The restriction that $P(E) \neq 0$ is needed so that we do not divide by zero. Notice that if S is a uniform space, then

$$P(A \cap E) = \frac{n(A \cap E)}{n(S)} \quad \text{and} \quad P(E) = \frac{n(E)}{n(S)}$$

so that

$$P(A|E) = \frac{P(A \cap E)}{P(E)} = \frac{n(A \cap E)}{n(E)}$$

agreeing with the conclusion we reached above.

Example 1 Suppose that two dice are tossed and we are told that their sum is 6. We wish to determine the conditional probability that one of the dice is a 2. If we let A denote this event, then we have

$$A = \begin{Bmatrix} (1, 2) & (2, 2) & (3, 2) & (4, 2) & (5, 2) & (6, 2) \\ & (2, 1) & (2, 3) & (2, 4) & (2, 5) & (2, 6) \end{Bmatrix}$$

so that the unconditional probability $P(A)$ of the event is $\frac{11}{36}$. If E denotes the event "sum is 6," then

$$E = \{(1, 5), (2, 4), (3, 3), (4, 2), (5, 1)\}$$

and so $P(E) = \frac{5}{36}$. Since $A \cap E = \{(2, 4), (4, 2)\}$, we have $P(A \cap E) = \frac{2}{36}$, and so

$$P(A|E) = \frac{P(A \cap E)}{P(E)} = \frac{\frac{2}{36}}{\frac{5}{36}} = \frac{2}{5}$$

Note that in the uniform conditional sample space E (S itself is uniform, so the subset E is also uniform), exactly two of the five elements have sum 6.

Example 2 Three fair coins are tossed, and we wish to calculate the probability of the event A "all are heads." Suppose that we have seen the first coin and know that it is a head. Let E_1 denote the event "first coin is a head." Then

$$E_1 = \{HHH, HHT, HTH, HTT\}$$

Since the original sample space (the usual one with eight elements) is uniform and since $A \cap E_1 = \{HHH\}$, we have $P(A|E_1) = \frac{1}{4}$. Surprisingly enough, if we had noticed only that one of the coins was a head and not necessarily that it was the first coin, the conditional probability of A is no longer $\frac{1}{4}$. For if we denote by E_2 the event "at least one head," then

$$E_2 = S - \{TTT\}$$

and $A \cap E_2 = \{HHH\}$ again, so that $P(A|E_2) = \frac{1}{7}$.

Example 3 As we are dealt a five-card poker hand, we pick up our cards one at a time. The first three cards we pick up are of the same type, and we are curious about our chances now of drawing a full house. We choose as our conditional sample space E the set of all hands containing three cards of the same type. Using the techniques of Section 1.5, we find that

$$n(E) = C_1{}^{13}C_3{}^4C_2{}^{49}$$

since we must choose

1. 1 of the 13 types;
2. 3 of the 4 cards of this type;
3. 2 of the remaining 49 cards.

Now, of these particular five-card hands, a total of

$$C_1{}^{13} \cdot C_3{}^4 \cdot C_1{}^{12} \cdot C_2{}^4$$

are full houses, since, instead of (3) above, we must choose

3. 1 of the 12 other types;
4. 2 of the 4 cards of this type.

As a result, if A denotes the event "full house," then

$$P(A|E) = \frac{C_1{}^{13} \cdot C_3{}^4 \cdot C_1{}^{12} \cdot C_2{}^4}{C_1{}^{13} \cdot C_3{}^4 \cdot C_2{}^{49}} = \frac{12 \cdot 6}{49 \cdot 24} = \frac{3}{49}$$

If we solve the equation defining the conditional probability $P(A|E)$ for $P(A \cap E)$, we get

$$P(A \cap E) = P(E)P(A|E)$$

Using mathematical induction, we can extend this to the following:

Theorem 3.1 The Multiplication Theorem

If $A_1, A_2, A_3, \ldots, A_n$ are events such that $P(\bigcap_{i=1}^{n-1} A_i) \neq 0$, then

$$P(A_1 \cap A_2 \cap \cdots \cap A_n) = P(A_1) \cdot P(A_2|A_1) \cdots P\left(A_n \left| \bigcap_{i=1}^{n-1} A_i\right.\right)$$

Proof: The case $n = 2$ follows from the definition; for the case $n = 3$, we have

$$P(A_1 \cap A_2 \cap A_3) = P(A_1 \cap (A_2 \cap A_3))$$
$$= P(A_1)P(A_2 \cap A_3|A_1)$$
$$= P(A_1)P(A_2|A_1)P(A_3|A_1 \cap A_2)$$

(Exercise 1). Assuming the result to be true for $n = k$, we have

$$P(A_1 \cap A_2 \cap \cdots \cap A_{k+1}) = P[A_1 \cap (A_2 \cap \cdots \cap A_{k+1})]$$
$$= P(A_1)P(A_2 \cap \cdots \cap A_{k+1}|A_1)$$
$$= P(A_1)P(A_2|A_1)P(A_3|A_1 \cap A_2) \cdots$$
$$P(A_{k+1}|A_1 \cap A_2 \cap \cdots \cap A_k)$$

hence, the result is true for $n = k + 1$.

The multiplication theorem is usually used to calculate the probability that *all* of a sequence of events occur. Event A_1 is presumed to be the first to occur (or not occur), followed in order by $A_2, A_3, \ldots, A_n$. As a result, conditional probabilities such as $P(A_k|A_1 \cap A_2 \cap \cdots \cap A_{k-1})$ make sense.

Example 4 A lot contains 12 items, of which 4 are defective. Three items are drawn at random from the lot, without replacement, and we wish to calculate the probability that all three are nondefective. If we let A_i denote the event "ith item is not defective," for $i = 1, 2, 3$, then we must find $P(A_1 \cap A_2 \cap A_3)$. By the multiplication theorem, we have

$$P(A_1 \cap A_2 \cap A_3) = P(A_1)P(A_2|A_1)P(A_3|A_1 \cap A_2)$$

(Note that the events A_1, A_2, and A_3 are arranged in the order of their occurrence. Item 1 is chosen before item 2, for example, so the probability $P(A_2|A_1)$ makes sense.) Now, $P(A_1) = \frac{8}{12}$, since 8 of the 12 items are nondefective. Given that A_1 has occurred, that is, that the first item chosen is nondefective, 7 of the remaining 11 items are nondefective. Therefore, $P(A_2|A_1) = \frac{7}{11}$. Similarly, the occurrence of $A_1 \cap A_2$ means that two nondefective items have already been chosen, leaving 6 nondefectives among the 10 remaining items so that $P(A_3|A_1 \cap A_2) = \frac{6}{10}$. Thus

$$P(A_1 \cap A_2 \cap A_3) = \left(\frac{8}{12}\right)\left(\frac{7}{11}\right)\left(\frac{6}{10}\right) = \frac{14}{55}$$

Example 5 Students in a summer school program took two courses, chemistry and history. Of these students, four percent failed chemistry, three percent failed history, and one percent failed both. We wish to determine what percentage passed chemistry and failed history. We let A be the event "passed chemistry" and B the event "passed history." Then we have $P(A) = 0.96$, $P(B) = 0.97$, and $P[C(A) \cap C(B)] = 0.01$. By applying De Morgan's law, $P(A \cup B) = 0.99$, then by Proposition 2.5,

$$P(A \cap B) = P(A) + P(B) - P(A \cup B)$$
$$= 0.96 + 0.97 - 0.99 = 0.94$$

Therefore,

$$P(B \mid A) = \frac{P(A \cap B)}{P(A)} = \frac{0.94}{0.96} = \frac{94}{96}$$

and so $P[C(B) \mid A] = \frac{2}{96}$. Hence,

$$P[A \cap C(B)] = P(A) \cdot P[C(B) \mid A]$$
$$= \frac{96}{100} \cdot \frac{2}{96} = 0.02$$

EXERCISES

1. Suppose that A, B, and C are events such that $P(C) \neq 0$ and $P(A \cap C) \neq 0$. Prove that $P(A \cap B|C) = P(A|C)P(B|A \cap C)$.
2. Suppose that A is an event such that $P(A) \neq 0$. Prove that $P[C(E)|A] = 1 - P(E|A)$.
3. Suppose that E is an event such that $P(E) \neq 0$. Prove that $P(A \cup B|E) = P(A|E) + P(B|E) - P(A \cap B|E)$.
4. Assume that the probability that a woman is a college graduate is 0.20, the probability that a woman is married is 0.35, and the probability that a female college graduate is married is 0.43. Suppose that a woman is chosen at random and found to be married. What are the chances that she is a college graduate?
5. Candidates A, B, and C have probabilities 0.5, 0.3, and 0.2, respectively, of winning an election. Just before the election, C withdraws. What are the probabilities now that A or B will win the election?
6. Thirty percent of the students on campus have cars and 25 percent have bikes. Of those with bikes, 90 percent don't have cars. What is the probability that a student has a car but not a bike?
7. A club consists of 50 couples. Six people are selected at random. Find the probability that 3 couples are selected given that 1 couple was selected.
8. A point (a, b) is chosen at random in the rectangle $0 \leq x \leq 2$, $0 \leq y \leq 1$. What is the probability that a < b, given that $a^2 + b^2 < 1$?

3.2 The Partition Theorem

We are all familiar with the ability of the television networks to predict the winners of political elections after only a small percentage of the vote has been reported. Perhaps you have wondered how they are able to make accurate predictions so quickly on the basis of what appears to be very little information. Quite similar in nature are the straw polls taken before an election and used to predict the outcome of the election. Both processes work in essentially the same manner. The voting public is divided into several disjoint categories. Most frequently these categories are geographic, although they may be chosen along lines that are ethnic or religious, for example. Election results for small, carefully selected groups from each category are then evaluated, and a prediction is made as soon as the network can be reasonably certain that its prediction will be accurate.

We shall explain this process in more detail later in this section. Now, we discuss the mathematical justification for this process, the partition theorem (also known as the theorem of total probability). We first define what we mean by a partition.

DEFINITION

A *partition* of a set S is a finite collection $A_1, A_2, \ldots, A_n$ of subsets of S which satisfy the following two conditions:

1. $S = A_1 \cup A_2 \cup \cdots \cup A_n$
2. $A_i \cap A_j = \varnothing$ for any choice of i and j.

Example 1 Given a set S, any subset A and its complement $C(A)$ partition S, since $S = A \cup C(A)$ and $A \cap C(A) = \varnothing$.

Example 2 The set S of all undergraduates at a college can be partitioned by year. If

$A_1 =$ freshmen

$A_2 =$ sophomores

$A_3 =$ juniors

$A_4 =$ seniors

then $S = A_1 \cup A_2 \cup A_3 \cup A_4$ and $A_i \cap A_j = \varnothing$ for any $i, j = 1, 2, 3, 4$.

Example 3 The 52 cards in the deck can be partitioned by color ($A_1 =$ red, $A_2 =$ black), by suit ($A_1 =$ hearts, $A_2 =$ clubs, $A_3 =$ diamonds, and $A_4 =$ spades), or by type ($A_1 =$ aces, $A_2 =$ deuces, $\ldots$, $A_{12} =$ queens, and $A_{13} =$ kings).

Theorem 3.2 The Partition Theorem

Suppose that $A_1, A_2, \ldots, A_n$ is a partition of a sample space S such that $P(A_i) \neq 0$ for any $i = 1, n$. If E is any event, then

$$P(E) = \sum_{i=1}^{n} P(A_i) \cdot P(E|A_i)$$

Proof: Since $A_1, A_2, \ldots, A_n$ partitions S, $E \cap A_1$, $E \cap A_2, \ldots, E \cap A_n$ partitions E (why?). Therefore, by the definition of a probability function,

$$P(E) = \sum_{i=1}^{n} P(E \cap A_i)$$

$$= \sum_{i=1}^{n} P(A_i) \cdot P(E \mid A_i)$$

with the latter following from the multiplication theorem.

In Figure 3.1, we have depicted a plot of land which has been divided into five subplots of unequal area denoted by A_1, A_2, A_3, A_4, and A_5. Each of the subplots contains part of a lake which lies on the land (shaded area in the picture). The owner of the land wishes to know what percentage of the plot is occupied by the lake, so he asks each of the individual tenants to determine the percentage of their land that is occupied by the lake. But since the five subplots have different areas, the owner cannot simply average the five percentages obtained from his tenants. Rather, he must weight these percentages by the percentage of the entire plot occupied by each tenant's subplot. If we let $P(A_i)$ denote the percentage of the entire plot occupied by subplot A_i and $P(L_i)$ the percentage of subplot A_i covered by the lake, then the percentage $P(L)$ of the entire plot that is occupied by the lake is

$$P(L) = P(A_1) \cdot P(L_1) + P(A_2) \cdot P(L_2) + \cdots + P(A_5) \cdot P(L_5)$$

Figure 3.1

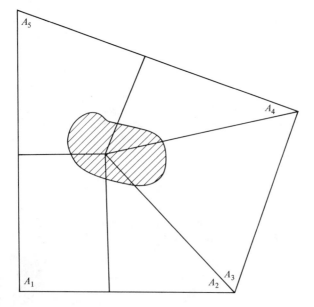

Actually, this expression is just the partition theorem, and is a good indication of what the partition theorem really means. The weights $P(A_i)$ are actually just the probabilities of the partition parts, while the percentages $P(L_i)$ are actually conditional probabilities $P(L|A_i)$.

In general, the probability of an event can be calculated by finding the conditional probabilities of the event relative to the parts of some partition and then taking a weighted average of these probabilities, the weights being the probabilities of the parts of the partition.

Note the alternate form of the partition theorem:

$$P(E) = \sum_{i=1}^{n} P(E \cap A_i)$$

With respect to the area of the lake in our example, this is saying that we need only determine the percentage of the entire plot occupied by the lake in each subplot $(L \cap A_i)$ and add.

The partition theorem can be better understood by means of a tree diagram. A partition having n parts gives rise to a tree with $2n$ branches (Figure 3.2). Of the $2n$ branches, exactly n correspond to the event E. The partition theorem says that $P(E)$ is the sum of the probabilities of these n branches; the multiplication theorem tells us how to calculate the probabilities of the branches—by multiplying the two probabilities along the

Figure 3.2

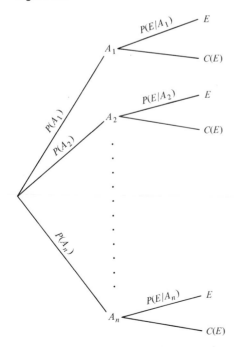

branch, the first being the probability of the partition part and the second being the conditional probability of the event relative to the partition part. These probabilities are usually indicated along the branches of the tree, as indicated in Figure 3.2.

Example 4 Suppose that the members of a college fraternity are distributed as follows: 30 percent are freshmen, 25 percent are sophomores, 25 percent are juniors, and 20 percent are seniors. If we are told that 50 percent of the freshmen, 30 percent of the sophomores, 10 percent of the juniors, and 5 percent of the seniors in the fraternity are enrolled in a math course, how would we determine the probability that a fraternity member chosen at random is enrolled in a math course? Due to the unequal distribution of the fraternity members by years, we could not simply average the four percentages 50, 30, 10, and 5, but rather must take a weighted average. Once the need for a weighted average is recognized, the partition theorem should suggest itself. The partition is usually suggested by the wording of the problem, and once a partition has been settled upon, a tree diagram should be drawn. In this problem, the following partition is suggested:

A_1 = the freshmen in the fraternity

A_2 = the sophomores in the fraternity

A_3 = the juniors in the fraternity

A_4 = the seniors in the fraternity

and we are given that $P(A_1) = 0.30$, $P(A_2) = 0.25$, $P(A_3) = 0.25$, and $P(A_4) = 0.20$. If E is the event "fraternity member is enrolled in a math course," then

Figure 3.3

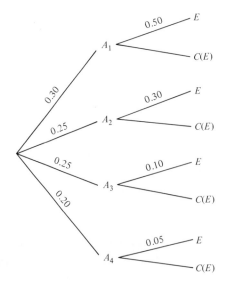

we are also given that $P(E|A_1) = 0.50$, $P(E|A_2) = 0.30$, $P(E|A_3) = 0.10$, and $P(E|A_4) = 0.05$. We therefore have the tree diagram in Figure 3.3. By the partition theorem, we have

$$P(E) = 0.30 \cdot 0.50 + 0.25 \cdot 0.30 + 0.25 \cdot 0.10 + 0.20 \cdot 0.05 = 0.254$$

thus, $P(E) = 0.254$. Notice that the four terms in the sum for $P(E)$ are the probabilities along the four branches of the tree that lead to E.

Example 5 Three guns, each containing five chambers, are placed in a box. The first gun has four empty chambers, the second gun three empty chambers, and the third gun two empty chambers. A man is blindfolded, forced to choose a gun from the box, place it to his head, and pull the trigger. What are the chances he will live to calculate the probability that he will live? The process we have defined is clear cut—choose a gun, shoot, and live or die. The partition is usually associated with the first step in the process, and in this case, is obvious. We let

$A_i = $ "choose gun i," $i = 1, 2, 3$

We are given that $P(A_i) = \frac{1}{3}$ for $i = 1, 2, 3$, since the gun is chosen at random, by the blindfolded person. If E is the event "live," then we are also given that $P(E|A_1) = \frac{4}{5}$, $P(E|A_2) = \frac{3}{5}$, and $P(E|A_3) = \frac{2}{5}$. By the partition theorem, we have

$$P(E) = (\tfrac{1}{3})(\tfrac{4}{5}) + (\tfrac{1}{3})(\tfrac{3}{5}) + (\tfrac{1}{3})(\tfrac{2}{5}) = \tfrac{3}{5}$$

Example 6 A deck of 52 cards is shuffled, and we wish to calculate the probability that the ace of spades is next to the queen of spades. That the partition theorem will be used in the solution of this problem is certainly not evident. Our first inclination might be to consider the sample space of all 52! permutations of the 52 cards in the deck. In trying to determine how frequently, in this sample space, the ace of spades is next to the queen of spades, we would soon notice that this depended on where in the deck the queen was situated. Whether or not the queen was one of the two end cards, rather than one of the 50 interior cards, would make a difference. Hence, we let

$A_1 = $ "queen of spades is an end card"

$A_2 = $ "queen of spades is an interior card"

partition the sample space. Clearly, $P(A_1) = \frac{2}{52}$ and $P(A_2) = \frac{50}{52}$. If E is the event "queen of spades is next to the ace of spades," then $P(E|A_1) = \frac{1}{51}$, since only one of the other 51 cards can be next to an end card, while $P(E|A_2) = \frac{2}{51}$, since two cards can be next to any interior card. Hence,

$$P(E) = (\tfrac{2}{52})(\tfrac{1}{51}) + (\tfrac{50}{52})(\tfrac{2}{51}) = \tfrac{1}{26}$$

Returning to the problem of election forecasting, suppose that $A_1, A_2, \ldots,$ A_m partitions the total voting public. If E is the event "candidate A wins the election," then the partition theorem gives

$$P(E) = \sum_{i=1}^{m} P(A_i) \cdot P(E|A_i)$$

Voting records from previous years yield an estimate as to what percentage of the voting public falls into each category of the partition. Thus, the values $P(A_i)$, $i = 1, 2, \ldots, m$, can be very closely approximated well in advance of election day. In order to approximate $P(E)$, we must be able to approximate each of the values $P(E|A_i)$. This is frequently done by selecting just a few precincts in each category of the partition, studying their past voting records, and interpreting their voting in the present election in light of their past preferences. Although it might be very difficult to obtain a single value for each $P(E|A_i)$, an interval of values which we could confidently say contained the true value $P(E|A_i)$ may be obtained, and the smallest (most pessimistic) value in the interval used for $P(E|A_i)$. If, upon using such values for the $P(E|A_i)$, we find that $P(E) > \frac{1}{2}$, then candidate A is projected a winner. We will have more to say about this process in Chapter 9.

A process consisting of a finite sequence of experiments, each of which has only a finite number of outcomes, is called a *finite stochastic process*. Probabilities relating to finite stochastic processes are usually calculated by the partition theorem and the multiplication theorem, the use of which is often facilitated by a tree diagram.

Example 7   Each of three boxes contains a number of items, some of which are defective, as indicated in the following table:

	Number of items	Number of defectives
Box 1	10	4
Box 2	6	1
Box 3	8	3

A box is selected at random, and an item is selected from the box. We wish to calculate the probability that the item is defective. We let A_i be the event "item is drawn from box *i*" and E the event "item is defective." The tree diagram in Figure 3.4 represents this stochastic process. The letters D and N represent defective and nondefective, respectively. By the partition theorem, we have

$$P(E) = (\tfrac{1}{3})(\tfrac{4}{10}) + (\tfrac{1}{3})(\tfrac{1}{6}) + (\tfrac{1}{3})(\tfrac{3}{8})$$
$$= \tfrac{113}{360}$$

the sum of the probabilities of all branches leading to D.

Example 8   An urn contains three red and seven white balls. A ball is drawn from the urn, and a ball of the opposite color is placed in the urn. A second ball is then drawn from the urn, and we wish to find the probability that it is red. This stochastic process is represented by the tree diagram in Figure 3.5.

Notice that the probabilities along the branches reflect the composition of the urn at the time of the draw. Hence, the probability that the second ball is red is either $\frac{2}{10}$ or $\frac{4}{10}$, depending on whether the first ball drawn was red or white. Since the first ball is more likely to be white than red, the $\frac{4}{10}$ must be weighted higher than the $\frac{2}{10}$. This is done by the partition theorem, with the weights being $\frac{7}{10}$ and $\frac{3}{10}$, respectively. If we let E be the event "second ball is red," then

$$P(E) = (\tfrac{3}{10})(\tfrac{2}{10}) + (\tfrac{7}{10})(\tfrac{4}{10}) = \tfrac{34}{100}$$

which is simply the sum of the probabilities along the two branches leading to R. If A is the event "both balls drawn have the same color," then

$$P(A) = P(R)P(R\,|\,R) + P(W)P(W\,|\,W)$$
$$= (\tfrac{3}{10})(\tfrac{2}{10}) + (\tfrac{7}{10})(\tfrac{6}{10}) = \tfrac{48}{100}$$

Figure 3.4

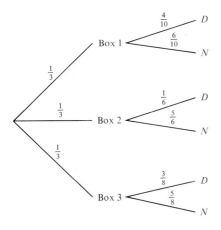

Figure 3.5

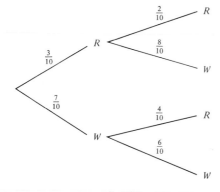

Example 9   Urn A contains three red and two white balls, while urn B contains two red and five white balls. An urn is selected at random, a ball is drawn from the urn and then placed in the other urn, and then a ball is drawn from the second urn. We wish to calculate the probability of the event "both balls drawn are of the same color." This can happen if either two red balls or two white balls are drawn from the two urns. In the tree diagram in Figure 3.6, the event E corresponds to the four branches ARR, AWW, BRR, and BWW. The probabilities along these branches reflect the distribution of the balls in the two urns at the time the two balls are drawn. In particular, the third segment of the branches in the upper half of the tree reflects the distribution of the balls in urn B at the time the second ball is drawn (the first ball having been drawn from urn A, since we are talking about the upper portion of the tree). To calculate $P(E)$ we need only add the probabilities of the four branches favorable to E. So

$$P(E) = (\tfrac{1}{2})(\tfrac{3}{5})(\tfrac{3}{8}) + (\tfrac{1}{2})(\tfrac{2}{5})(\tfrac{6}{8}) + (\tfrac{1}{2})(\tfrac{2}{7})(\tfrac{4}{6}) + (\tfrac{1}{2})(\tfrac{5}{7})(\tfrac{3}{6})$$

$$= \tfrac{901}{1680}$$

EXERCISES

1. Suppose that $A_1, A_2, \ldots, A_n$ is a partition of S and E is a subset of S. Prove that $E \cap A_1, E \cap A_2, \ldots, E \cap A_n$ is a partition of E.
2. Prove or disprove (with an example) each of the following:
   (a) $P(A|B) + P(A|C(B)) = 1$
   (b) $P(A|B) + P(C(A)|B) = 1$
   (c) $P(A|B) + P(C(A)|C(B)) = 1$
3. A rat going through a maze has a 50-50 chance of turning right the first time through. If he turns right the first time, he gets fed, so that his chances of turning right the

Figure 3.6

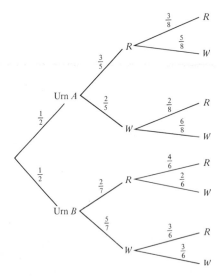

next time are increased to 60-40. If he turns left the first time, he receives a shock, so that his chances of turning left the next time are reduced to 30-70. What is the probability that the rat will turn right the second time through the maze?

4. An urn contains five red and three white balls. A ball is selected at random, discarded, and two balls of the opposite color are placed in the urn. A second ball is then selected at random from the urn. What is the probability that both balls have the same color?

5. Choose at random one of the numbers 1, 2, or 3 and then roll as many dice as indicated by the number chosen. What is the probability that the sum of the dice will be 5?

6. Roll a fair die, and then toss a coin the number of times indicated by the die. What is the probability that none of the coins will come up heads?

7. A store has five loaves of bread remaining, of which three are fresh. Three people come into the store to buy a loaf of bread. Who stands the best chance of getting a fresh loaf, the one who picks first, second, or third?

8. A subcontractor figures that there is a 40 percent chance of getting a particular job if company A wins the contract, a 50 percent if company B wins the contract, and a 70 percent chance if company C wins the contract. However, company A has a connection and stands a 50-50 chance of being awarded the contract, while the other two companies have equal chances. How likely is the subcontractor to get this job?

9. A student feels that there is an 80 percent chance of getting an A in course A, but only a 50 percent chance of getting an A in course B. However, the student feels that taking course B is twice as necessary as taking course A. What are the student's chances of getting an A?

10. The Democrats, Republicans, and Liberals all sponsored a candidate in a recent election. Fifty percent of the voters were Democrats, 40 percent were Republicans, and 10 percent were Liberals. Eighty percent of the Democrats voted for the Democratic candidate, 10 percent voted for the Republican, and 10 percent voted for the Liberal. Ninety percent of the Republicans voted for the Republican candidate, 5 percent for the Democratic, and 5 percent for the Liberal. Seventy percent of the Liberals voted for the Liberal candidate, 20 percent for the Republican, and 10 percent for the Democratic. Who won the election?

3.3 Bayes' Theorem

Another application of conditional probability is Bayes' formula, named after the British clergyman who discovered it. Bayes' formula is the starting point for an entire statistical philosophy known as Bayesian statistics.

Theorem 3.3 Bayes' Theorem

If $A_1, A_2, \ldots, A_n$ is a partition of the sample space S such that $P(A_i) \neq 0$ for any $i = 1, n$, and if E is an event for which $P(E) \neq 0$, then for each $k = 1, 2, \ldots, n$, we have

$$P(A_k|E) = \frac{P(A_k) \cdot P(E|A_k)}{\sum_{j=1}^{n} P(A_j) \cdot P(E|A_j)}$$

Proof: Applying the definition of conditional probability twice, we find

$$P(A_k|E) = \frac{P(E \cap A_k)}{P(E)} = \frac{P(A_k) \cdot P(E|A_k)}{P(E)}$$

By the partition theorem,

$$P(E) = \sum_{j=1}^{n} P(A_j) \cdot P(E|A_j)$$

and so the desired result follows.

The meaning of Bayes' theorem may seem confusing at first, because the left-hand side of the equation may seem backwards. In the examples we have seen thus far, the probabilities $P(A_k)$ of the partition parts were known, and conditional probabilities relative to the partition parts were calculated.

However, it is possible that the probabilities of the partition parts are unknown and that the very purpose of the experiment is to either approximate or obtain exact values for these probabilities. Before the experiment is performed, we may have a certain a priori idea (guess, estimate) as to what these probabilities might be. After the experiment has been run once, or perhaps several times, and we have observed the outcome(s), we will probably wish to revise our estimates for these probabilities and obtain what are called *a posteriori* probabilities. Bayes' formula is the tool that facilitates making these revisions.

The race track provides an excellent example of this process. The horse-player's task, basically, is to estimate, in his own head, each horse's chances (probability) of winning the race. After studying each horse's past performances, he comes up with his first guess as to what these probabilities might be. Then he observes the horses in the walking ring and post parade, noting which ones look good and which ones are bandaged up or are acting especially nervous. Using this information (consider it an experiment), he revises his probabilities. After the race has been run, of course, he has considerably more information at hand, and can once again make a revision in his estimate of the probabilities, information that he may be able to use at some future encounter between the same horses.

Perhaps it is helpful, in trying to understand Bayes' Theorem, to think of the relationship between the members of the partition and the event E as a cause-effect relationship. In the preceding two sections, we have calculated the probability of the effect, given the cause. But perhaps we know the probability of the effect, for various potential causes (the partition), and we are more interested in determining how likely each cause is, knowing the effect to have occurred. In the field of medical diagnosis, for example, an effect E, certain symptoms, has been observed in a patient, and the doctor must determine the cause C, the disease which is causing these symptoms to appear. Usually, the doctor will have information (statistics) regarding $P(E|C)$, how likely these symptoms are, given that the patient has a certain disease. But what the doctor really wants to know is the reverse probability

$P(C|E)$, how likely it is that it is a certain disease which is causing these symptoms to appear. It is here that Bayes' Theorem can be used.

The calculations involved in Bayes' formula can be made more meaningful by looking at the tree diagram discussed in the preceding section (Figure 3.2). If the event E has occurred, then only the branches leading to E remain relevant. These branches then constitute our conditional sample space. If we are calculating $P(A_k|E)$, then we are interested in the likelihood of the kth branch leading to E relative to all n branches that lead to E. The numerator in Bayes' formula is the probability along this (kth) branch, while the denominator is the sum of the probabilities along all branches in our conditional sample space.

Example 1 A rash and elevated temperature have been observed in a child, and the doctor has narrowed down the causes to chicken pox (CP), of which there is presently an epidemic, scarlet fever (SF), or possibly some other disease (D). Suppose that the symptoms (rash and elevated temperature) will certainly appear in a child suffering from scarlet fever, will almost certainly (85 percent of the time) appear in a child suffering from chicken pox, but will only 33 percent of the time appear in a child suffering from some other disease. In other words,

$$P(S|CP) = 0.85$$
$$P(S|SF) = 1.00$$
$$P(S|D) = 0.33$$

where S is the event "child has the given symptoms." The doctor also has statistics that indicate that the present frequencies of these diseases in his locale are as follows:

$$P(CP) = 0.150$$
$$P(SF) = 0.005$$
$$P(D) = 0.075$$

He can then use Bayes' formula in conjunction with the tree diagram in Figure 3.7 to estimate the likelihood that the child has any of the diseases mentioned. For example, the probability that the child does, in fact, have chicken pox is

$$P(CP|S) = \frac{P(CP) \cdot P(S|CP)}{P(CP) \cdot P(S|CP) + P(SF) \cdot P(S|SF) + P(D) \cdot P(S|D)}$$

$$= \frac{0.150 \cdot 0.85}{0.150 \cdot 0.85 + 0.005 \cdot 1.0 + 0.075 \cdot 0.33}$$

$$= 0.81$$

Notice that the sum of the probabilities of the three partition parts is 0.23, and not 1.00. Actually, we have eliminated the fourth, and by far the largest, partition part, the approximately 77 percent of the child population that is healthy, H. The reason that we can eliminate this group from our calculations is because it is reasonable to assume that $P(S|H) = 0$. Therefore, the H branch of the tree would make no contribution to any of our calculations.

Example 2   Three machines, A, B, and C, are responsible for 50, 30, and 20 percent, respectively, of the total number of items produced by a factory. The defective output from these machines are three, four, and five percent, respectively. The tree diagram in Figure 3.8 immediately suggests itself. If an

Figure 3.7

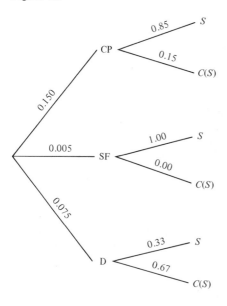

Figure 3.8

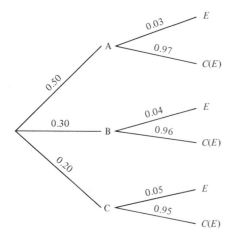

item is chosen at random from the produce of this factory, then we can calculate the probability that it is defective by means of the partition theorem. If E represents the event "item is defective," then

$$P(E) = 0.50 \cdot 0.03 + 0.30 \cdot 0.04 + 0.20 \cdot 0.05 = 0.037$$

Suppose, however, that we find the item to be defective, and we wish to calculate the probability that it was produced by machine A. Looking at the tree diagram, we see that we are given information about the second segment of the branches (E occurred), and are asked for a probability concerning the first segment of the branches (did A occur?). This inverted situation is the clue that Bayes' formula must be used. Our conditional sample space E consists of all three branches leading to a defective, and, as we have seen above, the sum of the probabilities along these three branches is 0.037. The event that the defective item was produced by machine A corresponds to only the first of these three branches, and since that branch has probability $0.50 \cdot 0.03 = 0.015$, we have, by Bayes' formula, that

$$P(A|E) = \frac{0.015}{0.037} = \frac{15}{37}$$

Example 3   Russell's Paradox   Three identical boxes each contain two coins. Box 1 contains two pennies, box 2 contains two nickels, and box 3 contains a penny and a nickel. A box is chosen at random, a coin is taken out and is found to be a penny. We wish to find the probability that the other coin in the box is also a penny, that is, that the first coin was taken from box 1. The solution appears simple: Either box 1 or box 3 was chosen, and so the probability that box 1 was chosen is $\frac{1}{2}$. However, Bayes' formula will tell us otherwise. Let E be the event "coin chosen was a penny" and A_i the event "coin was chosen from box i", $i = 1, 2, 3$. With the help of Bayes' formula and the tree diagram in Figure 3.9 we have

$$P(A_1|E) = \frac{\frac{1}{3} \cdot 1}{(\frac{1}{3} \cdot 1) + (\frac{1}{3} \cdot 0) + (\frac{1}{3} \cdot \frac{1}{2})} = \frac{2}{3}$$

We have calculated the probability of the branch representing drawing a penny from box 1, and divided by the sum of the probabilities of all branches leading to drawing a penny. Hence, $P(A_1|E) = \frac{2}{3}$ and not $\frac{1}{2}$ as reasoned above. The logic of the answer is quite simple: two of the three pennies in the three boxes are in box 1.

Example 4   A test for the detection of a particular disease is not quite foolproof. The test will correctly detect the disease 90 percent of the time, but will incorrectly detect the disease 1 percent of the time. From a large population of which an estimated 0.1 percent have the disease, a person is selected at random, given the test, and told he has the disease. What are the chances that the person actually does have the disease? Let us partition the population as follows:

A_1 = those people who have the disease

A_2 = those people who do not have the disease

We are given that $P(A_1) = 0.001$ and $P(A_2) = 0.999$, and we are also given the conditional probabilities $P(E|A_1) = 0.90$ and $P(E|A_2) = 0.01$, where E is the event "test is positive for the disease." We are given that a person having taken the test has been told that he has the disease. In other words, we are told that the event E has occurred for this person, and we wish to determine in which part of the partition he belongs. The actual probability that we are asked to calculate is $P(A_1|E)$, and this is clearly a case for Bayes' formula

$$P(A_1|E) = \frac{0.001 \cdot 0.90}{0.001 \cdot 0.90 + 0.999 \cdot 0.01} = 0.083$$

Therefore, only slightly more than eight percent of those whose test shows positive actually have the disease.

Example 5 At the beginning of the year, three economic theories are proposed concerning the American economy. At the time that they are proposed, they seem equally likely. At the end of the year, the actual state of the economy is evaluated in light of the three theories. It is determined that if the first theory were true, the probability that the economy would be in its present state is 0.6. Similar probabilities for the second and third theories are 0.4 and 0.2, respectively. With a year's hindsight, we can now, by means of Bayes' formula, reevaluate the likelihood that the three economic theories are valid. If A_i denotes the event "theory i is valid," for $i = 1, 2, 3$, then our initial guess is that $P(A_1) = P(A_2) = P(A_3) = \frac{1}{3}$. If E is the event "the economy is in its

Figure 3.9

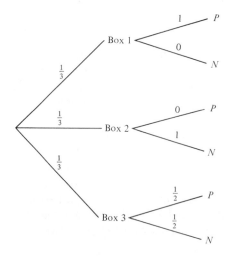

present state," then we calculate by the partition theorem that

$$P(E) = (\tfrac{1}{3} \cdot \tfrac{6}{10}) + (\tfrac{1}{3} \cdot \tfrac{4}{10}) + (\tfrac{1}{3} \cdot \tfrac{2}{10}) = \tfrac{4}{10}$$

Then, by Bayes' formula, we adjust our original guesses for the $P(A_i)$ as follows:

$$P(A_1|E) = (\tfrac{1}{3} \cdot \tfrac{6}{10})/\tfrac{4}{10} = \tfrac{1}{2}$$
$$P(A_2|E) = (\tfrac{1}{3} \cdot \tfrac{4}{10})/\tfrac{4}{10} = \tfrac{1}{3}$$
$$P(A_3|E) = (\tfrac{1}{3} \cdot \tfrac{2}{10})/\tfrac{4}{10} = \tfrac{1}{6}$$

Example 6 A crime has been committed by a person with a certain characteristic (hair color, blood type, etc). It is known that the probability that a person in the population will have this characteristic is p. A suspect has been apprehended and found to have the characteristic. Although he cannot be presumed guilty, the fact that he possesses the characteristic increases our estimate of the likelihood that he is guilty of the crime. Let G be the event "guilty," I the event "innocent," and C the event "has the characteristic in question." We are given, then, that $P(C|G) = 1$ and we assume that $P(C|I) = p$. Suppose that we felt before we knew that the suspect had the given characteristic that there was a 50-50 chance he was guilty. Then, with the added knowledge that he had the characteristic, we use Bayes' formula to revise our estimate of the chances of his guilt as follows:

$$P(G|C) = \frac{\tfrac{1}{2} \cdot 1}{\tfrac{1}{2} \cdot 1 + \tfrac{1}{2} \cdot p} = \frac{1}{1 + p} > \frac{1}{2}$$

EXERCISES

1. In Exercise 4 of Section 3.2, suppose that the second ball drawn was red. What is the probability that the first ball was also red?
2. In Exercise 6 of Section 3.2, suppose that none of the coins came up heads. What is the probability that only one coin was tossed?
3. In Exercise 8 of Section 3.2, suppose that the subcontractor did get the job. What is the probability that company A won the contract?
4. In Exercise 9 of Section 3.2, suppose that the student did get an A. What is the probability that he took course B?
5. Eighty percent of a college's students attended public high schools. Thirty percent of those from public high schools and 25 percent of those from private high schools graduate with honors. The valedictorian graduated with honors. What is the probability that he graduated from a public high school?
6. The probability that a male has tuberculosis is 0.05, while the probability that a female has tuberculosis is 0.10. From a group of 300 men and 150 women, a person is selected at random and found to have tuberculosis. What is the probability the person is female?
7. Four copies of an economics textbook and eight copies of a sociology textbook are in a box. Two books are chosen without replacement and the second is an economics book. What is the probability that the first book was also an economics book?

8. Suppose that we wish to estimate the probabilities of the partition members A_1, A_2, A_3, A_4 which at the outset seem equally likely. We will form our estimate of the $P(A_i)$ on the basis of 12 runs of an experiment which either does or does not result in an event E. Suppose we know that $P(E|A_1) = \frac{3}{4}$, $P(E|A_2) = \frac{3}{5}$, $P(E|A_3) = \frac{1}{2}$, and $P(E|A_4) = \frac{2}{5}$. Suppose that the 12 runs of the experiment result in $E, C(E), C(E), E, E, E, C(E), E, E, E, C(E)$, and $C(E)$. Write a computer program that will output $P(A_i)$, $i = 1, 4$, as the experiment develops.

9. A two-headed coin is placed in a box with three normal coins. A coin is selected from the box and tossed three times, with heads resulting each time. What is the probability that it was the two-headed coin?

10. The results of a questionnaire indicate that 40 percent of the men and 60 percent of the women employed by a particular company smoke. However, 75 percent of the people employed by the company are men. Suppose that a questionnaire is picked up and found to have been filled out by a smoker. What is the probability that the smoker is a woman?

3.4 Independent Events

A robbery was committed in Los Angeles in 1968* by an interracial couple consisting of a black man with a beard and mustache and a white girl with blond hair and a pony tail who drove a partly yellow car. A couple possessing all of these characteristics was convicted of this crime, partly on the basis of the following probability argument supplied by a local mathematics professor:

"Assuming the following individual probabilities for the given characteristics:

Interracial couple in a car	$\frac{1}{1000}$
Partly yellow car	$\frac{1}{10}$
Black man with a beard	$\frac{1}{10}$
Man with a mustache	$\frac{1}{4}$
Girl with blonde hair	$\frac{1}{3}$
Girl with ponytail	$\frac{1}{10}$

we multiply all six probabilities to find the probability that a couple will have all the characteristics to be 1/12,000,000. Since the defendants possessed all of these characteristics, the combination of which is so rare, they should be found guilty of the robbery." They were indeed found guilty, but later made a successful appeal based on a mathematical argument of their own, which we shall discuss in Chapter 6. Suppose that you were a lawyer, familiar with the concept of conditional probability, trying to defend them. Can you see a line of reasoning that might overturn their conviction?

What we shall be concerned with at the moment is the validity of the mathematical argument for the prosecution. The multiplication theorem allows us to calculate the probability that several events all occur by mul-

---

* *Trial by Mathematics*, Time Magazine, April 26, 1968.

tiplying probabilities, but the probabilities involved must be conditional probabilities. None of the probabilities listed above is a conditional probability.

In general, the occurrence of one event has an effect on the probability of occurrence of another event. $P(A)$ and $P(A|B)$ are, in general, not equal. There are times, however, when the occurrence of one event has no effect whatsoever on the occurrence of another event. This is the case in the situation above. The color of a person's car would not seem to have any effect on the color or style of a person's hair. Hence, although the probabilities that were multiplied above do not on the surface appear to be conditional probabilities, they are in fact conditional probabilities of a special type.

DEFINITION

An event A is said to be *independent* of an event B if $P(A) = P(A|B)$.

Implicit in this definition is the assumption that $P(B) \neq 0$. If A is independent of B, then

$$P(A \cap B) = P(B)P(A|B) = P(B) \cdot P(A) = P(A) \cdot P(B)$$

so the multiplication theorem takes the form

$$P(A \cap B) = P(A) \cdot P(B)$$

The multiplication theorem also says that

$$P(A \cap B) = P(A) \cdot P(B|A)$$

hence,

$$P(A) \cdot P(B|A) = P(A) \cdot P(B)$$

Thus, $P(B|A) = P(B)$ provided, of course, that $P(A) \neq 0$. Therefore, B is independent of A. Thus, if either A is independent of B or B is independent of A, then the other is also true and the multiplication rule $P(A \cap B) = P(A)P(B)$ holds, provided that neither $P(A) = 0$ nor $P(B) = 0$. In order that we might have a more general concept of independence, one that applies to *all* events, and not just those of nonzero probability, we restate our definition as follows:

DEFINITION

The events A and B are called *independent* if $P(A \cap B) = P(A) \cdot P(B)$.

The two previous rules for independence, $P(A|B) = P(A)$ and $P(B|A) = P(B)$, are simple consequences of the new definition and the multiplication theorem.

Proposition 3.4

If the events A and B are independent, then so are the events A and $C(B)$, $C(A)$ and B, and $C(A)$ and $C(B)$.

Proof: We prove the first case only, and leave the other two cases as an exercise. We are given that

$$P(A \cap B) = P(A) \cdot P(B)$$

by the independence of A and B. Since $\{B, C(B)\}$ is a partition of S, it follows that $\{A \cap B, A \cap C(B)\}$ is a partition of A, and so

$$P(A) = P(A \cap B) + P[A \cap C(B)]$$

Thus,

$$
\begin{aligned}
P[A \cap C(B)] &= P(A) - P(A \cap B) \\
&= P(A) - P(A) \cdot P(B) \quad \text{(independence)} \\
&= P(A) \cdot [1 - P(B)] \\
&= P(A) \cdot P[C(B)] \quad \text{(Proposition 2.2)}
\end{aligned}
$$

Hence, $P[A \cap C(B)] = P(A) \cdot P[C(B)]$, and so the multiplication theorem works for the events A and $C(B)$. As a result, A and $C(B)$ are independent.

Example 1 A fair coin is tossed three times, and we consider the events

A = "first toss is a head"

B = "second toss is a head"

C = "exactly two consecutive heads"

Using the usual sample space

$$S = \{\text{HHH, HHT, HTH, HTT, THH, THT, TTH, TTT}\}$$

we find that $P(A) = \frac{4}{8}$, $P(B) = \frac{4}{8}$, and $P(C) = \frac{2}{8}$. Since

$$A \cap B = \{\text{HHH, HHT}\}$$
$$A \cap C = \{\text{HHT}\}$$
$$B \cap C = \{\text{HHT, THH}\}$$

we have $P(A \cap B) = \frac{2}{8}$, $P(A \cap C) = \frac{1}{8}$, and $P(B \cap C) = \frac{2}{8}$. Therefore,

$$\tfrac{2}{8} = P(A \cap B) = P(A) \cdot P(B) = \tfrac{4}{8} \cdot \tfrac{4}{8} = \tfrac{1}{4}$$
$$\tfrac{1}{8} = P(A \cap C) = P(A) \cdot P(C) = \tfrac{4}{8} \cdot \tfrac{2}{8} = \tfrac{1}{8}$$
$$\tfrac{2}{8} = P(B \cap C) \neq P(B) \cdot P(C) = \tfrac{4}{8} \cdot \tfrac{2}{8} = \tfrac{1}{8}$$

so that the events A and B and the events A and C are independent, while the events B and C are not independent.

Example 2 The probability that A hits a certain target is $\frac{1}{4}$, while the probability that B hits the target is $\frac{2}{5}$. Suppose that they both shoot at the target. What is the probability that at least one of them hits the target? We are given that $P(A) = \frac{1}{4}$ and $P(B) = \frac{2}{5}$, and must calculate $P(A \cup B)$. By Proposition 2.5,

$$P(A \cup B) = P(A) + P(B) - P(A \cap B)$$

and so we must find $P(A \cap B)$. Ignoring the effect of any psychological factors, we assume that A's hitting the target is independent of B's hitting the target. Thus

$$P(A \cap B) = P(A) \cdot P(B) = \tfrac{1}{4} \cdot \tfrac{2}{5} = \tfrac{1}{10}$$

Hence

$$P(A \cup B) = P(A) + P(B) - P(A \cap B)$$
$$= \tfrac{1}{4} + \tfrac{2}{5} - \tfrac{1}{10} = \tfrac{11}{20}$$

DEFINITION

The events $A_1, A_2, \ldots, A_m$ are called *mutually independent* if A_i and A_j are independent for all $1 \leq i \neq j \leq n$. In other words, $P(A_i \cap A_j) = P(A_i) \cdot P(A_j)$ for any choice of i, j.

The definition of independence of events can be extended to three or more events as follows:

DEFINITION

Three events A, B, and C are called *independent* if

1. A, B, and C are mutually independent.
2. $P(A \cap B \cap C) = P(A) \cdot P(B) \cdot P(C)$

Condition 1 of the definition does *not* imply condition 2, as the following example shows:

Example 3 Two fair coins are tossed, and we consider the events

$A = $ "the first coin is a head"
$B = $ "the second coin is a head"
$C = $ "exactly one head occurs"

Using the usual sample space $S = \{HH, HT, TH, TT\}$, we see that

$$P(A) = P(B) = P(C) = \tfrac{1}{2}.$$

Since $A \cap B = \{HH\}$, $A \cap C = \{HT\}$, and $B \cap C = \{TH\}$, we have $P(A \cap B) = P(A \cap C) = P(B \cap C) = \tfrac{1}{4}$. Therefore,

$$\tfrac{1}{4} = P(A \cap B) = P(A) \cdot P(B) = \tfrac{1}{2} \cdot \tfrac{1}{2} = \tfrac{1}{4}$$
$$\tfrac{1}{4} = P(A \cap C) = P(A) \cdot P(C) = \tfrac{1}{2} \cdot \tfrac{1}{2} = \tfrac{1}{4}$$
$$\tfrac{1}{4} = P(B \cap C) = P(B) \cdot P(C) = \tfrac{1}{2} \cdot \tfrac{1}{2} = \tfrac{1}{4}$$

and so the events A, B, and C are mutually independent. However,

$$A \cap B \cap C = \varnothing, \text{ so } P(A \cap B \cap C) = 0,$$

while

$$P(A) \cdot P(B) \cdot P(C) = (\tfrac{1}{2})^3 = \tfrac{1}{8}.$$

Thus, (b) does not hold, so the events A, B, and C are not independent.

DEFINITION

The events $A_1, A_2, \ldots, A_n$ are said to be *independent* if the multiplication rule holds for all combinations of two or more events. (There are $2^n - n - 1$ such rules.)

Proposition 3.5

Suppose that A, B, and C are independent events. Then the event A is independent of any event which can be expressed in terms of B and C by means of unions, intersections, and complements. In particular, A is independent of $B \cap C$ and $B \cup C$.

Proof: To prove the independence of A and $B \cap C$, we argue

$$
\begin{aligned}
P[A \cap (B \cap C)] &= P(A \cap B \cap C) \\
&= P(A) \cdot P(B) \cdot P(C) \quad [\text{by (b)}] \\
&= P(A) \cdot P(B \cap C) \quad [B, C \text{ independent}]
\end{aligned}
$$

Therefore, A and $B \cap C$ are independent. To prove the independence of A and $B \cup C$, we argue

$$
\begin{aligned}
P[A \cap (B \cup C)] &= P[(A \cap B) \cup (A \cap C)] \quad (\text{distributive law}) \\
&= P(A \cap B) + P(A \cap C) - P(A \cap B \cap C) \\
&\qquad\qquad\qquad\qquad\qquad\qquad (\text{Proposition 2.5}) \\
&= P(A) \cdot P(B) + P(A) \cdot P(C) - P(A) \cdot P(B) \cdot P(C) \\
&\qquad\qquad\qquad\qquad\qquad\qquad [\text{by (a) and (b)}] \\
&= P(A) \cdot (P(B) + P(C) - P(B) \cdot P(C)) \\
&= P(A)P(B \cup C) \quad (\text{Proposition 2.5})
\end{aligned}
$$

The proof of the independence of A and any other logical combination of B and C is similar, with the assistance of Proposition 3.4.

Example 4 Each of A, B, and C shoot at a target, with respective probabilities $P(A) = 0.5$, $P(B) = 0.6$, and $P(C) = 0.8$ of hitting the target. We wish to calculate the probability of the event E "exactly one hits the target." Since

$$E = [A \cap C(B) \cap C(C)] \cup [C(A) \cap B \cap C(C)] \cup [C(A) \cap C(B) \cap C]$$

a union of three disjoint events, we have

$$P(E) = P[A \cap C(B) \cap C(C)] + P[C(A) \cap B \cap C(C)] + P[C(A) \cap C(B) \cap C]$$

We disregard psychological factors and assume that A, B, and C are independent. Applying Proposition 3.5, we find

$$P[A \cap C(B) \cap C(C)] = P(A) \cdot P[C(B)] \cdot P[C(C)]$$
$$= 0.5 \cdot 0.4 \cdot 0.2 = 0.04$$

$$P[C(A) \cap B \cap C(C)] = P[C(A)] \cdot P(B) \cdot P[C(C)]$$
$$= 0.5 \cdot 0.6 \cdot 0.2 = 0.06$$

$$P[C(A) \cap C(B) \cap C] = P[C(A)] \cdot P[C(B)] \cdot P(C)$$
$$= 0.5 \cdot 0.4 \cdot 0.8 = 0.16$$

Therefore, $P(E) = 0.04 + 0.06 + 0.16 = 0.26$.

EXERCISES

1. Prove that if $P(A) = 0$, then A and B are independent for any event B.
2. Complete the proof of Proposition 3.4.
3. If $P(A) = 0.4$ and $P(A \cup B) = 0.7$, what choice of $P(B)$ makes A and B independent?
4. A red and a green die are rolled. Are the events A "red die shows a 3" and B "total is even" independent?
5. A card is chosen from a standard deck. Are the events A "card is a club" and B "card is an ace" independent?
6. The probability that the husband's car starts on a cold morning is 0.2, while the probability that his wife's car starts is 0.1. On a given cold morning, what is the probability that (a) neither car starts, and (b) exactly one car starts.
7. Three people work independently at decoding a message with probabilities of success $\frac{1}{5}, \frac{1}{4}$, and $\frac{1}{3}$ respectively. What is the probability that one of them decodes the message?
8. If A and B are independent, prove that $P(A \cup B) = 1 - P[C(A)] \cdot P[C(B)]$.

3.5 Independent Trials

Many important probabilistic problems consist of the same experiment repeated several times, always under the same conditions. We come now to a formal definition of this situation.

DEFINITION

Let S be a probability space. By k *independent trials* of S we mean the sample space

$$T = \{(x_1, x_2, \ldots, x_k) : x_i \in S \qquad \text{for all } 1 \le i \le k\}$$

together with a probability function P defined by

$$P[(x_1, x_2, \ldots, x_k)] = P(x_1) \cdot P(x_2) \cdots \cdots P(x_k)$$

We sometimes write $T = S^k$ or $T = S \times S \times \cdots \times S$ (k times).

The sample elements of T are therefore ordered k-tuples of elements of S, with the individual coordinates of the k-tuple being the outcomes in S of the respective trials. Note that the probability function is defined according to the multiplication rule for independent events, and it is for this reason that the trials are called independent.

We must check, of course, that this definition of a probability function does, in fact, yield an acceptable assignment of probabilities. Assuming that $k = 2$ and that S is a finite set, we can argue

$$\sum_{i,j=1}^{n} P[(x_i, x_j)] = \sum_{i=1}^{n} \left\{ \sum_{j=1}^{n} P[(x_i, x_j)] \right\}$$

$$= \sum_{i=1}^{n} \left\{ \sum_{j=1}^{n} P(x_i) \cdot P(x_j) \right\}$$

$$= \sum_{i=1}^{n} P(x_i) \left\{ \sum_{j=1}^{n} P(x_j) \right\}$$

$$= \sum_{i=1}^{n} P(x_i) \cdot 1$$

$$= \sum_{i=1}^{n} P(x_i) = 1$$

The inductive argument for the case $k = m$ is quite similar and is left as an exercise for the reader.

Choosing a sample of size k *with* replacement is an example of k independent trials (of choosing a sample of size 1). When the sampling is done with replacement, each item chosen is taken from the entire population, and therefore, the multiplication law of the definition holds.

Example 1 A certain team wins with probability 0.6, loses with probability 0.3, and ties with probability 0.1. The team plays three games, and we wish to calculate the probability that they win at least two of the games, but do not lose. This is the event

$$E = \{(W, W, W), (W, W, T), (W, T, W), (T, W, W)\}$$

Ignoring psychological factors, we may assume that the outcomes of the three trials (games) are independent, and so

$$P[(W, W, W)] = 0.6 \cdot 0.6 \cdot 0.6 = 0.216$$

and

$$P[(W, W, T)] = P[(W, T, W)] = P[(T, W, W)]$$
$$= 0.6 \cdot 0.6 \cdot 0.1 = 0.036$$

Therefore, $P(E) = 0.216 + 3 \cdot 0.036 = 0.324$

The following example is a particular case of what is known as a *polya urn scheme*. Polya urn schemes can be used as mathematical models of the spread of contagious diseases.

Example 2 Suppose that an urn contains m red and k black balls. Each time a ball is drawn from the urn, it and one additional ball of the same color are placed back in the urn. For the sake of simplicity, we restrict ourselves to the first two selections from the urn. If we let B_i and R_i denote the events "drawing a black or red ball on the ith draw," $i = 1, 2$, then we have initially that

$$P(B_1) = \frac{k}{m + k} \quad \text{and} \quad P(R_1) = \frac{m}{m + k}$$

Using the partition theorem, we calculate

$$P(B_2) = P(R_1) \cdot P(B_2 | R_1) + P(B_1) \cdot P(B_2 | B_1)$$

$$= \frac{m}{m + k} \cdot \frac{k}{m + k + 1} + \frac{k}{m + k} \cdot \frac{k + 1}{m + k + 1}$$

$$= \frac{mk + k(k + 1)}{(m + k) \cdot (m + k + 1)} = \frac{k(m + k + 1)}{(m + k) \cdot (m + k + 1)} = \frac{k}{m + k}$$

Therefore, $P(B_2) = k/(m + k) = P(B_1)$, and in similar fashion, $P(R_2) = m/(m + k) = P(R_1)$. As a matter of fact, it can be shown that

$$P(B_n) = \frac{k}{m + k} \quad \text{and} \quad P(R_n) = \frac{m}{m + k}$$

for any $n = 1, 2, 3, \ldots$. Each trial of our experiment, then, has the same two possible outcomes, R and B, and they occur with the same probability on all trials. This might seem to indicate that the trials are independent, but they are not, since

$$P[(B_1, B_2)] = P(B_1) \cdot P(B_2 | B_1)$$

$$= \frac{k}{m + k} \cdot \frac{k + 1}{m + k + 1}$$

while

$$P(B_1) \cdot P(B_2) = \left(\frac{k}{m + k} \right)^2$$

Therefore, $P[(B_1, B_2)] \neq P(B_1) \cdot P(B_2)$

The following will give us a quick method to determine the independence (or dependence) of two events in a sample space of two independent trials.

DEFINITION

Given an experiment consisting of two independent trials of a sample space S, an event $A \subseteq S \times S$ is said to be *determined by the first trial* if there is a $C_1 \subseteq S$ such that $A = C_1 \times S$. An event $B \subseteq S \times S$ is said to be *determined by the second trial* if there is $C_2 \subseteq S$ such that $B = S \times C_2$.

Example 3 If a fair die is rolled twice, the sample space

$$S = \{(a, b) : a, b = 1, 2, 3, 4, 5, 6\}$$

is the sample space of two independent trials of rolling one die. If we let A be the event "first die is 6," then naturally A is determined by the first trial. If we choose $C_1 = \{6\}$, then $A = C_1 \times S$, the set of all ordered pairs with first coordinate 6.

Proposition 3.6

In the sample space $S \times S$ of two independent trials of S, if the events A and B are determined by the first and second trials, respectively, then A and B are independent events.

Proof: We prove first the following lemma.

Lemma 3.7

If $C_1 \subseteq S$ and $C_2 \subseteq S$, then $P(C_1 \times C_2) = P(C_1) \cdot P(C_2)$.

Proof: Suppose that $C_1 = \{x_1, x_2, \ldots, x_k\}$ and $C_2 = \{y_1, y_2, \ldots, y_m\}$. Then

$$C_1 \times C_2 = \{(x_i, y_j) : 1 \leq i \leq k, 1 \leq j \leq m\}$$

Hence,

$$
\begin{aligned}
P(C_1 \times C_2) &= \sum_{i=1}^{k} \sum_{j=1}^{m} P[(x_i, y_j)] \\
&= \sum_{i=1}^{k} \sum_{j=1}^{m} P(x_i) \cdot P(y_j) \\
&= \sum_{i=1}^{k} P(x_i) \sum_{j=1}^{m} P(y_j) \\
&= \sum_{i=1}^{k} [P(x_i) \cdot P(C_2)] \\
&= P(C_2) \cdot \sum_{i=1}^{k} P(x_i) \\
&= P(C_2) P(C_1)
\end{aligned}
$$

Returning now to the proof of Proposition 3.6, since the events A and B are determined by the first and second trials, respectively, there are events C_1 and C_2 in S such that

$$A = C_1 \times S \quad \text{and} \quad B = S \times C_2$$

By Lemma 3.7,

$$P(A) = P(C_1 \times S) = P(C_1) \cdot P(S) = P(C_1)$$
$$P(B) = P(S \times C_2) = P(S) \cdot P(C_2) = P(C_2)$$

so that $P(A) \cdot P(B) = P(C_1) \cdot P(C_2)$. On the other hand,

$$A \cap B = (C_1 \times S) \cap (S \times C_2) = C_1 \times C_2$$

so that, by Lemma 3.7,

$$P(A \cap B) = P(C_1 \times C_2) = P(C_1) \cdot P(C_2)$$

Therefore, $P(A \cap B) = P(A) \cdot P(B)$, so A and B are independent.

EXERCISES

1. A quiz has four multiple choice questions, each having three possible answers, only one of which is correct. Assuming that a student guesses at random, and that his successive guesses are independent, what is the probability that he gets more right than wrong answers?
2. A true-false exam consists of six questions, of which four must be answered correctly to pass. If you fail to answer the first two questions correctly, and are guessing, what are your chances of passing the exam?
3. If the probability that a ballplayer gets a hit is 0.250 (his batting average), the probability that he walks is 0.05, and the probability that he makes an out is 0.70, what is the probability that in four at-bats he gets exactly two hits and two walks?
4. If the two teams are equally matched, what is the probability that the World Series will go seven games, with the home team winning each game?
5. If the American League team has 60 percent probability of winning any game, but the Series starts in the National League city, what is the probability of the American League team winning in four games?
6. A system will fail only when each component fails. One particular component consists of three of the same item placed in triplicate so that the component will fail only if all three of the items fail. If any one of the items has a five percent chance of failing, what is the probability that that component will fail?

ADDITIONAL PROBLEMS

1. Solve the two problems posed by Chevalier de Mere that were mentioned in Section 1.1.
2. Each of two people toss three coins. What is the probability that they both get the same number of heads?
3. A sample of size 4 is drawn without replacement from an urn containing 12 balls, 8 of which are white. Given that the sample contains 3 white balls, what is the probability that the last ball drawn was white?
4. A penny is tossed twice and a nickel is tossed three times. What is the probability that the penny shows heads exactly as many times as the nickel shows heads?
5. Each of n people has his own deck and each draws one card from that deck. How large must n be so that nine times out of every ten the experiment is performed at least one of the people will draw an ace?

6. Twelve keys, of which only one fits, are tried one after another until the door opens. Find the probability that the door will be opened
   (a) on the twelfth try
   (b) on none of the first three tries
   (c) on either the first or twelfth try

7. Suppose that 80 percent of the population in a certain city is white, that 95 percent of the whites are employed, but only 75 percent of the blacks are employed. A person is chosen at random and found to be unemployed. What is the probability that the person is black?

8. It has been found that the workers in a certain factory are more likely to produce a defective item on Monday or Friday than on Tuesday, Wednesday, or Thursday. If these probabilities are 0.08 and 0.05, respectively, and if production is constant throughout the week, what percentage of the items produced by the factory are defective? What is the probability that a defective item was produced on Monday?

9. A player wins at "craps" if he rolls a 7 or 11, but loses if he rolls a 2, 3, or 12. Otherwise, he continues to roll the dice until he either wins or loses. What is the probability of winning at craps? Write a computer program to simulate 1000 games of craps and compare the results with the true probabilities.

10. A factory has a machine that tests each item that is produced with respect to a certain critical measurement. It is known that 96 percent of the items produced are acceptable. However, the test rejects acceptable items with probability 0.02 and accepts nonacceptable items with probability 0.05. What is the probability that an item which has passed this test is in fact acceptable?

4

Random Variables

4.1 Introduction

The sample spaces that we have been using fall into one of the following two categories:

1. qualitative, or descriptive;
2. quantitative, or numerical.

Examples of the first type were the 36 possible outcomes of the roll of two dice, the $C_5{}^{52}$ possible five-card poker hands, and the 365^k possible k-tuples of birthdays for a sample of k people. Qualitative sample spaces tend to be rather atomic, being very descriptive of all possible outcomes of the experiment.

Examples of the second type are the sample space indicating the number of heads occurring in three tosses of a fair coin and the sample space indicating the sum of two dice. Sample spaces of this type are not as atomic as those of the first type; a number is used to represent an outcome or a set of outcomes of the experiment. In order to calculate the probabilities associated with these numbers, it is usually necessary to work in a sample space of the first type.

The assignment of numerical values to the elements of a sample space can be thought of as a function from the sample space into the set of real numbers. Such functions are called random variables, and they are the main topic of discussion in the next three chapters.

Since random variables are functions, we shall briefly review some set-theoretic functional notation that we shall use. Suppose that f is a function from a set S to a set T (which we denote $f : S \to T$) and that $A \subseteq S$ and $B \subseteq T$. We shall use the notation

$$f(A) = \{f(x) : x \in A\}$$

to denote the *image* of the set A under the function f, and the notation

$$f^{-1}(B) = \{x \in S : f(x) \in B\}$$

to denote the *preimage*, or *inverse image* of the set B under the function f.

With these notational remarks out of the way, we now state the general definition of a random variable:

DEFINITION

Given a sample space S together with a set of events ξ, a *random variable* is a function $X : S \to \mathbb{R}$ such that $X^{-1}(I)$ is an event for every interval $I \subseteq \mathbb{R}$.

If the sample space S is either finite or countable, the condition $X^{-1}(I) \in \xi$ is superfluous because the set of all subsets of S can be chosen as the set of events. Then, since $X^{-1}(I)$ is a subset of S, it is necessarily an event.

A random variable is nothing more than a real-valued function defined on the elements of the sample space. It is random in the sense that the value it assumes depends on the random outcome of an experiment which specifies one element in the sample space, the domain of the random variable.

DEFINITION

A random variable $X:S \rightarrow \mathbb{R}$ is called *discrete* if any of the following is true:

1. S is a finite set.
2. $X(S)$ is a finite set, although S itself may be infinite.
3. $X(S)$ is a countable set.

We shall discuss only discrete random variables in this chapter, postponing our discussion of nondiscrete random variables until Chapter 8. We shall see that discrete random variables are quite different from nondiscrete random variables.

Example 1 A coin is tossed three times, giving rise to the sample space

$$S = \{\text{HHH, HHT, HTH, HTT, THH, THT, TTH, TTT}\}$$

Let X be the random variable which assigns to each element in S the largest number of successive heads which occur. Therefore,

$X(\text{HHH}) = 3 \qquad X(\text{THH}) = 2$

$X(\text{HHT}) = 2 \qquad X(\text{THT}) = 1$

$X(\text{HTH}) = 1 \qquad X(\text{TTH}) = 1$

$X(\text{HTT}) = 1 \qquad X(\text{TTT}) = 0$

Example 2 Suppose that two dice are rolled and we use the sample space

$$S = \{(a, b) : a, b = 1, 2, 3, 4, 5, 6\}$$

We define random variables X and Y by the rules

$$X[(a, b)] = a + b$$

and

$$Y[(a, b)] = \max\{a, b\}$$

That is, X is the sum of the two faces, and Y is the larger of the two faces.

Example 3 For each subset A of a sample space S, we can define a random variable I_A, called the *indicator* of the event A, as follows:

$$I_A(x) = \begin{cases} 1 & \text{if } x \in A \\ 0 & \text{if } x \notin A \end{cases} \quad \text{for any } x \in S$$

When the indicator of A takes the value 1, this indicates that the event A occurred. That is, the experiment resulted in a sample element $x \in A$.

We shall have occasion soon to speak of functions of a random variable. If $X:S \rightarrow \mathbb{R}$ is a random variable and $f:\mathbb{R} \rightarrow \mathbb{R}$ is a real-valued function defined on the set of real numbers, then we can define a random variable $f(X)$ as follows: For any $x \in S$, let $f(X)(x) = f(X(x))$. That is, we first apply the random variable X to the sample element x, obtaining a real number $X(x)$, to which we apply the function f, obtaining another real number $f(X(x))$.

Familiar functions that we shall use are $f(x) = x^2$, $f(x) = kx$, and $f(x) = x + k$, where k is a real constant. These functions give rise to random variables denoted X^2, kX, and $X + k$ which are defined as follows:

$$X^2(x) = [X(x)]^2$$
$$kX(x) = k \cdot X(x)$$
$$(X + k)(x) = X(x) + k$$

for any $x \in S$.

4.2 Probability and Distribution Functions

For the next three sections we will assume that our random variables are defined on a finite sample space

$$S = \{x_1, x_2, \ldots, x_n\}$$

We will use the notation

$$\{X = a\} = \{x \in S : X(x) = a\}$$

for the event that the random variable X assumes the value $a \in \mathbb{R}$. Since S is itself finite, it necessarily follows that the range of S under X must also be finite. We shall use the notation

$$X(S) = \{a_1, a_2, \ldots, a_m\}$$

for the range of the random variable X. Notice that $m = n[X(S)]$ is not necessarily equal to $n = n(S)$. It is possible that the random variable X may assign the same value to different sample elements.

DEFINITION

Given a random variable X defined on a probability space S with probability function P, the function $f : \mathbb{R} \to \mathbb{R}$ defined by

$$f(a) = P(\{X = a\})$$

is called the *probability* or *frequency* function of X.

The probability function f simply gives the probability that the random variable X assumes the value $a \in \mathbb{R}$.

Instead of writing $P(\{X = a\})$, we shall write simply $P(X = a)$.

Since $X(S)$ is a finite set, $\{X = a\} = \varnothing$, and consequently, $f(a) = 0$ for all but a finite number of real values. As a result, the probability function of X can be exhibited by means of a table, called a *frequency table*, as follows:

a_1	a_2	$\cdots$	a_m
$f(a_1)$	$f(a_2)$	$\cdots$	$f(a_m)$

The collection of sets $\{X = a_1\}, \{X = a_2\}, \ldots, \{X = a_m\}$ partitions the sample space. It is possible, however, that $P(X = a_k) = 0$ for some k, so care must be taken if conditional probabilities relative to any of these sets are used. As a direct consequence of this partition of S, we have

Proposition 4.1

$$\sum_{i=1}^{m} f(a_i) = 1$$

Proof: Since S is the union of the mutually disjoint sets $\{X = a_i\}$, we have

$$1 = P(S) = P\left(\bigcup_{i=1}^{m} X = a_i\right) = \sum_{i=1}^{m} P(X = a_i) = \sum_{i=1}^{m} f(a_i)$$

Example 1 Two coins are tossed, the first being weighted so as never to fall heads. In the sample space

$$S = \{HH, HT, TH, TT\}$$

we define the probabilities $P(HH) = P(HT) = 0$ and $P(TH) = P(TT) = \frac{1}{2}$. If X is the random variable that counts the number of heads occurring, then $X(HH) = 2$, $X(HT) = X(TH) = 1$ and $X(TT) = 0$. Therefore,

$f(0) = P(X = 0) = P(TT) = \frac{1}{2}$
$f(1) = P(X = 1) = P(HT, TH) = \frac{1}{2}$
$f(2) = P(X = 2) = P(HH) = 0$

The same information is presented in the following frequency table

a_i	0	1	2
$f(a_i)$	$\frac{1}{2}$	$\frac{1}{2}$	0

Notice that the partition part $\{X = 2\}$ has probability 0.

Example 2 Using the random variable X from Example 1 in Section 4.1, and assuming the coin is biased so that $P(H) = \frac{2}{3}$, we have

$f(0) = P(X = 0) = P(TTT) = \frac{1}{3} \cdot \frac{1}{3} \cdot \frac{1}{3} = \frac{1}{27}$

$f(1) = P(X = 1) = P(HTH, HTT, THT, TTH)$
$\qquad\qquad\qquad\quad = P(HTH) + P(HTT, THT, TTH)$
$\qquad\qquad\qquad\quad = (\frac{2}{3} \cdot \frac{1}{3} \cdot \frac{2}{3}) + 3 \cdot (\frac{2}{3} \cdot \frac{1}{3} \cdot \frac{1}{3})$
$\qquad\qquad\qquad\quad = \frac{4}{27} + \frac{6}{27} = \frac{10}{27}$

$f(2) = P(X = 2) = P(HHT, THH) = 2 \cdot (\frac{2}{3} \cdot \frac{2}{3} \cdot \frac{1}{3}) = \frac{8}{27}$

$f(3) = P(X = 3) = P(HHH) = \frac{2}{3} \cdot \frac{2}{3} \cdot \frac{2}{3} = \frac{8}{27}$

Therefore, we have the frequency table

a_i	0	1	2	3
$f(a_i)$	$\frac{1}{27}$	$\frac{10}{27}$	$\frac{8}{27}$	$\frac{8}{27}$

Notice that $f(0) + f(1) + f(2) + f(3) = 1$. This should always be checked to determine if an error has been made in calculating the probabilities $f(a_i)$.

Example 3 For the random variable X in Example 2 of Section 4.1, we have the following frequency table:

a_i	2	3	4	5	6	7	8	9	10	11	12
$f(a_i)$	$\frac{1}{36}$	$\frac{2}{36}$	$\frac{3}{36}$	$\frac{4}{36}$	$\frac{5}{36}$	$\frac{6}{36}$	$\frac{5}{36}$	$\frac{4}{36}$	$\frac{3}{36}$	$\frac{2}{36}$	$\frac{1}{36}$

To determine the $f(a_i)$, we must count the number of elements in the thirty-six–element uniform space $S = \{(a, b) : a, b = 1, 2, 3, 4, 5, 6\}$ favorable to the event $\{X = a_i\}$, the sum of the faces being a_i. If we picture this sample space in the form of a matrix, as indicated in Figure 4.1, then the subsets of the sample space corresponding to the events $\{X = a_i\}$ appear along the reverse diagonals, as indicated. Realizing this, it is then quite easy to calculate the probabilities $f(a_i)$.

Figure 4.1

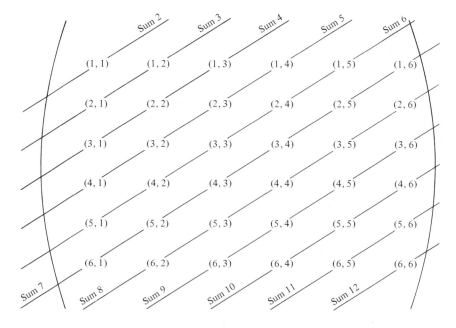

Example 4 For the random variable Y of Example 2 in Section 4.1, we have the frequency table:

a_i	1	2	3	4	5	6
$f(a_i)$	$\frac{1}{36}$	$\frac{3}{36}$	$\frac{5}{36}$	$\frac{7}{36}$	$\frac{9}{36}$	$\frac{11}{36}$

Looking at the sample space once again as a matrix, we see that the events $\{X = a_i\}$ occur in the form of backwards L's of length and height a_i, as indicated in Figure 4.2. Hence $n(X = a_i) = 2a_i - 1$, and so

$$f(a_i) = P(X = a_i) = \frac{2a_i - 1}{36}$$

It is often helpful to visualize the distribution of a random variable. We shall do this by means of a graph known as a *histogram*, which is quite similar in appearance to a bar graph. The values a_i are displayed on the x-axis, while the values $f(a_i)$ are displayed on the y-axis. Centered over each a_i is a bar whose height is chosen so that the area of the bar is $f(a_i)$. The base of the bar extends from the midpoint between a_{i-1} and a_i to the midpoint between a_i and a_{i+1}. The histogram of the random variable Y above appears in Figure 4.3.

Example 5 A sample of 3 items is selected at random from a box containing 12 items, 3 of which are defective. Let X be the random variable that counts the number of defective items in the sample. Then X can assume any of the

Figure 4.2

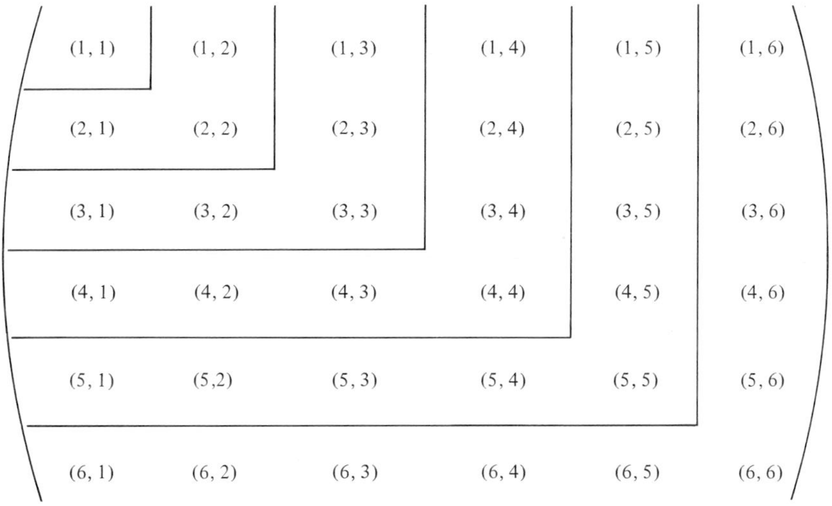

values $a_1 = 0$, $a_2 = 1$, $a_3 = 2$, and $a_4 = 3$, and

$$f(0) = P(X = 0) = \frac{C_0{}^3 \cdot C_3{}^9}{C_3{}^{12}} = \frac{84}{220}$$

$$f(1) = P(X = 1) = \frac{C_1{}^3 \cdot C_2{}^9}{C_3{}^{12}} = \frac{108}{220}$$

$$f(2) = P(X = 2) = \frac{C_2{}^3 \cdot C_1{}^9}{C_3{}^{12}} = \frac{27}{220}$$

$$f(3) = P(X = 3) = \frac{C_3{}^3 \cdot C_0{}^9}{C_3{}^{12}} = \frac{1}{220}$$

In the form of a frequency table, we have

a_i	0	1	2	3
$f(a_i)$	$\frac{84}{220}$	$\frac{108}{220}$	$\frac{27}{220}$	$\frac{1}{220}$

Example 6 Team A and team N compete in the World Series, and we make the assumption that the two teams are equally likely to win any particular game. Our sample space is the set of all possible game-by-game outcomes of the series. For example, the event $ANAANA$ represents a series won by team A in six games with team N winning only the second and fifth games. On this sample space, we define a random variable X to be the number of games in the series. Hence, $X(ANAANA) = 6$, for example. The random variable X can assume any of the values $a_1 = 4$, $a_2 = 5$, $a_3 = 6$, and $a_4 = 7$. The set $\{X = 5\}$, for example, consists of all five-tuples $(x_1, x_2, \ldots, x_5)$, where $x_i = A$ or $x_i = N$, and such that either A or N occurs exactly four times. If we denote by A_k the event that team A wins the series in k games,

Figure 4.3

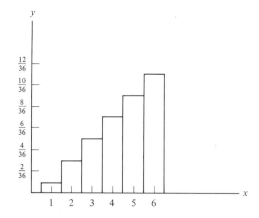

and by N_k the event that team N wins the series in k games, then $\{X = k\} = A_k \cup N_k$. Hence,

$$f(k) = P(X = k) = P(A_k \cup N_k) = P(A_k) + P(N_k)$$

Since the two teams are equally likely to win any game, $P(A_k) = P(N_k)$, and therefore, $f(k) = 2P(A_k)$. So the problem of determining the distribution of X has been reduced to that of determining the probabilities $P(A_k)$ for $k = 4, 5, 6$, and 7. Now each sample element in A_k is a k-tuple containing four A's, one of which must be the last (kth) coordinate. Each such k-tuple has probability $(\frac{1}{2})^4 \cdot (\frac{1}{2})^{k-4} = (\frac{1}{2})^k$, and there are a total of C_3^{k-1} such k-tuples, there being that many ways to place three A's among the first $k - 1$ coordinates. Thus,

$$P(A_k) = C_3^{k-1} \cdot (\tfrac{1}{2})^k$$

and so

$$f(k) = 2C_3^{k-1} \cdot (\tfrac{1}{2})^k = C_3^{k-1} \cdot (\tfrac{1}{2})^{k-1}$$

This gives us the frequency table

a_i	4	5	6	7
$f(a_i)$	$\frac{1}{8}$	$\frac{1}{4}$	$\frac{5}{16}$	$\frac{5}{16}$

Notice that the probabilities of six and seven game World Series are identical. Do you understand why this is so?

We will use the notation

$$\{X \le a\} = \{x \in S : X(x) \le a\}$$

for the set of those sample points whose functional value under X does not exceed a.

DEFINITION

The function $F : \mathbb{R} \to \mathbb{R}$ defined by $F(a) = P(X \le a)$, for any $a \in \mathbb{R}$, is called the *distribution (cumulative distribution)* function of the random variable X.

The (cumulative) distribution function is defined for all real values and is nonzero for all values which exceed the smallest value taken by the random variable. It is a rather simple task to calculate the probabilities of F once the distribution of X is known.

Example 7 Referring again to the random variable Y from Example 4 above, we construct the distribution table

a_i	1	2	3	4	5	6
$F(a_i)$	$\frac{1}{36}$	$\frac{4}{36}$	$\frac{9}{36}$	$\frac{16}{36}$	$\frac{25}{36}$	1

as follows: For each value $a_i = 1, 2, 3, 4, 5$, and 6, we express the event

$\{X \le a_i\}$ in terms of the events $\{X = a_i\}$. For example,

$$\{X \le 3\} = \{X = 1\} \cup \{X = 2\} \cup \{X = 3\}$$

which is simply a way of saying that X can take a value not exceeding 3 if and only if X takes one of the values 1, 2, or 3. Hence,

$$\begin{aligned} F(3) = P(X \le 3) &= P(X = 1) + P(X = 2) + P(X = 3) \\ &= f(1) + f(2) + f(3) \\ &= \tfrac{1}{36} + \tfrac{3}{36} + \tfrac{5}{36} = \tfrac{9}{36} \end{aligned}$$

Actually, the table given above is not complete—probabilities such as $F(-2)$, $F(3.5)$, and $F(10)$ are not included. The last two probabilities, in fact, are not zero. Since we have

$$\{X \le -2\} = \varnothing$$
$$\{X \le 3.5\} = \{X \le 3\}$$

and

$$\{X \le 10\} = S$$

we must have

$$F(-2) = P(\varnothing) = 0$$
$$F(3.5) = F(3) = \tfrac{9}{36}$$

and

$$F(10) = P(S) = 1$$

Figure 4.4

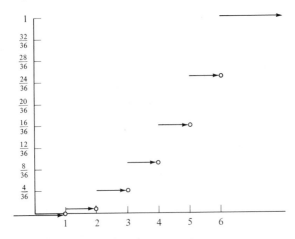

In general, if $a < 1$, then $\{X \leq a\} = \varnothing$, so $F(a) = 0$; if $a \geq 6$, then $\{X \leq a\} = S$, so $F(a) = 1$. On the other hand, the cumulative probability of any number that occurs between two consecutive values assumed by the random variable is equal to the cumulative probability of the smaller of those two values. The distribution function we have just obtained for the random variable Y is graphed in Figure 4.4. This is the graph of what is called a *step function*. The following proposition indicates, among other things, that the graph of the distribution function of a random variable with a finite range is always of this form.

Proposition 4.2

Let F be the distribution function of a discrete random variable X with finite range

$$X(S) = \{a_1 < a_2 < \cdots < a_m\}$$

Then

(a) there exist values L and M such that $F(a) = 0$ if $a < L$ and $F(a) = 1$ if $a \geq M$.
(b) F is a nondecreasing function; that is, if $a \geq b$, then $F(a) \geq F(b)$.
(c) F is a step function with a finite number of steps.
(d) If $b > a_1$, then $F(b) = \sum_{a_i \leq b} f(a_i)$.
(e) $P(a < X < b) = F(b) - F(a)$ for any pair of real numbers $a < b$.

Proof:

(a) We let $L = a_1$ and $M = a_m$. Then, if $a < L$, we have $\{X \leq a\} = \varnothing$, so that $F(a) = P(\varnothing) = 0$. On the other hand, if $a \geq M$, we have $\{X \leq a\} = S$, so that $F(a) = P(S) = 1$.
(b) If $a \geq b$, then $\{X \leq b\} \subseteq \{X \leq a\}$. For if $x \in \{X \leq b\}$, then $X(x) \leq b \leq a$, so that $x \in \{X \leq a\}$. Hence, by Proposition 2.3, $F(b) = P(X \leq b) \leq P(X \leq a) = F(a)$.
(c) If $a_i \leq a < a_{i+1}$, then $\{X \leq a\} = \{X \leq a_i\}$, and so $F(a) = P(X \leq a) = P(X \leq a_i) = F(a_i)$. Therefore, F is constant over each interval $[a_i, a_{i+1})$. However, since we can write

$$\{X \leq a_{i+1}\} = \{X \leq a_i\} \cup \{X = a_{i+1}\}$$

we have $F(a_{i+1}) = F(a_i) + f(a_{i+1})$, which means that F has a jump of $f(a_{i+1})$ at a_{i+1}, for $i = 0, 1, \ldots, m - 1$. That is, F is a step function.
(d) If $a_i \leq b < a_{i+1}$

$$F(b) = P(X \leq b) = P(X \leq a_i) = \sum_{j=1}^{i} P(X = a_j)$$

where the last equality follows from (c) by iteration. But

$$\sum_{j=1}^{i} P(X = a_j) = \sum_{a_i \leq b} P(X = a_i) = \sum_{a_i \leq b} f(a_i)$$

which is the desired form.

(e) If we assume that $a_{i-1} \leq a < a_i$ and $a_{i+k} \leq b < a_{i+k+1}$, where $k \geq 0$, then by (d) we have

$$F(a) = \sum_{j=1}^{i-1} f(a_j)$$

and

$$F(b) = \sum_{j=1}^{i+k} f(a_j)$$

Consequently,

$$F(b) - F(a) = \sum_{j=i}^{i+k} f(a_j)$$

$$= P(a_i \leq X \leq a_{i+k})$$
$$= P(a < X < b) \qquad \text{(why?)}$$

EXERCISES

1. If I_A is the indicator of the event A in S, prove that $P(I_A = 1) = P(A)$ and $P(I_A = 0) = P[C(A)]$.

2. Let X be the number of heads in four tosses of a fair coin. Determine the probability function of X.

3. Let X be the number of 6's in two rolls of a fair die. Determine the probability function of X.

4. Three balls are drawn without replacement from an urn containing eight red and eight black balls. A black ball is worth one point and a red ball is worth two points. If X denotes the total number of points for the three balls, determine the probability function of X.

5. Let X denote the number of spades in a thirteen-card bridge hand. Determine the probability function of X.

6. Let X denote the number of aces in a five-card poker hand. Find the probability function of X.

7. Let X represent the difference between the number of heads and the number of tails in n tosses of a fair coin. Find the probability function of X.

8. Urn A contains two white and four red balls, urn B eight white and four red balls, and urn C one white and three red balls. A ball is drawn from each urn, and we let X denote the number of whites drawn. Determine the probability function of X.

9. Box A contains nine $1 bills and one $10 bill, while box B contains five $1 bills, seven $5 bills, and three $10 bills. A box is selected at random and a bill is drawn from that box. If X represents the dollar value of the bill drawn, determine the probability function of X and find the probability that this dollar value exceeds $5.

10. Let X be the number of vowels in two-letter "words" formed from the letters of the alphabet with no restrictions imposed. Find the probability function of X. Repeat the problem, assuming that no letter may be repeated in the formation of these words.

4.3 Means, Modes, and Medians

The concept of average, which we are about to discuss, is one that is familiar to everybody. From the average grade on an examination to the day's average temperature to the Dow Jones' Index, the average number of sales for 30 industrial stocks, the term average has become part of our everyday life.

In order to motivate our formal definition of the average of a random variable, we take a look at how we might calculate the average of the following six numbers: 75, 90, 87, 75, 75, and 90. Obviously, we could simply calculate

$$\text{average} = \frac{75 + 90 + 87 + 75 + 75 + 90}{6} = 82$$

or we could rearrange this sum as

$$\text{average} = \frac{75 + 75 + 75 + 87 + 90 + 90}{6}$$

$$= \frac{(3 \cdot 75) + (1 \cdot 87) + (2 \cdot 90)}{6}$$

$$= (75 \cdot \tfrac{3}{6}) + (87 \cdot \tfrac{1}{6}) + (90 \cdot \tfrac{2}{6})$$

If we were to write down a relative frequency table for the six numbers in our example, we would get

a_i	75	87	90
$f(a_i)$	$\frac{3}{6}$	$\frac{1}{6}$	$\frac{2}{6}$

What we have in the last expression above is a weighted average of the three numbers 75, 87, and 90, with the weights being the relative frequencies of the three numbers. What we also have is the basic concept behind the definition of the average value of a random variable.

DEFINITION

Let X be a random variable with distribution

a_1	a_2	$\cdots$	a_m
$f(a_1)$	$f(a_2)$	$\cdots$	$f(a_m)$

The *average value* (*mean, expected value*) of X, denoted by either μ_x or $E(X)$, is defined to equal

$$a_1 \cdot f(a_1) + a_2 \cdot f(a_2) + \cdots + a_m \cdot f(a_m)$$

That is,

$$E(X) = \sum_{i=1}^{m} a_i \cdot f(a_i)$$

In other words, we calculate the mean of a random variable by multiplying the different values assumed by the random variable by their respective probability of occurrence, and adding.

Example 1 If X is the random variable from Example 3 in Section 4.2 (the sum of two dice), then

$$E(X) = (2 \cdot \tfrac{1}{36}) + (3 \cdot \tfrac{2}{36}) + \cdots + (12 \cdot \tfrac{1}{36}) = 7$$

Example 2 If Y is the random variable from Example 4 in Section 4.2 (the larger value on the two dice), then

$$E(Y) = (1 \cdot \tfrac{1}{36}) + (2 \cdot \tfrac{3}{36}) + \cdots + (6 \cdot \tfrac{11}{36}) = \tfrac{161}{36}$$

Note that $E(X) = 7$, a value that the random variable X actually assumes—the sum of two dice might possibly be 7. On the other hand, $E(Y) = \tfrac{161}{36} = 4.47$, which is not a value that Y may assume—the larger value of two dice can never be 4.47. In general, the mean of a random variable need not be a value actually assumed by the random variable.

Example 3 The average number of games in a World Series (Example 6, Section 4.2) equals

$$(4 \cdot \tfrac{1}{8}) + (5 \cdot \tfrac{1}{4}) + (6 \cdot \tfrac{5}{16}) + (7 \cdot \tfrac{5}{16}) = 5.81$$

The following result is a formula for calculating the mean of a random variable which is a function of another random variable.

Proposition 4.3

Let X be a random variable with distribution as above, $g: \mathbb{R} \to \mathbb{R}$ a function, and $Y = g(X)$. Then

$$E(Y) = E(g(X)) = \sum_{i=1}^{m} g(a_i) \cdot f(a_i)$$

Proof: If $Y(S) = \{b_1, b_2, \ldots, b_s\}$, then by definition

$$E(Y) = \sum_{j=1}^{s} b_j \cdot P(Y = b_j)$$

but

$$\{Y = b_j\} = \{x \in S : Y(x) = b_j\}$$
$$= \{x \in S : g(X(x)) = b_j\}$$
$$= \bigcup_{g(a_i)=b_j} \{x \in S : X(x) = a_i\}$$

In this last step, we have partitioned the set $\{x \in S : Y(x) = b_j\}$ by the intermediate values $a \in X(S)$ satisfying $g(a) = b_j$. It is possible that there are several values $a \in X(S)$ satisfying the condition $g(a) = b_j$, as indicated in Figure 4.5. As a result, we have

$$P(Y = b_j) = \sum_{g(a_i)=b_j} P(X = a_i)$$

and so

$$E(Y) = \sum_{j=1}^{s} b_j \left[\sum_{g(a_i)=b_j} P(X = a_i) \right]$$

$$= \sum_{j=1}^{s} \sum_{g(a_i)=b_j} g(a_i) \cdot P(X = a_i)$$

$$= \sum_{i=1}^{m} g(a_i) \cdot P(X = a_i)$$

the last step being valid because each a_i is mapped to a unique b_j, and therefore, each $g(a_i)$ occurs exactly once in the double sum.

Figure 4.5

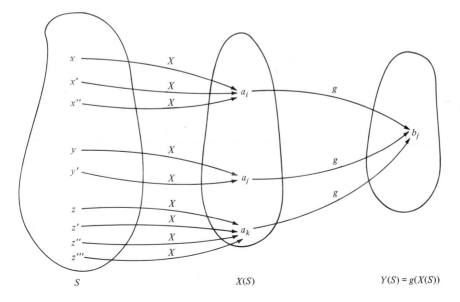

S $X(S)$ $Y(S) = g(X(S))$

Thus, we have a formula for evaluating $E(Y)$ without specifically having to construct the frequency table for Y.

Example 4 If X is the random variable recording the number of dots on the face of a die, then X has probability function $f(a_i) = \frac{1}{6}$, for $a_i = 1, 2, 3, 4, 5, 6$. If Y is the random variable X^2, then by Proposition 4.3 we have

$$E(Y) = (1^2 \cdot \tfrac{1}{6}) + (2^2 \cdot \tfrac{1}{6}) + \cdots + (6^2 \cdot \tfrac{1}{6}) = \tfrac{91}{6}$$

In this case, $g(a_i) = a_i^2$, and these terms are multiplied by the probabilities $f(a_i) = \frac{1}{6}$. The probability function of Y, which was not needed in these calculations, is $f(a_i) = \frac{1}{6}$, for $a_i = 1, 4, 9, 16, 25, 36$.

Corollary 4.4

If X is a random variable and a, b are real numbers, then

$$E(aX + b) = aE(X) + b$$

Proof: We apply Proposition 4.3 to the function $g(x) = ax + b$. Then $g(X) = aX + b$, and hence,

$$E(aX + b) = \sum_{i=1}^{m} g(a_i) \cdot f(a_i)$$

$$= \sum_{i=1}^{m} (aa_i + b) \cdot f(a_i)$$

$$= a \cdot \sum_{i=1}^{m} a_i \cdot f(a_i) + b \cdot \sum_{i=1}^{m} f(a_i)$$

$$= a \cdot E(X) + b$$

Corollary 4.5

If X is a random variable with distribution as stated above, then

(a) $E(X^2) = \sum_{i=1}^{m} a_i^2 \cdot f(a_i)$

(b) $E[X - E(X)] = \sum_{i=1}^{m} [a_i - E(X)] \cdot f(a_i)$

(c) $E\{[X - E(X)]^2\} = \sum_{i=1}^{m} [a_i - E(X)]^2 \cdot f(a_i)$

Proof: The proof is a simple application of Proposition 4.3, where in

(a) $g(x) = x^2$
(b) $g(x) = x - E(X)$
(c) $g(x) = [x - E(X)]^2$

The concept of mean arose, as did many concepts in probability theory, from the study of games of chance during the middle ages. Associated with each play of such a game is a random variable, the player's profit or loss. If the expected value of this random variable is positive, the game is said to be favorable to the player. If the expected value is zero, the game is said to be fair.

Example 5 A player rolls a fair die. If a prime number p (2, 3, 5) occurs on the top face, he wins p dollars. If a nonprime number q occurs, the player loses q dollars. We wish to determine if this game is favorable to the player, and to do so, we let

X = the player's net gain on one play

We find that X has the following distribution

a_i	-1	$+2$	$+3$	-4	$+5$	-6
$f(a_i)$	$\frac{1}{6}$	$\frac{1}{6}$	$\frac{1}{6}$	$\frac{1}{6}$	$\frac{1}{6}$	$\frac{1}{6}$

For example, the probability that $X = +3$ is the same as the probability that the die shows 3, which is $\frac{1}{6}$. Hence,

$$E(X) = (\tfrac{1}{6})(-1 + 2 + 3 - 4 + 5 - 6) = -\tfrac{1}{6}$$

and so the player would expect to lose one-sixth of every dollar that he bets.

Example 6 A roulette wheel consists of 38 positions numbered 0, 00, 1, 2, ..., 36. You bet \$1 on any number between 1 and 36, and get back \$36 if the wheel stops on your number. Otherwise, you lose your dollar. Using the same random variable X as above, we find the distribution in this case to be

a_i	$+35$	-1
$f(a_i)$	$\frac{1}{38}$	$\frac{37}{38}$

Hence,

$$E(X) = (+35 \cdot \tfrac{1}{38}) - (1 \cdot \tfrac{37}{38}) = -\tfrac{2}{38}$$

so the roulette player expects to lose one-nineteenth of every dollar that he bets. Notice that this game would be fair were it not for the two house numbers 0 and 00. Without them, X would have distribution

a_i	$+35$	-1
$f(a_i)$	$\frac{1}{36}$	$\frac{35}{36}$

and expected value

$$E(X) = (+35 \cdot \tfrac{1}{36}) - (1 \cdot \tfrac{35}{36}) = 0$$

Example 7 Pascal's famous wager with Chevalier de Méré on belief in the existence of God is based on the same type of reasoning as used in the preceding two examples. Pascal reasoned that, even if the probability p that God exists were rather small, the expected gain from believing in the existence of God, eternal happiness, is so large to make it a wise choice to believe in God. In other words,

$p \cdot$ eternal happiness $+ (1 - p) \cdot$ nothing

is extremely large, no matter how small you may wish to believe p.

The main purpose of the mean is as a measure of central tendency. It is the fulcrum, the center of gravity, the balance point of the distribution of probability on the real line. It is the average value one would expect if the experiment were conducted for a long series of trials.

However, the mean is not the only value that might be used to estimate the average or central value of a random variable. It is, however, the most widely used because it lends itself to far more theoretical applications than any other choice.

We shall now briefly mention two other measures of central tendency, the mode and the median.

DEFINITION

A possible value of a random variable X that occurs with probability at least as large as the probability of any other value of X is said to be *a mode* of X. That is,

$P(X = \text{a mode}) \geq P(X = \text{any other value})$

The mode is therefore the value taken most frequently by X, although it should be emphasized that several values might be equally "most frequent." Therefore, the mode of a random variable need not be unique.

Example 8   The mode of the random variable X (sum of the two faces) from Example 3 in Section 4.2 is 7 since

$P(X = 7) = \frac{6}{36} > P(X = \text{any other value})$

Example 9   If X denotes the number of games in a World Series, then the mode of X is either 6 or 7, since $P(X = 6) = P(X = 7) = \frac{5}{16}$, while both $P(X = 4)$ and $P(X = 5)$ are less than $\frac{5}{16}$.

DEFINITION

A value M taken by the random variable X that satisfies both $P(X \leq M) \geq \frac{1}{2}$ *and* $P(X \geq M) \geq \frac{1}{2}$ is called *a median* of X.

The median is the central value in the probability distribution and can best be calculated by means of the cumulative distribution function. The smallest value a_i such that $F(a_i) \geq \frac{1}{2}$ is always a median. By definition,

$F(a_i) \geq \frac{1}{2}$ implies that $P(X \leq a_i) \geq \frac{1}{2}$, and the fact that a_i is the smallest such value implies that $F(a_{i-1}) < \frac{1}{2}$. Therefore,

$$P(X \geq a_i) = 1 - P(X < a_{i-1}) = 1 - F(a_{i-1}) \geq \frac{1}{2}$$

Example 10   The median sum of two dice is 7 since

$$P(X \leq 7) = P(X \geq 7) = \frac{21}{36} > \frac{1}{2}$$

Example 11   The median number of games in a World Series is six since $P(X \leq 6) = \frac{11}{16}$ and $P(X \geq 6) = \frac{10}{16}$.

In both Examples 10 and 11 the median is unique. However, this is not necessarily true, as the following example shows.

Example 12   If X is a random variable with distribution

a_i	1	2	3	4
$f(a_i)$	0	$\frac{1}{2}$	$\frac{1}{2}$	0

then $P(X \leq 2) = \frac{1}{2}$ and $P(X \geq 2) = 1$, so that 2 is a median. However, $P(X \leq 3) = 1$ and $P(X \geq 3) = \frac{1}{2}$, so 3 is also a median.

Notice that in the case where X was the sum of two dice, the mean, median, and mode all equal 7. As the other examples indicate, it is not necessarily true that all three be equal. Finally, the possibility that there is more than one median or mode is another reason why the mean is the most preferred measure of the central value of a random variable.

EXERCISES

1. Calculate the means of the random variables in the exercises of the preceding section (except numbers 5, 6, 7).
2. A drunk wishes to open the front door of his home. He has five keys on his keychain. If he is sober enough to eliminate unsuccessful keys, what is the expected number of keys that he will try before opening the door?
3. An ice cream vendor can sell 100 ice cream bars on a warm day but only 35 on a cooler day. He makes a profit of 5¢ on every bar sold but loses 5¢ for every bar not sold. If the probability that a day will be warm is $\frac{2}{3}$, find his expected profit if he orders 35, 50, or 100 bars.
4. A coin is tossed at most three times. You win \$$k$ if the first head occurs on the kth toss. What is your expected profit?
5. On the first toss of a coin, you bet \$1 on heads. If a tail occurs, you then bet \$2 on heads on the second toss. It tails comes up again, you bet \$4 on heads on the third and final toss. Once you win, you stop. What is your expected profit?
6. One thousand lottery tickets are sold at \$1 per ticket. If there is one grand prize of \$500, four prizes of \$100, and five prizes of \$10, what is the expected value of an individual lottery ticket?
7. Write a computer program that will calculate the mean of a random variable.

4.4 Variance and Standard Deviation

We have seen that the mean is a weighted average of the values taken by a random variable. However, the random variables X and Y defined by the distributions below both have mean 2.75, and yet are clearly quite different types of random variables.

X:

	1	2	3	4
	$\frac{1}{8}$	$\frac{2}{8}$	$\frac{3}{8}$	$\frac{2}{8}$

Y:

	-20	1	10	30
	$\frac{2}{8}$	$\frac{2}{8}$	$\frac{3}{8}$	$\frac{1}{8}$

The values assumed by X are all very close to the mean, while the values assumed by Y are spread quite a bit both above and below the mean. Hence, the mean cannot stand alone as a description of a random variable. As a supplement to the mean, we seek some measure as to how much, or to what extent, the values assumed by the random variable spread out away from the mean. The first idea that might come to mind is to calculate the average of the values $a_i - \mu_X$, the differences between the values assumed by the random variable and its mean. This is equivalent to calculating the mean of the random variable $X - \mu_X$, which, according to Corollary 4.4, equals

$$E(X - \mu_X) = E(X) - \mu_X = 0$$

In other words, the signed differences $a_i - \mu_X$ cancel each other. Consequently, instead of considering the terms $a_i - \mu_X$, perhaps instead we should consider the terms $|a_i - \mu_X|$, the distances between the values assumed by X and the mean of X. In other words, we should consider the random variable $|X - \mu_X|$ rather than $X - \mu_X$. This is all very logical and would work very well were it not for the fact that $|X - \mu_X|$ does not lend itself very readily to the mathematical manipulations we shall need to make. As a substitute, we shall use the squared differences $(a_i - \mu_X)^2$, and make the following definition. We assume that X is a random variable with distribution as in the preceding section.

DEFINITION

The *variance* of the random variable X, denoted Var(X), is defined by

$$\mathrm{Var}(X) = E[(X - \mu_X)^2] = \sum_{i=1}^{m} (a_i - \mu_X)^2 f(a_i)$$

The *standard deviation* of X is defined to equal $+\sqrt{\mathrm{Var}(X)}$, and is denoted σ_X.

If the $\{a_i\}$, the values assumed by X, are in terms of a particular unit, such as dollars or pounds, for example, then Var(X) is in terms of the square of that unit, dollars squared or pounds squared, for example. For this reason, among others, we define the standard deviation, which is in terms of the same unit as is X.

We leave it as an exercise for the reader to prove that if $\text{Var}(X) = 0$, then the random variable X is almost a constant. By this we mean that X assumes the same value for all points in the sample space except possibly for some set of points whose probability is zero.

Example 1 Table 4.1 contains the calculations needed to determine the variance of the random variable X defined above. The mean of X is the sum of the numbers in the third column, while the variance of X is the sum of the numbers in the sixth column. The probability distribution of X is given in the first two columns.

Example 2 Table 4.2 contains the calculations needed to determine the variance of the random variable Y defined above. Since the variance of Y is 242.69, the standard deviation of Y is $+\sqrt{242.69} = 15.58$.

The larger the number of values a_i assumed by X becomes, the more tedious the above process becomes. And these calculations are most tedious when either $E(X)$ or some of the a_i are not whole numbers. In either situation, the following formula is of tremendous help:

Table 4.1
The Variance of X

a_i	$f(a_i)$	$a_i f(a_i)$	$(a_i - \mu_X)$	$(a_i - \mu_X)^2$	$(a_i - \mu_X)^2 f(a_i)$
1	$\frac{1}{8}$	$\frac{1}{8}$	$-\frac{7}{4}$	$\frac{49}{16}$	$\frac{49}{128}$
2	$\frac{2}{8}$	$\frac{4}{8}$	$-\frac{3}{4}$	$\frac{9}{16}$	$\frac{18}{128}$
3	$\frac{3}{8}$	$\frac{9}{8}$	$\frac{1}{4}$	$\frac{1}{16}$	$\frac{3}{128}$
4	$\frac{2}{8}$	$\frac{8}{8}$	$\frac{5}{4}$	$\frac{25}{16}$	$\frac{50}{128}$
		$\mu_X = \frac{22}{8}$			$\text{Var}(X) = \frac{120}{128}$

Table 4.2
The Variance of Y

a_i	$f(a_i)$	$a_i f(a_i)$	$(a_i - \mu_Y)$	$(a_i - \mu_Y)^2$	$(a_i - \mu_Y)^2 f(a_i)$
-20	$\frac{2}{8}$	$-\frac{40}{8}$	$-\frac{182}{8}$	517.56	129.39
1	$\frac{2}{8}$	$\frac{2}{8}$	$-\frac{14}{8}$	3.06	0.77
10	$\frac{3}{8}$	$\frac{30}{8}$	$\frac{58}{8}$	52.56	19.71
30	$\frac{1}{8}$	$\frac{30}{8}$	$\frac{218}{8}$	742.56	92.82
		$\mu_Y = \frac{22}{8}$			$\text{Var}(Y) = 242.69$

Proposition 4.6

$$\mathrm{Var}(X) = E(X^2) - E(X)^2$$

Proof: We simply expand the square $(a_i - \mu_X)^2$ in the definition of $\mathrm{Var}(X)$, and get:

$$\mathrm{Var}(X) = \sum_{i=1}^{m} (a_i - \mu_X)^2 \cdot f(a_i)$$

$$= \sum_{i=1}^{m} (a_i^2 - 2a_i\mu_X + \mu_x^2) \cdot f(a_i)$$

$$= \sum_{i=1}^{m} a_i^2 f(a_i) + \sum_{i=1}^{m} (-2\mu_X)a_i f(a_i) + \sum_{i=1}^{m} \mu_x^2 \cdot f(a_i)$$

$$= \sum_{i=1}^{m} a_i^2 f(a_i) - 2\mu_X \sum_{i=1}^{m} a_i f(a_i) + \mu_x^2 \sum_{i=1}^{m} f(a_i)$$

$$= \quad E(X^2) \quad - \quad 2E(X)^2 \quad + \quad E(X)^2$$

$$= \quad E(X^2) \quad - \quad E(X)^2$$

Example 1 (continued) Table 4.3 contains the calculations needed to determine the variance of X using Proposition 4.6 instead of the definition. Notice that there are only five columns this way, the third adding up to $E(X)$ and the fifth adding up to $E(X^2)$. Thus $E(X^2) = \frac{68}{8}$, and therefore, $\mathrm{Var}(X) = \frac{68}{8} - (\frac{22}{8})^2 = \frac{60}{64}$, which (as it should be) is the same answer that we found by the first method.

Example 3 If X represents the sum of two dice (Example 3, Section 4.2), we have seen that $E(X) = 7$, and since

$$E(X^2) = (2^2 \cdot \tfrac{1}{36}) + (3^2 \cdot \tfrac{2}{36}) + \cdots + (12^2 \cdot \tfrac{1}{36})$$

$$= \frac{1974}{36}$$

Table 4.3
The Variance of X by
Proposition 4.5

a_i	$f(a_i)$	$a_i f(a_i)$	a_i^2	$a_i^2 f(a_i)$
1	$\frac{1}{8}$	$\frac{1}{8}$	1	$\frac{1}{8}$
2	$\frac{2}{8}$	$\frac{4}{8}$	4	$\frac{8}{8}$
3	$\frac{3}{8}$	$\frac{9}{8}$	9	$\frac{27}{8}$
4	$\frac{2}{8}$	$\frac{8}{8}$	16	$\frac{32}{8}$
		$\mu_X = \frac{22}{8}$		$E(X^2) = \frac{68}{8}$

we have

$$\text{Var}(X) = \frac{1974}{36} - 7^2 = 54.8 - 49 = 5.8$$

Example 4 If Y represents the larger of two dice (Example 4, Section 4.2), then $E(Y) = 4.47$ and

$$E(Y^2) = (1^2 \cdot \tfrac{1}{36}) + (2^2 \cdot \tfrac{3}{36}) + \cdots + (6^2 \cdot \tfrac{11}{36})$$

$$= \frac{791}{36}$$

so that

$$\text{Var}(Y) = \frac{791}{36} - 4.47^2 = 21.94 - 19.98 = 1.99$$

Proposition 4.7

If a and b are real numbers, then

$$\text{Var}(aX + b) = a^2 \, \text{Var}(X) \quad \text{and} \quad \sigma_{aX+b} = |a|\,\sigma_X$$

Proof:

$$\begin{aligned}
\text{Var}(aX + b) &= E\{[(aX + b) - E(aX + b)]^2\} \\
&= E\{[aX + b - aE(X) - b]^2\} \\
&= E\{[aX - aE(X)]^2\} \\
&= E\{a^2 \cdot [X - E(X)]^2\} \\
&= a^2 \cdot E\{[X - E(X)]^2\} \\
&= a^2 \cdot \text{Var}(X)
\end{aligned}$$

Consequently,

$$\sigma_{aX+b} = +\sqrt{\text{Var}(aX + b)} = +\sqrt{a^2\text{Var}(X)} = |a|\,\sigma_X$$

Corollary 4.8

If a and b are real numbers, then

$$\text{Var}(aX) = a^2 \cdot \text{Var}(X) \quad \text{and} \quad \text{Var}(X + b) = \text{Var}(X)$$

The two results in Corollary 4.8 should seem logical. The values assumed by the random variable $X + b$ are simply the values assumed by X shifted b units to the right or left (depending on the sign of b). Their relative positions with respect to their mean $E(X + b) = E(X) + b$ are identical to the relative positions of the values of X about their mean $E(X)$. Consequently, one would expect the two random variables X and $X + b$ to have the same variance. On the other hand, the values assumed by aX are either spread out farther from $E(X)$ or concentrated much closer to $E(X)$, depending on whether $|a| > 1$ or $|a| < 1$. So one would expect the value of a to have some effect on the variance of aX.

By means of a simple transformation, we can transform any random variable into a random variable with mean 0 and variance 1. We shall see one example of the usefulness of such a transformation in Chapter 8 when we discuss the normal distribution.

DEFINITION

Let X be a random variable with mean μ_X and standard deviation σ_X. The random variable $X' = (X - \mu_X)/\sigma_X$ is called the *standardized* (or *normalized*) form of X.

Proposition 4.9

$E(X') = 0$ and $\text{Var}(X') = 1$

Proof: By Corollary 4.4,

$$E(X') = E\left(\frac{X - \mu_X}{\sigma_X}\right)$$

$$= \left(\frac{1}{\sigma_X}\right) E(X - \mu_X)$$

$$= \left(\frac{1}{\sigma_X}\right) [E(X) - \mu_X] = 0$$

By Proposition 4.7,

$$\text{Var}(X') = \text{Var}\left(\frac{X - \mu_X}{\sigma_X}\right)$$

$$= \left(\frac{1}{\sigma_X}\right)^2 \text{Var}(X - \mu_X)$$

$$= \left(\frac{1}{\sigma_X}\right)^2 \text{Var}(X) = 1$$

If we solve the equation $X' = (X - \mu_X)/\sigma_X$ for X, we find

$$X = \mu_X + (\sigma_X \cdot X')$$

Therefore, a standardized value of, say, $X' = 2$, means that the actual value of X is two standard deviations above the mean, while $X' = -3$ means that the value of X is three standard deviations below the mean. In general, the standardized value X' indicates how many standard deviations above or below the mean the X value lies.

The following result, due to the Russian mathematician Chebyshev, is of great importance in its own right. Also, a firm understanding of the concept involved greatly facilitates the study of statistics.

Theorem 4.10 Chebyshev's Theorem

Let X be a random variable with mean μ_X and standard deviation σ_X, and let c be any positive number. Then

$$P(|X - \mu_X| > c) < \frac{\text{Var}(X)}{c^2}$$

Proof: We start with the definition of $\text{Var}(X)$:

$$\text{Var}(X) = \sum_{i=1}^{m} (a_i - \mu_X)^2 \cdot f(a_i)$$

From this summation, we delete those terms (all of which are positive) for which $|a_i - \mu_X| \leq c$, and obtain

$$\text{Var}(X) \geq \sum_{|a_i - \mu_X| > c} (a_i - \mu_X)^2 \cdot f(a_i)$$

We further decrease this sum by replacing each $(a_i - \mu_X)^2$ by c^2, giving

$$\text{Var}(X) \geq \sum_{|a_i - \mu_X| > c} c^2 \cdot f(a_i) = c^2 \cdot \sum_{|a_i - \mu_X| > c} f(a_i)$$

But

$$\sum_{|a_i - \mu_X| > c} f(a_i) = P(|X - \mu_X| > c)$$

since we are adding in this summation the $P(X = a_i)$ for only those values of a_i satisfying $|a_i - \mu_X| > c$. Hence,

$$\text{Var}(X) \geq c^2 \cdot P(|X - \mu_X| > c)$$

from which Chebyshev's inequality follows.

What does Chebyshev's inequality tell us? The expression $P(|X - \mu_X| > c)$ is the probability that the random variable X will take a value that differs from its mean by more than c units, and we are told that this probability is bounded above by the value $\text{Var}(X)/c^2$. Note the effect that both c and $\text{Var}(X)$ have on this probability. If the value of c is large, the expression $\text{Var}(X)/c^2$ is correspondingly small. This is just another way of saying that the farther we venture from μ_X, that is, the larger we choose c, the less likely we are to find a value taken by X. On the other hand, if $\text{Var}(X)$ is small, the expression $\text{Var}(X)/c^2$ is also small. This is just another way of saying that a random variable with a small variance is less likely to assume a value beyond a given distance from its mean than is a random variable with a larger variance. Thus, the variance, in this sense, does control the dispersion of the values of the random variable about its mean, a characteristic we were hoping for when we made our definition of variance.

Chebyshev's inequality can be stated in its contrapositive form as follows:

Corollary 4.11

$$P(|X - \mu_X| \le c) \ge 1 - \frac{\text{Var}(X)}{c^2}$$

If we let $c = z\sigma_X$ in Corollary 4.11, we get

Corollary 4.12

$$P(|X - \mu_X| \le z\sigma_X) \ge 1 - \frac{1}{z^2}$$

That is, the probability that a random variable assumes a value within z standard deviations of its mean is at least $1 - (1/z^2)$. Choosing $z = 2$, for example, we find that

$$P(|X - \mu_X| \le 2\sigma_X) \ge 1 - (\tfrac{1}{2})^2 = \tfrac{3}{4}$$

Thus, at least three-fourths of the values taken by *any* random variable lie in the interval $[\mu_X - 2\sigma_X, \mu_X + 2\sigma_X]$. For $z = 3$, this probability is $\tfrac{8}{9}$; that is, at least eight-ninths of the values assumed by *any* random variable lie in the interval $[\mu_X - 3\sigma_X, \mu_X + 3\sigma_X]$. We should point out that since Chebyshev's inequality does apply to any random variable, the inequality is not very sharp. For example, although Chebyshev says that the probability that a random variable assumes a value within two standard deviations of its mean is at least 0.75, if the random variable happens to be normal (see Chapter 8), then this probability is actually 0.95.

Example 5   Suppose that X is a random variable with mean 50 and standard deviation 5. We wish to calculate the probability that X takes a value between 44 and 56. Since the difference between 44 and 50, or between 56 and 50, is 6, which is six-fifths of a standard deviation, we have $z = \tfrac{6}{5}$ and $P(44 \le X \le 56) \ge 1 - (1/z^2) = 1 - (\tfrac{5}{6})^2 = \tfrac{11}{36}$. Thus, the probability desired is at least $\tfrac{11}{36}$.

More examples on the use of Chebyshev's Theorem will come later. For one reference, see Example 7 in Section 5.2.

EXERCISES

1. Prove that if $\text{Var}(X) = 0$, then the random variable X is almost constant.
2. Calculate the variance of the random variables in the exercises to Section 4.2 (except numbers 5, 6, and 7).
3. If $E(X) = 50$ and $\text{Var}(X) = 4$, calculate $E(X^2)$, $\text{Var}(2X + 3)$, and $\text{Var}(-X)$.
4. An urn contains three balls numbered 1, two balls numbered 2, and one ball numbered 3. Two balls are drawn without replacement. Let X_i denote the number on the ith ball, $i = 1, 2$. Find $\text{Var}(X_1)$ and $\text{Var}(X_2)$.

5. Write a computer program that will calculate the variance of a random variable.
6. Suppose that X is a random variable with mean 100 and standard deviation 8. What is the probability that X takes a value between 90 and 110?

4.5 Discrete Random Variables

In this section, we remove the restriction that our random variables be defined on finite sample spaces and discuss discrete random variables in general. Any random variable which is defined on an infinite sample space, but has only a finite range, can be treated as if it were defined on a finite sample space. The only difference is that the probabilities $f(a_1), f(a_2), \ldots, f(a_m)$ are calculated in infinite sample spaces, using the techniques discussed in Sections 2.4 and 2.5.

The only remaining type of discrete random variable, then, is the one defined on an infinite sample space and having a countable range. So let us suppose that S is an infinite (countable or uncountable) sample space and that X is a random variable defined on S such that

$$X(S) = \{a_1, a_2, a_3, \ldots\}$$

a countable set of real numbers. The probabilities $f(a_i) = P(X = a_i)$, $i = 1, 2, 3, \ldots$, are calculated using the techniques of Sections 2.4 and 2.5, and we can form the countable distribution table

a_1	a_2	a_3	$\cdots$	a_k	$\cdots$
$f(a_1)$	$f(a_2)$	$f(a_3)$	$\cdots$	$f(a_k)$	$\cdots$

Imitating our definitions for the finite case, we define

$$E(X) = \sum_{i=1}^{\infty} a_i \cdot f(a_i)$$

$$\text{Var}(X) = \sum_{i=1}^{\infty} (a_i - \mu_X)^2 \cdot f(a_i)$$

when these series are absolutely convergent. Recall that an infinite series is absolutely convergent if the series of absolute values of the terms of the original series is convergent. Of course, the series defining $E(X)$ and $\text{Var}(X)$ need not be convergent at all, in which case the mean and variance would not exist. If these series are absolutely convergent, then they are necessarily convergent. This is a basic fact in the theory of infinite series. Why do we bother to impose the requirement of absolute convergence? The reason is to avoid a rather strange phenomenon that occurs with infinite series. It is possible that, should the terms of the series be added up in two different orders, two different sums might result. What is to prevent two different people from arranging the elements of $X(S)$ in two different orders? It would be unfortunate if the values of $E(X)$ and $\text{Var}(X)$ depended on the order in which the values of $X(S)$ were arranged. This is why absolute convergence is important. If a series is absolutely convergent, it does not matter in which order the terms are summed, the result will always be the same.

The reader may wish to review the basic facts about infinite series by consulting a good reference.*

The following formula will prove useful on many occasions when it is impossible to sum the series for $E(X)$ by means of the definition:

Proposition 4.13

If $X(S) = \{0, 1, 2, 3, \ldots\}$, then $E(X) = P(X > 0) + P(X > 1) + P(X > 2) + \ldots$. That is,

$$E(X) = \sum_{k=0}^{\infty} P(X > k)$$

Proof: Since the only possible values that X may assume are $0, 1, 2, 3, \ldots$, we have

$$P(X > 0) = P(X = 1) + P(X = 2) + P(X = 3) + \cdots$$
$$P(X > 1) = \qquad\qquad P(X = 2) + P(X = 3) + \cdots$$
$$P(X > 2) = \qquad\qquad\qquad\qquad P(X = 3) + \cdots$$

and so on. If we add the left-hand sides of these equations, we obtain $\sum_{k=0}^{\infty} P(X > k)$, and if we add the right-hand side, we obtain

$$P(X = 1) + 2 \cdot P(X = 2) + 3 \cdot P(X = 3) + \cdots$$

which equals $\sum_{k=0}^{\infty} k \cdot P(X = k)$, the term for $k = 0$ being zero. But since X assumes only the values $k = 0, 1, 2, 3, \ldots$, this second summation is actually $E(X)$.

Example 1 A fair coin is tossed until it comes up heads, and we let X denote the number of tosses required. Using the results of Example 1 in Section 2.4, we obtain the following probability distribution:

a_i	1	2	3	$\cdots$ k	$\cdots$
$f(a_1)$	$\frac{1}{2}$	$(\frac{1}{2})^2$	$(\frac{1}{2})^3$	$\cdots$ $(\frac{1}{2})^k$	$\cdots$

Therefore, X has mean

$$E(X) = 1 \cdot (\tfrac{1}{2}) + 2 \cdot (\tfrac{1}{2})^2 + 3 \cdot (\tfrac{1}{2})^3 + \cdots$$

$$= \sum_{k=1}^{\infty} k \cdot (\tfrac{1}{2})^k$$

$$= \frac{\frac{1}{2}}{(1 - \frac{1}{2})^2} = 2$$

---

* G. B. Thomas, *Calculus and Analytic Geometry* (Reading, Massachusetts: Addison-Wesley, 1968), chap. 18.

Hence, on the average, you would expect to toss the coin two times to get the first head. The series for the variance is

$$\text{Var}(X) = \sum_{k=1}^{\infty} (k-2)^2 \cdot (\tfrac{1}{2})^k$$

which is much more difficult to sum.

Example 2 An honest die is rolled until a 6 appears, and we let X denote the number of rolls required. Using the results from Example 2 in Section 2.4, we obtain the following probability distribution:

a_i	1	2	3	$\cdots$	k
$f(a_i)$	$\tfrac{1}{6}$	$\tfrac{5}{6} \cdot \tfrac{1}{6}$	$(\tfrac{5}{6})^2 \cdot \tfrac{1}{6}$	$\cdots$	$(\tfrac{5}{6})^{k-1} \cdot \tfrac{1}{6}$

In order to calculate $E(X)$, we must sum the series

$$1(\tfrac{1}{6}) + 2(\tfrac{5}{6})(\tfrac{1}{6}) + 3(\tfrac{5}{6})^2(\tfrac{1}{6}) + \cdots$$

That is,

$$E(X) = \sum_{k=1}^{\infty} k(\tfrac{5}{6})^{k-1}(\tfrac{1}{6})$$

$$= \tfrac{1}{6} \sum_{k=1}^{\infty} k(\tfrac{5}{6})^{k-1}$$

which is almost in the form of the derivative of a geometric series. So we multiply each term of the series by $1 = \tfrac{6}{5} \cdot \tfrac{5}{6}$, then factor out the $\tfrac{6}{5}$, to get

$$E(X) = (\tfrac{1}{6})(\tfrac{6}{5}) \sum_{k=1}^{\infty} k \cdot (\tfrac{5}{6})^k$$

$$= \tfrac{1}{5} \sum_{k=1}^{\infty} k \cdot (\tfrac{5}{6})^k$$

$$= \tfrac{1}{5} \cdot \frac{\tfrac{5}{6}}{(1 - \tfrac{5}{6})^2} = 6$$

Therefore, we would have to wait, on the average, six rolls of the die before a 6 appears. The solution to this problem would have been much easier numerically had we noticed that the event

$$\{X > k\} = \text{``first 6 occurs after the } k\text{th roll''}$$

is equivalent to the event

"no 6's occur on the first k rolls"

Therefore, we have $P(X > k) = (\tfrac{5}{6})^k$, for $k = 0, 1, 2, 3, \ldots$. Note that for $k = 0$, the event $\{X > k\}$ is a sure event. Using Proposition 4.13, we have

$$E(X) = \sum_{k=0}^{\infty} P(X > k) = \sum_{k=0}^{\infty} (\tfrac{5}{6})^k = \frac{1}{1 - \tfrac{5}{6}} = 6$$

Example 3 An urn contains one white ball and one black ball. A ball is drawn at random from the urn. If it is white, the process stops, but if it is black, it and one additional black ball are placed back in the urn. Then another ball is drawn from the urn. This process continues until a white ball is drawn, and we let X denote the number of draws required. X is a discrete random variable with range $\{1, 2, 3, \ldots\}$, and

$P(X = 1) = \frac{1}{2}$

$P(X = 2) = (\frac{1}{2})(\frac{1}{3})$

$P(X = 3) = (\frac{1}{2})(\frac{2}{3})(\frac{1}{4}) = (\frac{1}{3})(\frac{1}{4})$

$P(X = 4) = (\frac{1}{2})(\frac{2}{3})(\frac{3}{4})(\frac{1}{5}) = (\frac{1}{4})(\frac{1}{5})$

In general, the event $\{X = k\}$ means drawing $k - 1$ black balls before finally drawing a white ball on the kth draw. Each succeeding (black) draw is made from an urn containing exactly one white ball, and one more black ball than for the previous draw. Hence, we have

$$P(X = k) = (\tfrac{1}{2})(\tfrac{2}{3})(\tfrac{3}{4}) \cdots \left(\frac{k - 1}{k}\right)\left(\frac{1}{k + 1}\right)$$

$$= \frac{1}{k} \cdot \frac{1}{k + 1}$$

$$= \frac{1}{k} - \frac{1}{k + 1}$$

We have therefore obtained the probability distribution

a_i	1	2	3	$\cdots$
$f(a_i)$	$1 - \frac{1}{2}$	$\frac{1}{2} - \frac{1}{3}$	$\frac{1}{3} - \frac{1}{4}$	$\cdots$

Checking that our assignment of probabilities is valid, we find

$$\sum_{k=1}^{\infty} f(a_k) = \sum_{k=1}^{\infty}\left[\frac{1}{k} - \left(\frac{1}{k + 1}\right)\right]$$

$$= (1 - \tfrac{1}{2}) + (\tfrac{1}{2} - \tfrac{1}{3}) + (\tfrac{1}{3} - \tfrac{1}{4}) + \cdots$$

$$= 1 + (\tfrac{1}{2} - \tfrac{1}{2}) + (\tfrac{1}{3} - \tfrac{1}{3}) + \cdots$$

Since the nth partial sum

$$s_n = \sum_{k=1}^{n}\left[\frac{1}{k} - \left(\frac{1}{k + 1}\right)\right]$$

$$= 1 - \left(\frac{1}{n + 1}\right)$$

the $\lim_{n \to \infty} s_n = 1$. Therefore, $\sum_{k=1}^{\infty} f(a_k) = 1$, so our assignment of probabilities is valid. Reverting back to the formula $f(k) = (1/k) \cdot (1/(k + 1))$,

we have

$$E(X) = \sum_{k=1}^{\infty} k \cdot \left(\frac{1}{k}\right) \cdot \left(\frac{1}{k+1}\right)$$

$$= \sum_{k=1}^{\infty} \left(\frac{1}{k+1}\right)$$

$$= \sum_{k=2}^{\infty} \left(\frac{1}{k}\right)$$

Thus, $E(X) = \frac{1}{2} + \frac{1}{3} + \frac{1}{4} + \cdots$, an infinite series which diverges—the sum of these terms is ∞. Hence, $E(X) = \infty$. On the average, you would not expect to ever draw a white ball.

Absolute convergence is necessary to prove the following extension of Proposition 4.6 to discrete random variables:

Proposition 4.14

Suppose that X is a discrete random variable such that $E(X^2)$ converges. Then both $E(X)$ and $\mathrm{Var}(X)$ converge absolutely and $\mathrm{Var}(X) = E(X^2) - E(X)^2$.

Proof: Our assumption is that

$$\sum_{i=1}^{\infty} a_i^2 f(a_i) < \infty$$

and since all terms of this series are positive, convergence is equivalent to absolute convergence. We divide the proof into three parts:

(a) Since $|x| \leq x^2 + 1$ for any value of x, we have

$$\sum_{i=1}^{\infty} |a_i| \cdot f(a_i) \leq \sum_{i=1}^{\infty} (a_i^2 + 1) \cdot f(a_i)$$

The absolute convergence of $E(X^2)$ and $\sum_{i=1}^{\infty} f(a_i)$ allows us to make the following rearrangement in the right-hand summation:

$$\sum_{i=1}^{\infty} (a_i^2 + 1) \cdot f(a_i) = \sum_{i=1}^{\infty} a_i^2 f(a_i) + \sum_{i=1}^{\infty} f(a_i)$$

Therefore,

$$\sum_{i=1}^{\infty} |a_i| f(a_i) \leq E(X^2) + 1 < \infty$$

so the series defining $E(X)$ is absolutely convergent.

(b) Since $(x - y)^2 \leq 2(x^2 + y^2)$ for all x, y, we have

$$\mathrm{Var}(X) = \sum_{i=1}^{\infty} (a_i - \mu_X)^2 \cdot f(a_i)$$

$$\leq \sum_{i=1}^{\infty} 2(a_i^2 + \mu_X^2) \cdot f(a_i)$$

$$= 2 \sum_{i=1}^{\infty} a_i^2 \cdot f(a_i) + 2\mu_X^2 \sum_{i=1}^{\infty} f(a_i)$$

with the last equality again following because of absolute convergence. Therefore, $\text{Var}(X) \leq 2 \cdot E(X^2) + 2 \cdot E(X)^2$, and since both $E(X^2)$ and $E(X)$ are absolutely convergent [the latter by (a)], we have $\text{Var}(X) < \infty$.

(c) Finally, the proof that $\text{Var}(X) = E(X^2) - E(X)^2$ is identical to the original proof (see Proposition 4.6). The square $(a_i - \mu_X)^2$ is expanded term by term, and then the absolute convergence of $E(X^2)$, $E(X)$, and $\sum_{i=1}^{\infty} f(a_i)$ allows the rearrangement into three sums.

We also note that the convergence of $E(X^2)$ allows the manipulations involved in the proof of Chebyshev's theorem to be carried over to the discrete case.

EXERCISES

1. Prove the two inequalities asserted in the proof of Proposition 4.14.
2. What is the expected number of rolls of a fair die before an even number occurs?
3. What is the expected number of rolls of two fair dice before sum 7 occurs? Before sum 12 occurs?
4. On the average, how many cards must be drawn from a deck, with replacement, before a queen is drawn?
5. How many poker hands must be dealt on the average before a full house is dealt?
6. An urn contains 20 balls, only one of which is red. How many balls must be drawn with replacement, on the average, before the red ball is drawn?

4.6 Moments and Moment-generating Functions

The concepts we are about to discuss are generalizations of the formal definitions of mean and variance. As we shall see, their main use is theoretical. They will prove time and again to be powerful tools in the proofs of important theorems presented later in the book.

The mean and variance are two examples of what are called the moments of a random variable. There are two different types of moments, both of which will be studied in this section. First, we have

DEFINITION

The *kth moment about the origin* of a discrete random variable X is defined to equal $E(X^k)$, denoted μ'_k. That is,

$$\mu'_k = \sum_i a_i^k \cdot f(a_i)$$

when this summation (if infinite) is absolutely convergent.

Note that $\mu'_1 = E(X)$ and $\mu'_2 = E(X^2)$. Therefore $\text{Var}(X) = \mu'_2 - (\mu'_1)^2$.

DEFINITION

The *kth moment about the mean* of a discrete random variable X is defined to equal $E[(X - \mu_X)^k]$, denoted μ_k. That is,

$$\mu_k = \sum_i (a_i - \mu_X)^k \cdot f(a_i)$$

when this summation (if infinite) is absolutely convergent.

Note here that $\mu_1 = 0$ and $\mu_2 = \text{Var}(X)$.

Moments about the mean are of particular importance because they give some idea of how a random variable is distributed about its mean. We have already seen the significance of the variance in this respect. The third moment about the mean (actually, μ_3/σ^3) is used to measure how (if at all) symmetric a random variable is with respect to its mean. Recall that the signed differences $(a_i - \mu_X)$ tend to cancel each other, and therefore, the first moment about the mean is always zero. The third moment about the mean then becomes the lowest moment that reflects differences in the shape of the distribution from one side of the mean to the other. The term *skewness* is used as a synonym for lack of symmetry. Also, the fourth moment about the mean (actually μ_4/σ^4, which is called the *kurtosis* of the distribution) is used to measure the flatness or peakedness of a distribution. The kurtosis is usually compared to the value 3, the kurtosis of the standard normal distribution (Chapter 8), which is used as a point of reference.

Although the moment about the origin and the moment about the mean can be evaluated directly from the definition by adding the terms of the corresponding sum, the arithmetic involved does become rather tedious. There is, however, an alternate method which often simplifies the work considerably. This method is based on the moment-generating function of the random variable, which we now define.

DEFINITION

The *moment-generating function* $M_X(t)$ of a discrete random variable X is defined to equal $E(e^{tX})$, where t is an arbitrary real number. That is,

$$M_X(t) = \sum_i e^{a_i t} \cdot f(a_i)$$

when this summation (if infinite) is absolutely convergent.

Proposition 4.15

$$M_X(t) = \sum_{k=0}^{\infty} \mu'_k \left(\frac{t^k}{k!} \right)$$

Proof: Using the Maclaurin expansion of e^{tx},

$$e^{tx} = \sum_{k=0}^{\infty} \frac{(tx)^k}{k!} = 1 + tx + \frac{t^2 x^2}{2} + \frac{t^3 x^3}{6} + \cdots$$

we obtain

$$M_X(t) = \sum_i \left(1 + a_i t + \frac{a_i^2 t^2}{2} + \frac{a_i^3 t^3}{6} + \cdots\right) \cdot f(a_i)$$

$$= \sum_i \left(f(a_i) + a_i f(a_i)t + \frac{a_i^2 f(a_i)t^2}{2} + \frac{a_i^3 f(a_i)t^3}{6} + \cdots\right)$$

$$= \sum_i f(a_i) + \left(\sum_i a_i f(a_i)\right) \cdot t + \left(\sum_i a_i^2 f(a_i)\right) \cdot \frac{t^2}{2} + \left(\sum_i a_i^3 f(a_i)\right) \cdot \frac{t^3}{6} + \cdots$$

That is,

$$M_X(t) = \sum_{k=0}^{\infty} \left(\sum_{i=1}^{\infty} a_i^k f(a_i)\right) \cdot \frac{t^k}{k!}$$

That is, if we expand $M_X(t)$ as a Taylor series about $t = 0$, the coefficient of the term $t^k/k!$ is the kth moment about the origin. But this coefficient is known to be the kth derivative of the function at the origin, giving

Corollary 4.16

$$\mu_k' = \frac{d^k[M_X(t)]}{dx^k}\bigg|_{t=0}$$

Example 1 Suppose that X is a discrete uniform random variable with probability function $f(x) = 1/n$ if $x = 1, 2, \ldots, n$ and $f(x) = 0$ otherwise. That is, $a_k = k$, for $k = 1, 2, \ldots, n$, and $f(a_k) = 1/n$. Then

$$M_X(t) = \sum_{k=1}^{n} e^{kt}\left(\frac{1}{n}\right) = \left(\frac{1}{n}\right)\sum_{k=1}^{n} e^{kt}$$

If we let

$$s_n = 1 + e^t + e^{2t} + \cdots + e^{(n-1)t}$$

then

$$e^t s_n = e^t + e^{2t} + e^{3t} + \cdots + e^{nt}$$

Therefore,

$$s_n - e^t s_n = 1 - e^{nt}$$

so that

$$s_n = \frac{1 - e^{nt}}{1 - e^t}$$

and so

$$e^t s_n = \frac{e^t \cdot (1 - e^{nt})}{1 - e^t}$$

Consequently,

$$M_X(t) = \frac{1}{n} \cdot e^t s_n = \frac{e^t \cdot (1 - e^{nt})}{n(1 - e^t)}$$

is the moment-generating function of X.

The moment-generating function is so-called because all the moments of a random variable can be generated by successively differentiating the moment generating function of the random variable.

Example 2   If X is the indicator of the event E, then X assumes two values, 0 and 1, with probabilities

$$f(0) = P(X = 0) = 1 - P(E) \quad \text{and} \quad f(1) = P(X = 1) = P(E)$$

Therefore,

$$\begin{aligned} M_X(t) &= e^{0t}[1 - P(E)] + e^{1t}P(E) \\ &= 1 - P(E) + e^t P(E) \\ &= 1 + (e^t - 1)P(E) \end{aligned}$$

Conveniently, all derivatives of this function equal $P(E)e^t$, and therefore, all moments about the origin of an indicator random variable equal $P(E)$.

Our work with moment-generating functions in later chapters will be simplified by the following formulas, all of which are simple consequences of the definition.

Proposition 4.17

If X is a discrete random variable and a and b are real constants, then

(a) $M_{X+a}(t) = e^{at} \cdot M_X(t)$
(b) $M_{bX}(t) = M_X(bt)$
(c) $M_{(X+a)/b}(t) = e^{(a/b)t} \cdot M_X(t/b)$

Proof: Exercise

EXERCISES

1. Prove Proposition 4.17.
2. Write a computer program that will calculate the moments of a random variable.
3. Prove that

$$\mu_k = \sum_{i=0}^{k} (-1)^k C_i^k \mu'_{k-i} \mu^k$$

4. Find the moment-generating function and the first five moments about the origin of the random variable X that counts the number of heads in two tosses of a fair coin.

5. Find the moment-generating function and the first five moments about the origin of the random variable X that counts the number of 6's in two rolls of a fair die.

6. Find the moment-generating function of the random variable X that counts the number of aces in a poker hand.

ADDITIONAL PROBLEMS

1. Prove that

$$E[(aX + b)^k] = \sum_{i=0}^{k} C_i^k \cdot a^{k-i} \cdot b^i \cdot E(X^{k-i})$$

2. A cattleman buys 100 calves in the fall. Each represents a profit of $50 if it survives the winter, but a loss of $500 if it dies. If the probability that a calf will survive the winter is 0.95, what is the cattleman's expected profit?

3. An urn contains three green and two red balls. Let X denote the number of balls drawn until getting a red. Find the distribution and mean of X if
   (a) the draw is with replacement, and
   (b) the draw is without replacement.

4. The ball-producing machine currently used by a company produces ten percent defectives, and the process costs 50¢ per ball. A new machine will produce only five percent defectives, but at a cost of 75¢ per ball. If the company sells the non-defective balls for $1 each, and does not sell the defectives, should the company buy the new machine?

5. Two urns contain five balls each. An urn is chosen at random and a ball is removed from that urn. This process is repeated until one of the two urns is empty. Let X denote the number of balls remaining in the other urn. Find the probability function and mean of X.

6. An urn contains 18 white and 2 black balls. Suppose that $m \le 18$ balls are drawn without replacement. Find the probability that at least one black ball is drawn. What is the smallest value of m for which this probability is greater than $\frac{1}{2}$?

7. Suppose that the series defining $E(X^r)$ is absolutely convergent. Prove that the series defining $E(X^k)$ is also absolutely convergent, for any $1 \le k < r$.

8. The probability that a 55 year old person lives for at least one more year is 0.98. What premium should an insurance company charge such a person for a $10,000 term life insurance policy if the company wishes to make a $50 profit on the transaction?

9. A circle of radius 1 is divided into n concentric disks of radii $1/n, 2/n, \ldots, 1$. A point is chosen at random in the circle, and is worth one point if it falls in the disk of outer radius $1/n$, two points if it falls in the disk of outer radius $2/n$, and so forth. Let X denote the point value assigned. Find $E(X)$.

10. Suppose that n balls are distributed at random into r urns. Let X_i indicate whether or not urn i is empty, for $i = 1, 2, \ldots, r$. Find $E(X_i)$.

Several Random
Variables

5.1 Joint Probability

In many cases, more than one outcome of a random experiment is of interest. The very purpose of the experiment may be to determine if one or several variables can be used to explain or predict the outcome of another variable (or variables). For example, we may be interested in the relationship between the number of cigarettes a person smokes per day and the results of a test for lung cancer that the person has recently taken, or we may be interested in how a student's high school grades can be used to predict his achievement in college, or we may wish to determine if there is any relationship between the behavior of three key stocks and the behavior of the stock market as a whole.

It is for the purpose of studying the relationship between different random variables, the effect that one might have on another, that we now introduce the concept of the joint probability function of two random variables. We shall restrict ourselves, for the most part, to discrete random variables which have a finite range, although what we shall do can easily be extended to discrete random variables in general. We shall indicate, where relevant, how proofs can be extended to discrete random variables that have a countable range. And in Chapter 8 we will discuss how these concepts carry over to nondiscrete random variables.

DEFINITION

Let X and Y be random variables, defined on a sample space S, having probability functions f and g, respectively. The function $h: \mathbb{R} \times \mathbb{R} \to \mathbb{R}$ defined by

$$h[(a, b)] = P(\{X = a\} \cap \{Y = b\})$$

that is,

$$h[(a, b)] = P(\{x \in S : X(x) = a \quad \text{and} \quad Y(x) = b\})$$

is called the *joint probability function* of X and Y. We shall abbreviate $h[(a, b)]$ to $h(a, b)$.

Table 5.1
Joint Probability Table of X and Y

X \ Y	b_1	b_2	$\cdots$	b_k
a_1	$h(a_1, b_1)$	$h(a_1, b_2)$	$\cdots$	$h(a_1, b_k)$
a_2	$h(a_2, b_1)$	$h(a_2, b_2)$	$\cdots$	$h(a_2, b_k)$
a_3	$h(a_3, b_1)$	$h(a_3, b_2)$	$\cdots$	$h(a_3, b_k)$
$\vdots$				
a_m	$h(a_m, b_1)$	$h(a_m, b_2)$	$\cdots$	$h(a_m, b_k)$

If $X(S) = \{a_1, a_2, \ldots, a_m\}$ and $Y(S) = \{b_1, b_2, \ldots, b_k\}$, then the joint probability function h can be displayed by means of the two-dimensional Table 5.1. Each row of the table contains the probabilities that $X = a_i$, for some fixed $i = 1, 2, \ldots, m$, while Y in turn takes each of the values $b_1, b_2, \ldots, b_k$.

The joint table not only contains information about how the two random variables X and Y behave together, but also contains information about how each of the two random variables behave individually. Should we calculate the sum of the values across any row of the table, we would find

$$\sum_{j=1}^{k} h(a_i, b_j) = \sum_{j=1}^{k} P(\{X = a_i\} \cap \{Y = b_j\})$$

$$= P\left(\bigcup_{j=1}^{k} \{X = a_i\} \cap \{Y = b_j\} \right) \quad \text{(a disjoint union)}$$

$$= P\left(\{X = a_i\} \cap \bigcup_{j=1}^{k} \{Y = b_j\} \right)$$

$$= P(\{X = a_i\} \cap S)$$

$$= P(X = a_i) = f(a_i)$$

Thus, the sum of the probabilities across the row corresponding to the X-value a_i equals $f(a_i)$, the probability that X assumes the value a_i. Likewise, we have

$$\sum_{i=1}^{m} h(a_i, b_j) = g(b_j)$$

that is, the sum of the probabilities down the column corresponding to the Y-value b_j equals $g(b_j)$, the probability that Y assumes the value b_j.

The values $f(a_1), f(a_2), \ldots, f(a_m)$ are usually displayed in the margin to the right of the table, while the values $g(b_1), g(b_2), \ldots, g(b_k)$ are usually displayed in the margin below the table. For this reason, the probability functions f and g are sometimes referred to as the *marginal probabilities*.

Note that since the sums of the m rows of the table are $f(a_1), f(a_2), \ldots, f(a_m)$, and since $\sum_{i=1}^{m} f(a_i) = 1$, the sum of all the probabilities in the two-dimensional joint probability table is 1, as is the case for one-dimensional probability tables.

Example 1 A fair coin is tossed three times. We define random variables X and Y as follows:

$$X = \begin{cases} 0 & \text{if the first coin is a head} \\ 1 & \text{if the first coin is a tail} \end{cases}$$

and

$Y = $ number of heads in the three tosses

These random variables have probability distributions

$$X: \quad \begin{array}{c|c|c} a_i & 0 & 1 \\ \hline f(a_i) & \frac{1}{2} & \frac{1}{2} \end{array}$$

and

$$Y: \quad \begin{array}{c|c|c|c|c} b_j & 0 & 1 & 2 & 3 \\ \hline g(b_j) & \frac{1}{8} & \frac{3}{8} & \frac{3}{8} & \frac{1}{8} \end{array}$$

In order to calculate the values in the joint probability table, we note that the first line of the table corresponds to the event

$$\{X = 0\} = \{HHH, HHT, HTH, HTT\}$$

and the second line of the table corresponds to the event

$$\{X = 1\} = \{THH, THT, TTH, TTT\}$$

It is then just a matter of counting the number of heads in each of these sample elements to obtain the joint table

X \ Y	0	1	2	3	$f(a_i)$
0	0	$\frac{1}{8}$	$\frac{2}{8}$	$\frac{1}{8}$	$\frac{1}{2}$
1	$\frac{1}{8}$	$\frac{2}{8}$	$\frac{1}{8}$	0	$\frac{1}{2}$
$g(b_j)$	$\frac{1}{8}$	$\frac{3}{8}$	$\frac{3}{8}$	$\frac{1}{8}$	1

For example, to calculate $h(1, 2) = P(\{X = 1\} \cap \{Y = 2\})$, we look at the event $X = 1$ and determine which of its elements have two heads. Since just the event THH has two heads, we have

$$P(\{X = 1\} \cap \{Y = 2\}) = P(THH) = \frac{1}{8}$$

Note that the rows of the table add up to the probability function of X and the columns add up to the probability function of Y.

Example 2 In the sample space of C_{13}^{52} different possible bridge hands, we define random variables X and Y as follows:

X = number of aces in the hand

Y = number of kings in the hand

Then $X(S) = Y(S) = \{0, 1, 2, 3, 4\}$. Instead of exhibiting the entire 5×5 joint probability table, we simple exhibit a formula for the joint probability function. Using the concept of stratified sampling, we find

$$h(a, b) = \begin{cases} (C_a^4 C_b^4 C_{13-(a+b)}^{44})/C_{13}^{52} & \text{if } a, b = 0, 1, 2, 3, 4 \\ 0 & \text{otherwise} \end{cases}$$

The numerator above is simply the number of ways we can choose a of the four aces, b of the four kings, and $13 - (a + b)$ of the 44 other cards.

Example 3   Five balls are randomly distributed among three urns numbered 1, 2, and 3. We let

X = number of balls in urn 1

and

Y = number of balls in the first two urns

Our sample space is the set

$$S = \{(a_1, a_2, a_3, a_4, a_5) : a_i = 1, 2, \text{ or } 3\}$$

where each coordinate of the five-tuple is associated with one of the five balls, and indicates into which urn that ball was placed. Since each of the five balls can be placed into any of the three urns, there are 3^5 different elements in S. Since the event $\{X = k\}$ is the subset of S consisting of those five-tuples in which exactly k of the coordinates are 1's and the other $5 - k$ coordinates are either 2 or 3, and since there are C_k^5 ways to choose the k coordinates in which to place the k 1's, and then 2^{5-k} ways to place either a 2 or 3 in the other $5 - k$ coordinates, we have the following probability function for X:

$$f(k) = \begin{cases} C_k^5 \cdot 2^{5-k}/3^5 & \text{if } 0 \le k \le 5 \\ 0 & \text{otherwise} \end{cases}$$

In order to calculate the probabilities $g(k)$, we note that the number of ways that we can place k balls into the first two urns is the same as the number of ways we can place $5 - k$ balls into urn 3. Since the distribution of balls in urn 3 is identical to the distribution of balls in urn 1 (use 3's instead of 1's in the argument for $f(k)$ above), we find that

$$g(k) = f(5 - k) = C_{5-k}^5 \cdot 2^{5-(5-k)}/3^5 \qquad \text{for } k = 0, 1, \ldots, 5$$

that is,

$$g(k) = \begin{cases} C_k^5 \cdot 2^k/3^5 & \text{if } 0 \le k \le 5 \\ 0 & \text{otherwise} \end{cases}$$

which is the number of ways to choose k of the five coordinates in which to place a 1 or a 2, and then place either a 1 or a 2 in those coordinates. In order to calculate a joint probability $h(a, b)$, we must determine how many of the $C_b^5 \cdot 2^b$ different ways to place b balls in the first two urns correspond to having a of these b balls in urn 1. This immediately sets the restriction that a cannot exceed b. Instead of randomly choosing a 1 or a 2 for each of the b coordinates chosen, we must choose a of the b coordinates to contain 1's, leaving the other $b - a$ coordinates to contain 2's. The other $5 - b$ coordinates, remember, already contain 3's. Hence,

$$h(a, b) = \begin{cases} C_b^5 \cdot C_a^b/3^5 & \text{if } 0 \le a \le b \le 5 \\ 0 & \text{otherwise} \end{cases}$$

This formula gives the joint probability Table 5.2. Notice that the frequencies of f and g occur correctly in the margins.

Each row and each column of the joint probability table can be used to define what are called conditional probability functions.

DEFINITION

For each $b_j \in Y(S)$, the function $f(.\,|b_j): \mathbb{R} \to \mathbb{R}$ defined by

$$f(a\,|\,b_j) = \frac{h(a, b_j)}{g(b_j)}$$

is called the *conditional probability function of X given that* $Y = b_j$. Similarly, given $a_i \in X(S)$, the function $g(.\,|a_i): \mathbb{R} \to \mathbb{R}$ defined by

$$g(a_i\,|\,b) = \frac{h(a_i, b)}{f(a_i)}$$

is called the *conditional probability function of Y given that* $X = a_i$.

Note that $f(.\,|b_j)$ is nonzero only (possibly) for the values in $X(S)$, while $g(.\,|a_i)$ is nonzero only (possibly) for the values in $Y(S)$.

The conditional probability functions are, in fact, probability functions. The one property that must be checked is that the sum of the probabilities over the domain of the function is, in fact, 1. For fixed b_j, then, we have

$$\sum_{i=1}^{m} f(a_i\,|\,b_j) = \sum_{i=1}^{m} \frac{h(a_i, b_j)}{g(b_j)}$$

$$= \frac{1}{g(b_j)} \sum_{i=1}^{m} h(a_i, b_j)$$

$$= \frac{1}{g(b_j)} g(b_j) = 1$$

A similar argument proves that $\sum_{j=1}^{k} g(b_j\,|\,a_i) = 1$ for any fixed a_i.

Table 5.2
Joint Probability Table of X and Y

X＼Y	0	1	2	3	4	5	$f(a_i)$
0	$1/3^5$	$5/3^5$	$10/3^5$	$10/3^5$	$5/3^5$	$1/3^5$	$32/3^5$
1	0	$5/3^5$	$20/3^5$	$30/3^5$	$20/3^5$	$5/3^5$	$80/3^5$
2	0	0	$10/3^5$	$30/3^5$	$30/3^5$	$10/3^5$	$80/3^5$
3	0	0	0	$10/3^5$	$20/3^5$	$10/3^5$	$40/3^5$
4	0	0	0	0	$5/3^5$	$5/3^5$	$10/3^5$
5	0	0	0	0	0	$1/3^5$	$1/3^5$
$g(b_j)$	$1/3^5$	$10/3^5$	$40/3^5$	$80/3^5$	$80/3^5$	$32/3^5$	1

Using the joint probability table, it is a rather simple task to calculate the probabilities in the distribution table of one of the conditional probability functions. For example, the function $f(.|b_j)$ presumes that the event $\{Y = b_j\}$ has occurred, and so the only probabilities in the joint table that remain relevant are those in the jth column of the table. The sum of the probabilities in this column is

$$\sum_{i=1}^{m} h(a_i, b_j) = g(b_j)$$

and the definition indicates that we obtain the values for $f(.|b_j)$ by dividing the individual probabilities in the jth column by the value $g(b_j)$, the column sum.

Example 4   Referring back to Example 3, we shall determine the probability distribution of $f(.|b_4)$. The relevant column of the joint table is the fourth column

	$b_4 = 3$
$a_1 = 0$	$10/3^5$
$a_2 = 1$	$30/3^5$
$a_3 = 2$	$30/3^5$
$a_4 = 3$	$10/3^5$
$a_5 = 4$	0
$a_6 = 5$	0
	$g(b_4) = 80/3^5$

Since by definition

$$f(a_i|b_4) = \frac{h(a_i, b_4)}{g(b_4)} = \frac{h(a_i, b_4)}{80/3^5}$$

we have

$$f(0|b_4) = \frac{10/3^5}{80/3^5} = \frac{1}{8}$$

$$f(1|b_4) = \frac{30/3^5}{80/3^5} = \frac{3}{8}$$

$$f(2|b_4) = \frac{30/3^5}{80/3^5} = \frac{3}{8}$$

$$f(3|b_4) = \frac{10/3^5}{80/3^5} = \frac{1}{8}$$

$$f(4|b_4) = \frac{0}{80/3^5} = 0$$

$$f(5|b_4) = \frac{0}{80/3^5} = 0$$

and so $f(.|b_4)$ has probability distribution

a_i	0	1	2	3	4	5
$f(a_i|b_4)$	$\frac{1}{8}$	$\frac{3}{8}$	$\frac{3}{8}$	$\frac{1}{8}$	0	0

As a result, if we knew that there were three balls in the first two urns ($b_4 = 3$), then the probability is $\frac{3}{8}$ that two of these are in urn 1.

It is clear from the example above that the probabilities that X will assume any of its possible values will be affected by the fact that Y has assumed a particular value. Using the random variables X and Y from Example 4 above, we see that $P(X = 3) = 40/3^5$, while $P(X = 3 | Y = 3) = \frac{1}{8}$. Thus, in general,

$$f(a_i|b_j) = P(X = a_i | Y = b_j) \neq P(X = a_i) = f(a_i)$$

That is, the events $\{X = a_i\}$ and $\{Y = b_j\}$ are not necessarily independent.

We are now ready to extend the concept of independent events to random variables.

DEFINITION

The random variables X and Y are called *independent* if, for any pair (a_i, b_j),

$$h(a_i, b_j) = f(a_i) \cdot g(b_j)$$

that is,

$$P(\{X = a_i\} \cap \{Y = b_j\}) = P(X = a_i)P(Y = b_j)$$

This means, of course, that the events $\{X = a_i\}$ and $\{Y = b_j\}$ must be independent. Since

$$f(a_i|b_j) = \frac{h(a_i, b_j)}{g(b_j)}$$

the fact that $h(a_i, b_j) = f(a_i) \cdot g(b_j)$ allows the necessary cancellation to prove that

$$f(a_i|b_j) = f(a_i)$$

or

$$P(X = a_i | Y = b_j) = P(X = a_i)$$

whenever X and Y are independent. Likewise, we have

$$g(b_j|a_i) = g(b_j)$$

or

$$P(Y = b_j | X = a_i) = P(Y = b_j)$$

whenever X and Y are independent.

Given the joint probability table, the definition affords us a simple method for determining whether or not the random variables X and Y are independent. If they are independent, then *all* values $h(a_i, b_j)$ can be obtained by multiplying the corresponding marginal probabilities $f(a_i)$ and $g(b_j)$. If even one product $f(a_i) \cdot g(b_j)$ fails to equal the corresponding $h(a_i, b_j)$, then X and Y are dependent. In each of the Examples 1–3 above, the random variables X and Y were dependent, as a quick glance at the joint tables will show.

Example 5   An urn contains three red and two green balls. Two balls are drawn with replacement from the urn, and we let

$$X = \begin{cases} 0 & \text{if the first ball is green} \\ 1 & \text{if the first ball is red} \end{cases}$$

and

$$Y = \begin{cases} 0 & \text{if the second ball is green} \\ 1 & \text{if the second ball is red} \end{cases}$$

Since we draw the two balls with replacement, the contents of the urn are the same for both draws, and so we would expect X and Y to be independent. Both X and Y have probability distribution

0	1
$\frac{2}{5}$	$\frac{3}{5}$

and their joint distribution is

$$Y$$

X	0	1	$f(a_i)$
0	$\frac{4}{25}$	$\frac{6}{25}$	$\frac{10}{25}$
1	$\frac{6}{25}$	$\frac{9}{25}$	$\frac{15}{25}$
$g(b_j)$	$\frac{10}{25}$	$\frac{15}{25}$	1

For example,

$$\begin{aligned} h(0, 1) &= P(\{X = 0\} \cap \{Y = 1\}) \\ &= P(X = 0) \cdot P(Y = 1 \mid X = 0) \\ &= P(X = 0) \cdot P(Y = 1) \quad (\textit{with} \text{ replacement}) \\ &= \tfrac{2}{5} \cdot \tfrac{3}{5} = \tfrac{6}{25} \end{aligned}$$

It is easy to see that the joint table could have been obtained by multiplying the corresponding marginal probabilities, and so X and Y are independent.

Example 6 If the events A and B are independent, then the indicator random variables I_A and I_B of these events are also independent. These random variables have probability distributions

$$I_A: \quad \frac{0 \quad | \quad 1}{P[C(A)] \quad | \quad P(A)} \quad \text{and} \quad I_B: \quad \frac{0 \quad | \quad 1}{P[C(B)] \quad | \quad P(B)}$$

By Proposition 3.4, we have

$$h(0, 0) = P(I_A = 0 \cap I_B = 0) = P[C(A) \cap C(B)] = P[C(A)] \cdot P[C(B)]$$
$$h(0, 1) = P(I_A = 0 \cap I_B = 1) = P[C(A) \cap B] = P[C(A)] \cdot P(B)$$
$$h(1, 0) = P(I_A = 1 \cap I_B = 0) = P(A \cap C(B)] = P(A) \cdot P[C(B)]$$
$$h(1, 1) = P(I_A = 1 \cap I_B = 1) = P(A \cap B) = P(A) \cdot P(B)$$

Notice, for example, that

$$\{I_A = 0\} \cap \{I_B = 1\} = \{x \in S : I_A(x) = 0 \quad \text{and} \quad I_B(x) = 1\}$$
$$= \{x \in S : x \notin A \quad \text{and} \quad x \in B\}$$
$$= \{x \in S : x \in C(A) \cap B\} = C(A) \cap B$$

So we have

$$h(0, 0) = I_A(0) \cdot I_B(0)$$
$$h(0, 1) = I_A(0) \cdot I_B(1)$$
$$h(1, 0) = I_A(1) \cdot I_B(0)$$
$$h(1, 1) = I_A(1) \cdot I_B(1)$$

and therefore, I_A and I_B are independent random variables.

Proposition 5.1

If X and Y are independent random variables and $u, v : \mathbb{R} \to \mathbb{R}$, then $u(X)$ and $v(Y)$ are also independent random variables.

Proof: We must show that

$$P(\{u(X) = a\} \cap \{v(Y) = b\}) = P[u(X) = a] \cdot P[v(Y) = b]$$

Since

$$\{u(X) = a \cap v(Y) = b\} = \{x \in S : u(X)(x) = a \quad \text{and} \quad v(Y)(x) = b\}$$
$$= \{x \in S : u(X)(x) = a\} \cap \{x \in S : v(Y)(x) = b\}$$
$$= \bigcup_{u(a_i) = a} \{x \in S : X(x) = a_i\} \cap \bigcup_{v(b_j) = b} \{x \in S : Y(x) = b_j\}$$

which, by the distributive law,

$$= \bigcup_{\substack{u(a_i) = a \\ v(b_j) = b}} \{x \in S : X(x) = a_i\} \cap \{x \in S : Y(x) = b_j\}$$

Therefore,

$$P(u(X) = a \cap v(Y) = b) = P\left(\bigcup_{\substack{u(a_i)=a \\ v(b_j)=b}} \{X = a_i\} \cap \{Y = b_j\}\right)$$

$$= \sum_{\substack{u(a_i)=a \\ v(b_j)=b}} P(\{X = a_i\} \cap \{Y = b_j\})$$

$$= \sum_{\substack{u(a_i)=a \\ v(b_j)=b}} P(X = a_i) \cdot P(Y = b_j) \qquad \begin{array}{l}\text{(since } X \text{ and } Y \\ \text{are independent)}\end{array}$$

$$= \sum_{u(a_i)=a} P(X = a_i) \sum_{v(b_j)=b} P(Y = b_j)$$

(distributive law)

$$= P[u(X) = a] \cdot P[v(Y) = b]$$

Corollary 5.2

Suppose that X and Y are independent random variables, then so are

(a) X^2 and Y^2
(b) $X - \mu_X$ and $Y - \mu_Y$

The concepts of joint distributions and independent random variables can be extended to the case of several random variables. If $X_1, X_2, \ldots, X_n$ are discrete random variables with finite ranges and probability functions $f_1, f_2, \ldots, f_n$, respectively, then we can define the joint probability function f as follows:

$$f(a_1, a_2, \ldots, a_n) = P\left(\bigcap_{i=1}^{n} \{X_i = a_i\}\right)$$

We say that these random variables are independent if the events $\{X_1 = a_1\}$, $\{X_2 = a_2\}, \ldots, \{X_n = a_n\}$ are independent. This is equivalent to the rule

$$f(a_1, a_2, \ldots, a_n) = f_1(a_1) \cdot f_2(a_2) \cdot \cdots \cdot f_n(a_n)$$

EXERCISES

1. Two fair dice are rolled, and we let

   X = number of 6's that occur

   Y = sum of the top faces

   Z = larger value of the top faces

   Find the joint probability function of X and Y, X and Z, and Y and Z. Are any of these pairs of random variables independent?

2. Three balls are placed at random into three urns, and we let

   X = number of balls in urn 1

   Y = number of balls in urn 3

   Z = number of occupied urns

Find the joint probability function of X and Y, X and Z, and Y and Z. Are any of these pairs of random variables independent?

3. Two balls are drawn at random from an urn containing five balls which are numbered 1, 1, 2, 2, 3. Let X denote the sum of the two balls and Y the larger value of the two balls. Find the joint probability function of X and Y. Are X and Y independent?

4. A number is drawn at random from the set $\{1, 2, 3, \ldots, n\}$, and should that number be k, another number is drawn from the set $\{k, k + 1, \ldots, n\}$. Let X denote the first number drawn and Y the second number drawn. Find the joint probability function of X and Y. Are X and Y independent?

5. Let X be the number of aces and Y the number of kings in a poker hand. Find the joint probability function of X and Y. Are X and Y independent?

6. If the two balls are drawn without replacement in Example 5, are the random variables X and Y still independent?

7. Suppose that a box contains eight operable, six defective, and ten semioperable items. If two items are drawn at random from the box, and X denotes the number of operable items and Y the number of defective items in this sample, find the joint probability function of X and Y. Are X and Y independent?

8. An urn contains three balls numbered 1, two balls numbered 2, and one ball numbered 3. Two balls are drawn without replacement from the urn. Let X_i denote the number on the ith ball, $i = 1, 2$. Find the joint probability function of X_1 and X_2. Are X_1 and X_2 independent?

5.2 Functions of Several Random Variables

In Chapter 4 we discussed functions of one random variable. For example, we studied random variables of the form $X + b$, where b is a constant. In this section, we shall go one step farther and study random variables like $X + Y$, where both X and Y are random variables. In addition, we shall study, in general, functions of two random variables and functions of several random variables. Once again, we restrict ourselves to discrete random variables.

DEFINITION

If $z: \mathbb{R} \times \mathbb{R} \to \mathbb{R}$ is a real-valued function of two real variables, and if X and Y are discrete random variables defined on a sample space S, then we can define a random variable $z(X, Y)$ by

$$z(X, Y)(x) = z(X(x), Y(x)) \qquad \text{for any } x \in S$$

That is, we first apply both X and Y to the sample element x, obtaining a pair of real values $X(x)$ and $Y(x)$, to which we then apply the function z, which is a function of two real arguments, obtaining a third real value $z(X(x), Y(x))$, which we define to be the image of x under the random variable $z(X, Y)$.

Our definition extends to several random variables as follows:

DEFINITION

If $z: \mathbb{R}_n \to \mathbb{R}$ is a real-valued function of n real variables, and if $X_1, X_2, \ldots, X_n$ are discrete random variables defined on a sample space S, then we can define a random variable $z(X_1, X_2, \ldots, X_n)$ by

$$z(X_1, X_2, \ldots, X_n)(x) = z(X_1(x), X_2(x), \ldots, X_n(x))$$

Example 1 Let X and Y be the random variables defined in Example 1 of the preceding section. We have already constructed the joint probability table

X	Y 0	1	2	3
0	0	$\frac{1}{8}$	$\frac{2}{8}$	$\frac{1}{8}$
1	$\frac{1}{8}$	$\frac{2}{8}$	$\frac{1}{8}$	0

Suppose that we use the function $z(x, y) = x + y$ to define a new random variable $z(X, Y)$, which is simply $X + Y$. Clearly, the values that $X + Y$ may assume range from 0 to 4. In order to construct the probability table of $X + Y$, we must relate events of the form $\{X + Y = k\}$ to events of the form $\{X = a\} \cap \{Y = b\}$. This must be done for all values $k = 0, 1, 2, 3, 4$:

$$\{X + Y = 0\} = \{X = 0\} \cap \{Y = 0\}$$
$$\{X + Y = 1\} = \{X = 0\} \cap \{Y = 1\} \cup \{X = 1\} \cap \{Y = 0\}$$
$$\{X + Y = 2\} = \{X = 0\} \cap \{Y = 2\} \cup \{X = 1\} \cap \{Y = 1\}$$
$$\{X + Y = 3\} = \{X = 0\} \cap \{Y = 3\} \cup \{X = 1\} \cap \{Y = 2\}$$
$$\{X + Y = 4\} = \{X = 1\} \cap \{Y = 3\}$$

In order that the sum $X + Y$ be 2, for example, since X assumes only the values 0 and 1 and since Y assumes only the values 0, 1, 2, and 3, we must have either $\{X = 0\}$ and $\{Y = 2\}$ or $\{X = 1\}$ and $\{Y = 1\}$. Now the probabilities of the events $\{X = a\} \cap \{Y = b\}$ listed above can be found in the joint probability table, and so we must make use of this table to form the probability table of $X + Y$. With practice, this can be done by scanning the joint table for the values of X and Y that correspond to the value of $z(X, Y)$, and then adding these probabilities. This gives the following probability table for $X + Y$:

$X + Y$:	0	1	2	3	4
	0	$\frac{1}{4}$	$\frac{1}{2}$	$\frac{1}{4}$	0

For example,

$$P(X + Y = 2) = P(X = 0 \cap Y = 2) + P(X = 1 \cap Y = 1) = \frac{2}{8} + \frac{2}{8} = \frac{1}{2}$$

From the marginal probabilities, we find that $E(X) = \frac{1}{2}$ and $E(Y) = \frac{3}{2}$, and from the probability table for $X + Y$, we find that $E(X + Y) = 2$. Thus, at least in this case, $E(X + Y) = E(X) + E(Y)$. We shall see soon that this formula is true for any choice of random variables X and Y.

Example 2 Using the same X and Y as above, had we defined $z(x, y) = xy$ instead, we would have found the probability distribution

$$XY: \begin{array}{c|c|c|c|c} 0 & 1 & 2 & 3 \\ \hline \frac{5}{8} & \frac{1}{4} & \frac{1}{8} & 0 \end{array}$$

for the random variable $z(X, Y) = XY$. For example,

$$\{XY = 0\} = \{X = 0\} \cup \{Y = 0\}$$

and so

$$P(XY = 0) = P(X = 0) + P(Y = 0) - P(X = 0 \cap Y = 0)$$
$$= \frac{1}{2} + \frac{1}{8} - 0 = \frac{5}{8}$$

Also,

$$\{XY = 2\} = \{X = 1\} \cap \{Y = 2\}$$

and so

$$P(XY = 2) = P(X = 1 \cap Y = 2) = \frac{1}{8}$$

From the probability distribution of XY we see that $E(XY) = \frac{1}{2}$. Since $E(X) \cdot E(Y) = (\frac{1}{2})(\frac{3}{2}) = \frac{3}{4}$, we see that $E(XY) \neq E(X)E(Y)$ for this choice of X and Y. Hence, the product rule for means will not be true in general.

The next result will allow us to bypass the construction of the probability table for $z(X, Y)$ should we be interested only in calculating $E[z(X, Y)]$.

Proposition 5.3

If $z : \mathbb{R} \times \mathbb{R} \to \mathbb{R}$, $X(S) = \{a_1, a_2, \ldots, a_m\}$ and $Y(S) = \{b_1, b_2, \ldots, b_k\}$, then

$$E[z(X, Y)] = \sum_{\substack{i = 1, 2, \ldots, m \\ j = 1, 2, \ldots, k}} z(a_i, b_j) \cdot h(a_i, b_j)$$

Proof: Suppose that $z(X, Y)(S) = \{c_1, c_2, \ldots, c_p\}$. Then, by definition, we have

$$E[z(X, Y)] = \sum_{q=1}^{p} c_q \cdot P[z(X, Y) = c_q]$$

but $c_q = z(a_i, b_j)$ for at least one pair (a_i, b_j), and so

$$\{z(X, Y) = c_q\} = \{x \in S : z(X, Y)(x) = c_q\}$$
$$= \bigcup_{z(a_i, b_j) = c_q} \{x \in S : X(x) = a_i \quad \text{and} \quad Y(x) = b_j\}$$

As a result, we have

$$P(z(X, Y) = c_q) = \sum_{z(a_i, b_j) = c_q} P(X = a_i \cap Y = b_j)$$

Therefore,

$$E[z(X, Y)] = \sum_{q=1}^{p} c_q \sum_{z(a_i, b_j) = c_q} P(X = a_i \cap Y = b_j)$$

$$= \sum_{q=1}^{p} \sum_{z(a_i, b_j) = c_q} c_q \cdot P(X = a_i \cap Y = b_j)$$

$$= \sum_{q=1}^{p} \sum_{z(a_i, b_j) = c_q} z(a_i, b_j) P(X = a_i \cap Y = b_j)$$

$$= \sum_{\substack{i = 1, 2, \ldots, m \\ j = 1, 2, \ldots, k}} z(a_i, b_j) \cdot h(a_i, b_j)$$

the last step being valid because each pair (a_i, b_j) occurs in exactly one of the sums

$$\sum_{z(a_i, b_j) = c_q} P(X = a_i \cap Y = b_j)$$

that sum which corresponds to the value $c = z(a_i, b_j)$.

What Proposition 5.3 is saying is really quite simple. In order to calculate $E[z(X, Y)]$, we can form a sum consisting of one term for each probability in the joint table for X and Y. Each of these terms is the product of that probability $h(a_i, b_j)$ with the corresponding functional value $z(a_i, b_j)$.

Example 1 (*continued*)   Referring to the joint table above, we use Proposition 5.3 to recalculate $E(X + Y)$. We write our calculations in the form of a table, with the value of $z(a_i, b_j) \cdot h(a_i, b_j)$ occurring in the ith row and jth column:

$(0 + 0)(0)$ $(0 + 1)(\frac{1}{8})$ $(0 + 2)(\frac{2}{8})$ $(0 + 3)(\frac{1}{8})$

$(1 + 0)(\frac{1}{8})$ $(1 + 1)(\frac{2}{8})$ $(1 + 2)(\frac{1}{8})$ $(1 + 3)(0)$

Therefore, $E(X + Y) = 0 + (\frac{1}{8}) + (\frac{4}{8}) + (\frac{3}{8}) + (\frac{1}{8}) + (\frac{4}{8}) + (\frac{3}{8}) + 0 = 2$, the sum of the eight values in the table above.

Example 2 (*continued*)   We proceed in the same manner, using Proposition 5.3, to recalculate $E(XY)$:

$(0 \cdot 0)(0)$ $(0 \cdot 1)(\frac{1}{8})$ $(0 \cdot 2)(\frac{2}{8})$ $(0 \cdot 3)(\frac{1}{8})$

$(1 \cdot 0)(\frac{1}{8})$ $(1 \cdot 1)(\frac{2}{8})$ $(1 \cdot 2)(\frac{1}{8})$ $(1 \cdot 3)(0)$

Therefore, $E(XY) = 0 + 0 + 0 + 0 + 0 + \frac{2}{8} + \frac{2}{8} + 0 = \frac{1}{2}$.

Before we proceed any farther, a few comments are in order concerning absolute convergence and discrete random variables that have countable ranges. First of all, implicit in our discussion of the joint probability function is the assumption that if either X or Y has a countable range, then the series $\sum_{i,j} h(a_i, b_j)$ be (absolutely) convergent. (Since it is a series of positive terms, absolute convergence is equivalent to convergence.) This carries along with it the additional assumptions that the series $\sum_i f(a_i)$ and $\sum_j g(b_j)$ also be (absolutely) convergent. It is the absolute convergence of these latter two series that justifies the use of the distributive law in the last step of the proof of Proposition 5.1.

Proposition 5.3 can also be extended, in a straightforward manner, to the case of discrete random variables with countable ranges, provided we make the assumption that the series $\sum_{i,j} z(a_i, b_j) \cdot h(a_i, b_j)$ be absolutely convergent.

We now proceed to derive a series of corollaries to Proposition 5.3.

Corollary 5.4

$E(X + Y) = E(X) + E(Y)$

Proof: In this case, $z(x, y) = x + y$, so that $z(a_i, b_j) = a_i + b_j$. Therefore,

$$E(X + Y) = \sum_{i,j} (a_i + b_j) \cdot h(a_i, b_j)$$

$$= \sum_{i,j} a_i \cdot h(a_i, b_j) + \sum_{i,j} b_j h(a_i, b_j)$$

$$= \sum_{i=1}^{m} a_i \left[\sum_{j=1}^{k} h(a_i, b_j) \right] + \sum_{j=1}^{k} b_j \left[\sum_{i=1}^{m} h(a_i, b_j) \right]$$

$$= \sum_{i=1}^{m} a_i \cdot f(a_i) + \sum_{j=1}^{k} b_j \cdot g(b_j)$$

$$= \quad E(X) \quad + \quad E(Y)$$

If X and Y were allowed to have countable ranges, we would require the absolute convergence of both $E(X)$ and $E(Y)$. This would justify the separation into two sums in the first step and the use of the distributive law in the second step of the proof above.

Corollary 5.5

(a) $E(aX + bY) = aE(X) + bE(Y)$
(b) If $X_1, X_2, \ldots, X_n$ are all discrete random variables, then

$$E\left(\sum_{i=1}^{n} a_i X_i \right) = \sum_{i=1}^{n} a_i \cdot E(X_i)$$

Proof: By Corollary 5.4, $E(aX + bY) = E(aX) + E(bY)$. By Corollary 4.4, $E(aX) = aE(X)$ and $E(bY) = bE(Y)$, which proves (a). Then (b) follows from (a) by mathematical induction on n.

Corollary 5.6

If $X_1, X_2, \ldots, X_n$ are independent discrete random variables with moment-generating functions $M_{X_1}(t), M_{X_2}(t), \ldots, M_{X_n}(t)$, respectively, and if $Y = X_1 + X_2 + \cdots + X_n$, then

$$M_Y(t) = \prod_{i=1}^{n} M_{X_i}(t)$$

Proof: By definition,

$$M_Y(t) = E(e^{Yt}) = E\{\exp[(X_1 + X_2 + \cdots + X_n)t]\}$$

and if we let a_i denote an arbitrary value assumed by the random variable X_i, we have

$$M_Y(t) = \sum_{a_i} \exp[(a_1 + a_2 + \cdots + a_n)t] \cdot f(a_1, a_2, \ldots, a_n)$$

Using the independence of the X_i, we have

$$M_Y(t) = \sum_{a_i} \exp[(a_1 + a_2 + \cdots + a_n)t] \cdot f(a_1) \cdot f(a_2) \cdots f(a_n)$$

Combining terms in a_i, we obtain

$$M_Y(t) = \sum_{a_i} [e^{a_1 t} \cdot f(a_1)][e^{a_2 t} \cdot f(a_2)] \cdots [e^{a_n t} \cdot f(a_n)]$$

Then, using either the simple distributive law or the absolute convergence of the series for the $M_{X_i}(t)$, we get

$$M_Y(t) = \left[\sum_{a_1} e^{a_1 t} \cdot f(a_1) \right] \cdots \left[\sum_{a_n} e^{a_n t} \cdot f(a_n) \right]$$

But

$$M_{X_i}(t) = \sum_{a_i} e^{a_i t} \cdot f(a_i)$$

and so the desired result follows.

In many cases, the calculation of the mean of a random variable can be greatly simplified by writing the random variable as a sum of several indicator random variables and then applying Corollary 5.5. Care must be taken in choosing the proper indicators, so that the sum of the indicators does equal the original random variable. The following three examples illustrate this technique.

Example 3 Five balls numbered 1, 2, 3, 4, and 5 are placed at random into five urns marked 1, 2, 3, 4, and 5, so that one ball is placed into each urn. A *match* occurs when the ball numbered k is placed into the urn marked k, $k = 1, 2, 3, 4, 5$. We let X denote the number of matches, and wish to calculate the mean of X. We could work in the sample space S of all 5! permutations of the five numbers 1, 2, 3, 4, and 5, and try to construct the probability function of X. Instead we will express X as a sum of several indicator random variables. The indicators that we choose must have two nice properties: They must add up to X, and they must be considerably easier to work with than X. In this case, we choose indicators I_k that indicate whether or not ball k is in urn k. That is, we let I_k be the random variable

$$I_k = \begin{cases} 1 & \text{if ball } k \text{ is in urn } k \\ 0 & \text{otherwise} \end{cases}, \, k = 1, 2, 3, 4, 5$$

Each match forces the corresponding indicator to be 1; otherwise (no match), that indicator would be 0. Therefore, $X = I_1 + I_2 + I_3 + I_4 + I_5$. So adding the indicators does, in fact, count the number of matches. By Corollary 5.5, then, we have

$$E(X) = E\left(\sum_{k=1}^{5} I_k \right) = \sum_{k=1}^{5} E(I_k)$$

and since the five indicators are essentially the same, we have $E(X) = 5 \cdot E(I_k)$. Thus, the problem of calculating $E(X)$ has been reduced to one of calculating $E(I_k)$ for any one k. Since I_k is an indicator random variable, we know that

$$E(I_k) = P(\text{ball } k \text{ is in urn } k)$$

In the sample space of $5! = 120$ permutations of the numbers (numbered balls) 1, 2, 3, 4, and 5, exactly $4! = 24$ correspond to the event that number k is in position k. This is actually just the number of ways of permuting the other four numbers in the other four positions. Therefore,

$$P(\text{ball } k \text{ is in urn } k) = \frac{24}{120} = \frac{1}{5}$$

and hence, $E(X) = 5 \cdot E(I_k) = 5 \cdot (\frac{1}{5}) = 1$. Thus, on the average, we would expect one match. (As a matter of fact, if there had been n balls and n urns, we would still expect just one match.)

Example 4 Suppose that n indistinguishable balls are randomly distributed among three urns numbered 1, 2, and 3. We let X denote the number of empty urns, and we wish to calculate $E(X)$. We could work in the sample space

$$S = \{(a_1, a_2, \ldots, a_n) : a_i = 1, 2, \text{ or } 3\}$$

and determine which of these 3^n n-tuples correspond to zero, one, or two empty urns. Instead, we define indicator random variables

$$I_k = \begin{cases} 1 & \text{if urn } k \text{ is empty} \\ 0 & \text{otherwise} \end{cases}$$

for $k = 1, 2, 3$. Each indicator tells whether or not the corresponding urn is empty, and so their sum gives the number of empty urns. Therefore, $X = I_1 + I_2 + I_3$. As a result,

$$E(X) = E(I_1) + E(I_2) + E(I_3) = 3 \cdot E(I_1)$$

since the three indicators are essentially the same. Since we are once again dealing with indicators, we have

$$E(I_1) = P(\text{urn 1 is empty})$$

Since the event "urn 1 is empty" corresponds to the 2^n n-tuples (as described above) whose coordinates $a_i = 2$ or 3, we have $P(\text{urn 1 is empty}) = 2^n/3^n = (\frac{2}{3})^n$, and so,

$$E(X) = 3 \cdot E(I_1) = 3 \cdot \left(\frac{2}{3}\right)^n = \frac{2^n}{3^{n-1}}$$

Example 5   We let X denote the number of aces in a poker hand, and wish to find $E(X)$, the average number of aces in a poker hand. Using the techniques of Chapter 1, we find that the probability function f of X is

$$f(k) = \begin{cases} C_k^4 \cdot C_{5-k}^{48}/C_5^{52} & \text{for } k = 0, 1, 2, 3, 4 \\ 0 & \text{otherwise} \end{cases}$$

and therefore,

$$E(X) = \sum_{k=0}^{4} kC_k^4 \cdot C_{5-k}^{48}/C_5^{52}$$

Unfortunately, the arithmetic involved in calculating $E(X)$ this way is tedious, and so we try an alternative approach using indicators. We define four indicators as follows:

$I_1 = $ "the ace of spades is in the hand"
$I_2 = $ "the ace of hearts is in the hand"
$I_3 = $ "the ace of clubs is in the hand"
$I_4 = $ "the ace of diamonds is in the hand"

For example,

$$I_1 = \begin{cases} 1 & \text{if the ace of spades is in the hand} \\ 0 & \text{otherwise} \end{cases}$$

Now $X = I_1 + I_2 + I_3 + I_4$, and so $E(X) = E(I_1) + E(I_2) + E(I_3) + E(I_4) = 4E(I_1)$, since the four indicators are essentially the same. Because the I_k are indicators, we have

$$E(I_1) = P(\text{ace of spades is in the hand})$$

By the formula derived in Example 8 of Section 2.3, we have

$P($ace of spades is in the hand$) = \frac{5}{52}$

the specified item being the ace of spades. Therefore,

$E(X) = 4 \cdot E(I_1) = 4 \cdot (\frac{5}{52}) = \frac{5}{13}$

We present now one final corollary to Proposition 5.3.

Corollary 5.7

$E(XY) - E(X)E(Y) = E[(X - \mu_X)(Y - \mu_Y)]$

Proof: We expand the right-hand side of this equation:

$E[(X - \mu_X)(Y - \mu_Y)] = E(XY - \mu_Y X - \mu_X Y + \mu_X \mu_Y)$

which by Corollary 5.5

$$= E(XY) + E(-\mu_Y X) + E(-\mu_X Y) + E(\mu_X \mu_Y)$$

which, in turn, by Corollary 4.4

$$= E(XY) - \mu_Y E(X) - \mu_X E(Y) + \mu_X \mu_Y$$
$$= E(XY) - E(X)E(Y)$$

Thus, in general, the multiplication rule for means, $E(XY) = E(X) \cdot E(Y)$, does not hold, and the amount by which the rule fails is $E[(X - \mu_X)(Y - \mu_Y)]$. We shall study this term in more detail in the next section. The following result gives sufficient conditions that the multiplication rule holds. We shall see that these conditions are not necessary.

Proposition 5.8

If X and Y are independent discrete random variables, then $E(XY) = E(X) \cdot E(Y)$.

Proof: Since we are using the function $z(x, y) = xy$, we have $z(a_i, b_j) = a_i b_j$, and so, by Proposition 5.3, we have

$E(XY) = \sum_{i, j} a_i \cdot b_j \cdot h(a_i b_j)$

$\quad = \sum_{i, j} a_i \cdot b_j \cdot f(a_i) \cdot g(b_j)$ (by the independence of X and Y)

$\quad = \sum_{i=1}^{m} a_i \cdot f(a_i) \cdot \sum_{j=1}^{k} b_j \cdot g(b_j)$ (distributive law)

$\quad = \quad E(X) \quad \cdot \quad E(Y)$

In the case of discrete X and Y with countable ranges, the absolute convergence of $E(X^2)$ and $E(Y^2)$ is sufficient to guarantee the absolute convergence of $E(XY)$. For, since $2|xy| \le x^2 + y^2$ for any x, y, we have

$$E(XY) = \sum_{i,j=1}^{\infty} |a_i b_j| h(a_i, b_j)$$

$$\le \sum_{i,j=1}^{\infty} a_i^2 \cdot h(a_i, b_j) + \sum_{i,j=1}^{\infty} b_j^2 \cdot h(a_i, b_j)$$

$$= \sum_{i,j=1}^{\infty} a_i^2 f(a_i) g(b_j) + \sum_{i,j=1}^{\infty} b_j^2 f(a_i) g(b_j)$$

$$= \sum_{i=1}^{\infty} a_i^2 f(a_i) \left[\sum_{j=1}^{\infty} g(b_j) \right]$$

$$+ \sum_{j=1}^{\infty} b_j^2 g(b_j) \left[\sum_{i=1}^{\infty} f(a_i) \right] \qquad \text{(by the independence of } X \text{ and } Y \text{)}$$

$$= \sum_{i=1}^{\infty} a_i^2 f(a_i) + \sum_{j=1}^{\infty} b_j^2 g(b_j) \qquad \text{(distributive law)}$$

$$= \quad E(X^2) \quad + \quad E(Y^2)$$

The original proof, then, is valid for the infinite case, with the third step, the separation into a product of two sums, being valid because of the absolute convergence of $E(XY)$.

The converse of Proposition 5.8 is false. The fact that the multiplication rule $E(XY) = E(X) \cdot E(Y)$ holds does *not* imply the independence of X and Y, as the following example illustrates.

Example 6 Suppose that X is a random variable with probability distribution

a_i	-1	0	1
$f(a_i)$	$\frac{1}{4}$	$\frac{1}{2}$	$\frac{1}{4}$

and that $Y = X^2$. Since Y is a function of X, we would not expect X and Y to be independent. This is evident from the joint probability table

X Y	0	1	$f(a_i)$
-1	0	$\frac{1}{4}$	$\frac{1}{4}$
0	$\frac{1}{2}$	0	$\frac{1}{2}$
1	0	$\frac{1}{4}$	$\frac{1}{4}$
$g(b_j)$	$\frac{1}{2}$	$\frac{1}{2}$	1

Obviously, the multiplication rule $h(a_i, b_j) = f(a_i) \cdot g(b_j)$ does not hold, so X and Y are not independent. But $E(X) = 0$, $E(Y) = \frac{1}{2}$, and $E(XY) = 0$ and so $E(XY) = E(X) \cdot E(Y)$.

Notice the special form of the joint table in the preceding example. Each row of the table contains exactly one nonzero probability. This is a strong indication that the values assumed by Y are very much dependent on the values assumed by X. In other words, Y is a function of X.

Corollary 5.9

If X and Y are independent discrete random variables, then

$$E[(X - \mu_X)(Y - \mu_Y)] = 0$$

Proof: By Corollary 5.7,

$$E(XY) - E(X)E(Y) = E[(X - \mu_X)(Y - \mu_Y)]$$

and by Proposition 5.8,

$$E(XY) - E(X)E(Y) = 0$$

since X and Y are independent, proving the desired result.

Proposition 5.10

If X and Y are independent discrete random variables, then

$$\text{Var}(X + Y) = \text{Var}(X) + \text{Var}(Y)$$

Proof: By definition,

$$
\begin{aligned}
\text{Var}(X + Y) &= E\{[(X + Y) - E(X + Y)]^2\} \\
&= E\{[X + Y - E(X) - E(Y)]^2\} \\
&= E\{[(X - E(X)) + (Y - E(Y))]^2\} \\
&= E\{[X - E(X)]^2 + 2[X - E(X)][Y - E(Y)] + [Y - E(Y)]^2\} \\
&= E\{[X - E(X)]^2\} + 2E\{[X - E(X)][Y - E(Y)]\} \\
&\quad + E\{[Y - E(Y)]^2\} \\
&= \text{Var}(X) + 2E\{[X - E(X)][Y - E(Y)]\} + \text{Var}(Y)
\end{aligned}
$$

Since X and Y are independent, the middle term is 0, and so $\text{Var}(X + Y) = \text{Var}(X) + \text{Var}(Y)$.

Corollary 5.11

If X and Y are independent discrete random variables, then $\text{Var}(aX + bY) = a^2 \, \text{Var}(X) + b^2 \, \text{Var}(Y)$.

Proof: The independence of X and Y implies the independence of aX and bY (Proposition 5.1). Applying Proposition 5.10 to aX and bY, we have

$$\text{Var}(aX + bY) = \text{Var}(aX) + \text{Var}(bY)$$

The result now follows from Corollary 4.8.

We note that the independence of $X_1, X_2, \ldots, X_n$ implies the independence of $a_1 X_1 + a_2 X_2 + \cdots + a_{n-1} X_{n-1}$ and X_n (Exercise 1). We use this result to prove

Corollary 5.12

If $X_1, X_2, \ldots, X_n$ are independent discrete random variables, then

$$\text{Var}\left(\sum_{i=1}^{n} a_i X_i \right) = \sum_{i=1}^{n} a_i^2 \cdot \text{Var}(X_i)$$

Proof: The proof is by induction on n. The case $n = 2$ is Corollary 5.11. We have

$$\text{Var}\left(\sum_{i=1}^{n} a_i X_i \right) = \text{Var}\left(\sum_{i=1}^{n-1} a_i X_i + a_n X_n \right)$$

$$= \text{Var}\left(\sum_{i=1}^{n-1} a_i X_i \right) + \text{Var}(a_n X_n)$$

by Proposition 5.10 and the remark above. By induction, this

$$= \sum_{i=1}^{n-1} a_i^2 \cdot \text{Var}(X_i) + a_n^2 \cdot \text{Var}(X_n)$$

$$= \sum_{i=1}^{n} a_i^2 \cdot \text{Var}(X_i)$$

The following example exhibits the use of Chebyshev's theorem, as well as the preceding results.

Example 7 An honest die is rolled 300 times and we wish to estimate the probability that the top face shows a 1 or a 2 fewer than 70 times. We define a random variable X by

$X =$ number of 1's or 2's in 300 rolls

and decompose X into a sum of 300 indicators I_k, where

$$I_k = \begin{cases} 1 & \text{if the } k\text{th roll results in a 1 or a 2} \\ 0 & \text{otherwise} \end{cases}$$

We therefore have $X = \sum_{k=1}^{300} I_k$, and so $E(X) = \sum_{k=1}^{300} E(I_k)$ and, assuming that the 300 rolls of the die are independent, $\text{Var}(X) = \sum_{k=1}^{300} \text{Var}(I_k)$. Therefore, in order to calculate $E(X)$ and $\text{Var}(X)$, we need only calculate $E(I_k)$ and $\text{Var}(I_k)$ for one indicator I_k, the 300 indicators being essentially the same. But I_k has probability distribution

a_i	0	1
$f(a_i)$	$\frac{2}{3}$	$\frac{1}{3}$

and so $E(I_k) = \frac{1}{3}$ and $\mathrm{Var}(I_k) = \frac{2}{9}$. Thus

$$E(X) = 300 \cdot E(I_k) = 300 \cdot \tfrac{1}{3} = 100$$

and

$$\mathrm{Var}(X) = 300 \cdot \mathrm{Var}(I_k) = 300 \cdot \tfrac{2}{9} = \tfrac{600}{9}$$

But why have we bothered to calculate $E(X)$ and $\mathrm{Var}(X)$? For the simple reason that we want to relate the event $\{X < 70\}$ to $E(X)$. If we consider the following picture of the event $\{X < 70\}$

we might be able to see a potential application of Chebyshev's inequality. Since the event $\{X < 70\}$ is a subset of the event

$$\{X < 70\} \cup \{X > 130\} = \{|X - 100| > 30\}$$

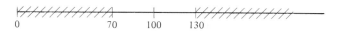

we have $P(X < 70) \le P(|X - 100| > 30)$. [Actually, it is true that $P(X < 70) = \frac{1}{2}P(|X - 100| > 30)$.] This last probability can be given an upper bound by means of Chebyshev's inequality:

$$P(|X - 100| > 30) \le \frac{\mathrm{Var}(X)}{30^2}$$

$$= \frac{\frac{600}{9}}{30^2} = \frac{2}{27}$$

Thus, according to Chebyshev's inequality, the probability that there will be fewer than 70 1's or 2's in 300 rolls of a fair die is at most 2/27. (Using the fact that the event $X < 70$ is exactly half of the event $|X - 100| > 30$, we can reduce this estimate in half, to 1/27.)

EXERCISES

1. Suppose that $X_1, X_2, \ldots, X_n$ are independent discrete random variables. Prove that $a_1X_1 + a_2X_2 + \cdots + a_{n-1}X_{n-1}$ and X_n are independent.
2. Determine the probability distribution of the random variables $X + Y$, $X + Z$, $Y + Z$, XY, XZ, YZ where X, Y, and Z are the random variables from Exercises 1 and 2 of the preceding section.
3. Let X and Y be the random variables from Exercise 8 of the preceding section. Find the probability distribution of $X + Y$ and XY. Find $E(X + Y)$, $E(XY)$, $\mathrm{Var}(X + Y)$, and $\mathrm{Var}(XY)$.
4. A cube has four sides with one spot and two sides with two spots. Let X denote the total number of spots shown in ten rolls of the cube. Find $E(X)$ and $\mathrm{Var}(X)$.
5. How many different faces should be expected to show in five rolls of a fair die?

6. A fair die is rolled (independently) 900 times. Let T denote the total number of dots for the 900 rolls. Find a lower bound for $P(2900 \leq T \leq 3400)$.

7. A newspaper prints the pictures of ten famous people and, in scrambled order, a baby picture of each. If you are just guessing, how many of these people should you expect to match to their baby pictures?

8. A fair die is rolled (independently) 1000 times and we let X denote the number of even outcomes (2, 4, 6). Estimate $P(480 \leq X \leq 520)$.

9. What is the expected number of hearts in a poker hand?

10. A box contains 1000 items, 20 of which are defective. What is the expected number of defectives in a sample of 50 items chosen from the box?

5.3 Covariance and Correlation

The term $E[(X - \mu_X)(Y - \mu_Y)]$ has already occurred several times in this chapter. We have seen that it takes the value zero when the random variables X and Y are independent and, in any case, is in some sense a measure of the independence of the two random variables. The term thus seems to have some intrinsic value, so we shall give it a name and some further study.

DEFINITION

The *covariance* of two discrete random variables X and Y, denoted Cov(X, Y), is defined to equal $E[(X - \mu_X)(Y - \mu_Y)]$.

As the name suggests, the covariance is also a measure of how the two random variables vary together. If, for a given sample element, both random variables take values larger than their respective means, or both take values smaller than their means, then the value of $(X - \mu_X)(Y - \mu_Y)$ will be positive for that sample element. If, on the other hand, one of the random variables assumes a value larger than its mean while the other assumes a value smaller than its mean, then the value of $(X - \mu_X)(Y - \mu_Y)$ will be negative for that sample element. Thus, if the random variables X and Y tend to simultaneously assume values on the same side of their respective means, the value of their covariance will be positive. But if they tend to simultaneously assume values on the opposite sides of their means, the value of their covariance will be negative.

The facts already known about the covariance, together with some results that are direct consequences of the definition, are included in the following:

Proposition 5.13

If X and Y are discrete random variables, then

(a) $\text{Cov}(X, Y) = E(XY) - E(X) \cdot E(Y)$
(b) $\text{Var}(X + Y) = \text{Var}(X) + \text{Var}(Y) + 2 \cdot \text{Cov}(X, Y)$
(c) $\text{Cov}(Y, X) = \text{Cov}(X, Y)$

(d) $\text{Cov}(X - \mu_X, Y - \mu_Y) = \text{Cov}(X, Y)$
(e) $\text{Cov}(aX, bY) = ab \cdot \text{Cov}(X, Y)$
(f) If X and Y are independent, then $\text{Cov}(X, Y) = 0$

Proof: Exercise

Example 1 Suppose that X and Y are the random variables defined in Example 1 of Section 5.1. We have seen that $E(X) = \frac{1}{2}$, $E(Y) = \frac{3}{2}$, and $E(XY) = \frac{1}{2}$. Therefore, $\text{Cov}(X, Y) = \frac{1}{2} - (\frac{1}{2} \cdot \frac{3}{2}) = -\frac{1}{4}$. Recall that X is the indicator of "tails on the first toss," while Y counts the number of heads in the three tosses. Notice that a high value of X (a 1) naturally results in a lower value for Y. If the first toss results in tails, there will naturally be fewer heads for the three tosses, so there is a kind of reverse relationship between X and Y, and this explains the negative covariance.

Example 2 Suppose that X and Y are the random variables defined in Example 5 in Section 5.1. We calculate that $E(X) = E(Y) = \frac{3}{5}$ and that $E(XY) = \frac{9}{25}$. Therefore, $\text{Cov}(X, Y) = \frac{9}{25} - (\frac{3}{5} \cdot \frac{3}{5}) = 0$. Recall that these random variables are independent.

Example 3 Suppose that X and Y are the random variables defined in Example 6 of Section 5.2. We already know that $E(X) = 0$, $E(Y) = \frac{1}{2}$, and $E(XY) = 0$. Therefore, $\text{Cov}(X, Y) = 0 - 0(\frac{1}{2}) = 0$, even though the random variable Y is very much dependent on X (as a matter of fact, Y is defined to be X^2).

Proposition 5.14

If X, Y, and Z are discrete random variables, then

$$\text{Cov}(X + Y, Z) = \text{Cov}(X, Z) + \text{Cov}(Y, Z)$$

Proof: By definition,

$$
\begin{aligned}
\text{Cov}(X + Y, Z) &= E\{[(X + Y) - E(X + Y)][Z - E(Z)]\} \\
&= E(\{[X - E(X)] + [Y - E(Y)]\} [Z - E(Z)]) \\
&= E\{[X - E(X)][Z - E(Z)]\} + E\{[Y - E(Y)][Z - E(Z)]\} \\
&= \qquad \text{Cov}(X, Z) \qquad\qquad + \text{Cov}(Y, Z)
\end{aligned}
$$

Proposition 5.15

Suppose that $X_1, X_2, \ldots, X_n$ are discrete random variables. Then

(a) $\text{Var}\left(\sum_{i=1}^{n} X_i\right) = \sum_{i=1}^{n} \text{Var}(X_i) + 2 \sum_{i<j} \text{Cov}(X_i, X_j)$

(b) $\text{Var}\left(\sum_{i=1}^{n} a_i X_i\right) = \sum_{i=1}^{n} a_i^2 \, \text{Var}(X_i) + 2 \sum_{i<j} a_i a_j \, \text{Cov}(X_i, X_j)$

(c) If $Y_1 = \sum\limits_{i=1}^{n} a_i X_i$ and $Y_2 = \sum\limits_{i=1}^{n} b_i X_i$, then

$$\text{Cov}(Y_1, Y_2) = \sum_{i=1}^{n} a_i b_i \, \text{Var}(X_i) + \sum_{i<j} (a_i b_j + a_j b_i)\text{Cov}(X_i, X_j)$$

(d) If the $\{X_i\}$ are independent, then

$$\text{Cov}(Y_1, Y_2) = \sum_{i=1}^{n} a_i b_i \, \text{Var}(X_i)$$

Proof:

(a) This follows by induction on n, using Proposition 5.13(b) and Proposition 5.14.
(b) This follows from (a), using Proposition 5.13(e).
(c) By definition,

$$\text{Cov}(Y_1, Y_2) = E(Y_1 Y_2) - E(Y_1)E(Y_2)$$

$$= E\left(\sum_{i=1}^{n} a_i X_i \cdot \sum_{j=1}^{n} b_j X_j\right) - E\left(\sum_{i=1}^{n} a_i X_i\right)E\left(\sum_{j=1}^{n} b_j X_j\right)$$

$$= E\left(\sum_{i,j=1}^{n} a_i b_j X_i X_j\right) - E\left(\sum_{i=1}^{n} a_i X_i\right)E\left(\sum_{j=1}^{n} b_j X_j\right)$$

$$= \sum_{i,j=1}^{n} a_i b_j E(X_i X_j) - \sum_{i=1}^{n} a_i E(X_i) \sum_{j=1}^{n} b_j E(X_j)$$

$$= \sum_{i,j=1}^{n} a_i b_j E(X_i X_j) - \sum_{i,j=1}^{n} a_i b_j E(X_i)E(X_j)$$

$$= \sum_{i,j=1}^{n} a_i b_j [E(X_i X_j) - E(X_i) \cdot E(X_j)]$$

Separating the $i = j$ and $i \neq j$ parts of this summation, we obtain

$$\text{Cov}(Y_1, Y_2) = \sum_{i=1}^{n} a_i b_i [E(X_i^2) - E(X_i)^2]$$

$$+ \sum_{i \neq j} a_i b_j [E(X_i X_j) - E(X_i)E(X_j)]$$

$$= \sum_{i=1}^{n} a_i b_i \, \text{Var}(X_i) + \sum_{i<j} a_i b_j \, \text{Cov}(X_i, X_j)$$

$$+ \sum_{i>j} a_i b_j \, \text{Cov}(X_i, X_j)$$

According to Proposition 5.13(c), we can reverse the roles of i and j in the third summand, and this gives the desired result.
(d) Since the $\{X_i\}$ are independent, we have $\text{Cov}(X_i, X_j) = 0$ for any pair X_i, X_j. The desired result now follows from (c).

The covariance of two random variables has a major shortcoming—it is affected by the size of the random variables. The following example illustrates this point.

Example 4 Suppose that X_1 and Y_1 are discrete random variables, and that $X_2 = 2X_1$ and $Y_2 = 2Y_1$. Then $E(X_2) = 2E(X_1)$, $E(Y_2) = 2E(Y_1)$, and $E(X_2 Y_2) = 4E(X_1 Y_1)$. Therefore,

$$
\begin{aligned}
\text{Cov}(X_2, Y_2) &= E(X_2 Y_2) - E(X_2) \cdot E(Y_2) \\
&= 4E(X_1 Y_1) - [2E(X_1)] \cdot [2E(Y_1)] \\
&= 4\,\text{Cov}(X_1, Y_1)
\end{aligned}
$$

Therefore, even though the random variables X_1 and Y_1, and X_2 and Y_2, behave in the exact same manner with respect to each other, the second pair have a covariance four times as large as the first pair.

The covariance, as a matter of fact, is a boundless function. We can find random variables X and Y to make $\text{Cov}(X, Y)$ any number we choose. For if we start with an arbitrary pair of random variables X and Y, then any real number c can be written in the form $c = ab\,\text{Cov}(X, Y)$, for a suitable choice of a and b. It then follows from Proposition 5.13(e) that $\text{Cov}(aX, bY) = ab\,\text{Cov}(X, Y) = c$.

It is more convenient, however, to have a measure of the relationship between two random variables that cannot vary so widely. It is for this reason that we make the following definition.

DEFINITION

The *correlation coefficient* of the random variables X and Y, denoted $\rho(X, Y)$, is defined to equal the covariance of the standardized random variables X' and Y'. That is, $\rho(X, Y) = \text{Cov}(X', Y')$. If either $\sigma_X = 0$ or $\sigma_Y = 0$, we define $\rho(X, Y) = 0$. If $\rho(X, Y) = 0$, then X and Y are said to be *uncorrelated*; otherwise, X and Y are said to be *correlated*.

Proposition 5.16

$$
\rho(X, Y) = \frac{\text{Cov}(X, Y)}{\sigma_X \sigma_Y}
$$

Proof: By definition,

$$
\begin{aligned}
\rho(X, Y) = \text{Cov}(X', Y') &= \text{Cov}\left(\frac{X - \mu_X}{\sigma_X}, \frac{Y - \mu_Y}{\sigma_Y}\right) \\
&= \left(\frac{1}{\sigma_X \sigma_Y}\right) \text{Cov}(X - \mu_X, Y - \mu_Y) \qquad [\text{Prop. 5.13(e)}] \\
&= \left(\frac{1}{\sigma_X \sigma_Y}\right) \text{Cov}(X, Y) \qquad\qquad\quad [\text{Prop. 5.13(d)}]
\end{aligned}
$$

We shall use this formula, rather than the definition, to calculate the correlation coefficient of two random variables. The values σ_X, σ_Y, and $\text{Cov}(X, Y)$ are relatively simple to calculate, and quite likely have already been calculated for other purposes.

Example 1 (continued)   Since $E(X) = E(X^2) = \frac{1}{2}$, we have $\text{Var}(X) = \frac{1}{2} - (\frac{1}{2})^2 = \frac{1}{4}$. Also, $E(Y) = \frac{3}{2}$ and

$$E(Y^2) = (1 \cdot \tfrac{3}{8}) + (4 \cdot \tfrac{3}{8}) + (9 \cdot \tfrac{1}{8}) = 3$$

so that $\text{Var}(Y) = 3 - (\frac{3}{2})^2 = \frac{3}{4}$. We have already calculated $\text{Cov}(X, Y) = -\frac{1}{4}$, and therefore,

$$\rho(X, Y) = \frac{-\frac{1}{4}}{\sqrt{\tfrac{1}{4}}\sqrt{\tfrac{3}{4}}} = \frac{-1}{\sqrt{3}} = -0.58$$

Example 2 (continued)   Since the random variables X and Y in this example were independent, we found $\text{Cov}(X, Y) = 0$. Therefore, $\rho(X, Y) = 0$.

Example 3 (continued)   In this example, even though the random variables X and Y were not independent, we found that $\text{Cov}(X, Y) = 0$, and so $\rho(X, Y) = 0$.

The following result justifies our use of the correlation coefficient rather than the covariance.

Proposition 5.17

$$|\rho(X, Y)| \leq 1$$

Proof: We begin the proof by applying Proposition 5.13(b) to the random variables X' and Y':

$$\text{Var}(X' + Y') = \text{Var}(X') + \text{Var}(Y') + 2\,\text{Cov}(X', Y')$$
$$= \quad 1 \quad + \quad 1 \quad + 2\rho(X, Y)$$

Thus,

$$\text{Var}(X' + Y') = 2 + 2\rho(X, Y) = 2[1 + \rho(X, Y)]$$

but the variance of any random variable is positive, and so

$$2(1 + \rho(X, Y)) \geq 0$$

which is equivalent to $1 + \rho(X, Y) \geq 0$, or $\rho(X, Y) \geq -1$. In similar fashion, we can show that

$$\text{Var}(X' - Y') = 2[1 - \rho(X, Y)]$$

and since $\text{Var}(X' - Y') \geq 0$, we have

$$2[1 - \rho(X, Y)] \geq 0$$

which is equivalent to $1 - \rho(X, Y) \geq 0$, or $\rho(X, Y) \leq 1$. Thus $-1 \leq \rho(X, Y) \leq 1$, which is equivalent to $|\rho(X, Y)| \leq 1$.

Thus, the correlation coefficient is indeed a bounded function. It would be helpful if the value of $\rho(X, Y)$ told us something about the relationship between X and Y. However, Examples 2 and 3 above have indicated that both independent and highly dependent random variables X and Y can have $\rho(X, Y) = 0$, so it would seem unlikely that much sense can be made out of the fact that $\rho(X, Y) = 0$. On the other hand, recall that a large (either positive or negative) value for the covariance indicates that the random variables are behaving quite similarly with respect to their means, or else quite the opposite with respect to their means. One might suspect that there is a functional relationship between the random variables in this case. The following two results indicate that relationship:

Proposition 5.18

Suppose that $\sigma_X = 0$ and that $Y = mX + b$, where $m \neq 0$. Then $\rho(X, Y) = 1$ if $m > 0$ and $\rho(X, Y) = -1$ if $m < 0$.

Proof: If $Y = mX + b$, then $E(Y) = mE(X) + b$ and $\text{Var}(Y) = m^2 \text{Var}(X)$. Therefore,

$$
\begin{aligned}
\text{Cov}(X, Y) &= E[(X - \mu_X)(Y - \mu_Y)] \\
&= E\{(X - \mu_X)[mX + b - mE(X) - b]\} \\
&= E\{(X - \mu_X)[mX - mE(X)]\} \\
&= E[m(X - \mu_X)^2] \\
&= m \cdot E[(X - \mu_X)^2] \\
&= m \cdot \text{Var}(X)
\end{aligned}
$$

and so

$$
\begin{aligned}
\rho(X, Y) &= \frac{\text{Cov}(X, Y)}{\sigma_X \sigma_Y} = \frac{m \,\text{Var}(X)}{\sigma_X \sqrt{m^2 \,\text{Var}(X)}} \\
&= \frac{m \sigma_X{}^2}{\sigma_X |m| \sigma_X} \\
&= \frac{m}{|m|}
\end{aligned}
$$

Thus, $\rho(X, Y) = m/|m|$. If $m > 0$, then $m/|m| = +1$, while $m < 0$ implies that $m/|m| = -1$. Hence, $\rho(X, Y) = \pm 1$, depending on the sign of m.

The following is the converse of Proposition 5.18:

Proposition 5.19

If $|\rho(X, Y)| = 1$, then there exists $m \neq 0$ and b such that $Y = mX + b$ except possibly for a set of sample values whose probability is 0.

Proof: We consider the case $\rho(X, Y) = 1$; the case $\rho(X, Y) = -1$ is similar and is left as an exercise. Since

$$\text{Var}(X' - Y') = 2[1 - \rho(X, Y)]$$

$\rho(X, Y) = +1$ implies that $\text{Var}(X' - Y') = 0$. This implies (Exercise 1 of Section 4.4) that $X' - Y'$ is constant except possibly on a set of probability 0. We write this as

$$X' - Y' = k \qquad \text{ae} \qquad \text{(almost everywhere)}$$

This implies that

$$\frac{X - \mu_X}{\sigma_X} - \frac{Y - \mu_Y}{\sigma_Y} = k \qquad \text{ae}$$

which implies that

$$Y = \frac{\sigma_Y}{\sigma_X} \cdot X + \frac{\mu_Y - k\sigma_Y - (\sigma_Y \mu_X)}{\sigma_X} \qquad \text{ae}$$

So we choose $m = \sigma_Y/\sigma_X$ (which is nonzero since $\sigma_Y \neq 0$) and $b = \mu_Y - k\sigma_Y - (\sigma_Y \mu_X)/\sigma_X$, and we have $Y = mX + b$ (ae).

EXERCISES

1. Prove Proposition 5.13.
2. Prove Proposition 5.15(a).
3. Show that $\rho(X_1, Y_1) = \rho(X_2, Y_2)$ for the random variables X_1, Y_1, X_2, and Y_2 in Example 4.
4. Verify the result of Proposition 5.19 for the case $\rho(X, Y) = -1$.
5. Calculate the covariance and correlation coefficient for the following pairs of random variables:
   (a) X and Y in Example 3, Section 5.1.
   (b) X and Y, X and Z, and Y and Z in Exercise 1 from Section 5.1.
   (c) X and Y, X and Z, and Y and Z in Exercise 2 from Section 5.1.
   (d) X_1 and X_2 in Exercise 8 from Section 5.1.
6. If $U = aX + b$ and $V = cY + d$, prove that $\rho(U, V) = \rho(X, Y)$.

5.4 The Sample Mean

When we discuss statistical inference in Chapters 9 and 10, one of the major problems that will confront us will be to say something meaningful about the mean of a population based on the mean of a small sample taken from the population. In this section, we will lay the foundation that will allow us not only to make an estimate concerning the mean of a population, but also to determine how reliable that estimate might be.

As we mentioned in Chapter 1, statistical reasoning proceeds from knowledge of the sample to knowledge of the population, whereas probabilistic reasoning proceeds in exactly the opposite direction. However, the statistical reasoning process must be preceded by a little probabilistic reasoning. In order to make a reasonable statement about the unknown population based on the information contained in just one (known) sample, we must have some idea about the kinds of samples that we might possibly obtain. To do this, in this preliminary stage, we assume that we have knowledge about the structure of the population, and deduce information about the structure of samples. This information will allow us later to qualify the accuracy of our estimate of the population mean based on the mean of our sample.

So let us suppose that we have a population whose makeup is specified by the probability function f of a random variable X:

a_1	a_2	a_3	$\cdots$	a_m
$f(a_1)$	$f(a_2)$	$f(a_3)$	$\cdots$	$f(a_m)$

That is to say, each member of the population has been assigned a numerical value a_i, which may represent his height to the nearest inch, or his salary to the nearest \$100, or his answer, in coded form, to a particular question, either YES (coded 0), UNDECIDED (coded 1) or NO (coded 2), to give just three examples. The probabilities $f(a_i)$ then indicate what percentages of the population fall into each of these numerical categories.

To make this all more concrete, suppose that we have an urn containing N balls that are numbered either $a_1, a_2, \ldots,$ or a_m. A ball is drawn at random from the urn, and X is defined to be the number on the ball. If among the N balls there are f_i balls numbered a_i, then the probability $f(a_i)$ is simply the relative frequency f_i/N, for $i = 1, 2, \ldots, m$. Using this terminology, we have

$$E(X) = \left(\frac{1}{N}\right) \sum_{i=1}^{m} a_i f_i$$

and

$$\text{Var}(X) = \left(\frac{1}{N}\right) \sum_{i=1}^{m} [a_i - E(X)]^2 \cdot f_i$$

$E(X)$ and $\text{Var}(X)$ are called the *population mean* and *population variance*, respectively.

Suppose that we choose n balls, with replacement, from the urn (that is, choose a sample of size n, with replacement, from our population), and define n random variables $X_1, X_2, \ldots, X_n$ as follows:

$X_k =$ the number on the kth ball drawn

for $k = 1, 2, \ldots, n$. Since we are sampling with replacement, these random variables are independent and have the same probability distribution as X.

DEFINITION

The random variable $\bar{X} = (X_1 + X_2 + \cdots + X_n)/n$ is called the *sample mean* of X for samples of size n.

Note that $\bar{X}$ is, indeed, a random variable. It is defined on the sample space of all possible samples of size n taken with replacement from the urn (population). The value that $\bar{X}$ assigns to such a sample is simply the average of the sample—the average of the numbers on the n balls drawn from the urn.

In statistics, we will have available this one value of $\bar{X}$, the average of the one sample that we have taken, and from it will try to deduce an estimate for $E(X)$, the population average. For now, we will think of $\bar{X}$ as a random variable, whose domain is all possible samples of size n and whose range is the set of all possible averages for these samples, and our interest will be to gain knowledge about the probability distribution of this random variable. Of course, this is a purely theoretical pursuit, because in practice it would be both physically and economically impossible to investigate each such sample. However, the information to be gained from our study will allow us to put the results of our one sample in the proper perspective.

Example 1 Suppose that all the citizens of a certain town have been asked whether or not they support a certain piece of legislation and that a YES answer is graded as $+1$, a NO answer is graded as -1, while an UNDE-CIDED answer is graded as 0. Suppose that it is known that the population is divided as follows:

a_i	-1	0	$+1$
$f(a_i)$	$\frac{3}{10}$	$\frac{2}{10}$	$\frac{5}{10}$

As a model for this situation, we can imagine an urn that contains ten balls, three numbered -1, two numbered 0, and five numbered $+1$. If X denotes the number on a ball drawn at random from the urn, then X has the probability distribution above. From the probability table of X, we calculate

$$E(X) = (-1 \cdot \tfrac{3}{10}) + (0 \cdot \tfrac{2}{10}) + (1 \cdot \tfrac{5}{10}) = \tfrac{2}{10}$$

and

$$E(X^2) = (1 \cdot \tfrac{3}{10}) + (0 \cdot \tfrac{2}{10}) + (1 \cdot \tfrac{5}{10}) = \tfrac{8}{10}$$

Therefore,

$$\text{Var}(X) = \tfrac{8}{10} - (\tfrac{2}{10})^2 = \tfrac{76}{100} = 0.76$$

Thus, the population mean is 0.2, and the population variance is 0.76. Suppose that we now draw a sample of size 2, with replacement, from the urn. In Table 5.3, we have listed all possible samples of size 2 (order considered), their probabilities of occurring, and their averages. Since the two draws from the urn are independent, the probability that any particular sample occurs is simply the product of the probabilities of the outcomes of the two draws. Note that the sum of the probabilities of all possible samples is 1, which can be checked by adding the values in the second column of the table. If $\bar{X}$ is the sample mean for samples of size 2, then the possible values that $\bar{X}$ might take can be found in the third column of the table. Using both the second and third columns of the table, we can construct the following probability distribution table for $\bar{X}$:

-1	$-\frac{1}{2}$	0	$\frac{1}{2}$	1
0.09	0.12	0.34	0.20	0.25

For example,

$$P(\bar{X} = \tfrac{1}{2}) = P[(0, 1), (1, 0)] = 0.10 + 0.10 = 0.20$$

and

$$P(\bar{X} = 0) = P[(0, 0), (-1, 1), (1, -1)]$$
$$= 0.04 + 0.15 + 0.15 = 0.34$$

From the probability distribution of $\bar{X}$, we can calculate

$$E(\bar{X}) = -1 \cdot (0.09) - \tfrac{1}{2} \cdot (0.12) + 0 \cdot (0.34) + \tfrac{1}{2} \cdot (0.20) + 1 \cdot (0.25)$$
$$= 0.20$$

and

$$E(\bar{X}^2) = 1 \cdot (0.09) + \tfrac{1}{4} \cdot (0.12) + 0 \cdot (0.34) + \tfrac{1}{4} \cdot (0.20) + 1 \cdot (0.25)$$
$$= 0.42$$

Therefore,

$$\text{Var}(\bar{X}) = E(\bar{X}^2) - E(\bar{X})^2 = 0.42 - (0.20)^2 = 0.38$$

Notice that, at least in this one case, $E(\bar{X}) = E(X)$ and $\text{Var}(\bar{X}) = \text{Var}(X)/2$. The following result indicates that this was not just a fluke.

Table 5.3
Probability and Averages of Samples of Size 2

Sample	Probability	Average
$(-1, -1)$	0.09	-1
$(-1, 0)$	0.06	$-\frac{1}{2}$
$(-1, 1)$	0.15	0
$(0, -1)$	0.06	$-\frac{1}{2}$
$(0, 0)$	0.04	0
$(0, 1)$	0.10	$\frac{1}{2}$
$(1, -1)$	0.15	0
$(1, 0)$	0.10	$\frac{1}{2}$
$(1, 1)$	0.25	1

Proposition 5.20

If $\bar{X}$ is the sample mean for samples of size n taken with replacement from a population specified by a random variable X, then

(a) $E(\bar{X}) = E(X)$
(b) $\text{Var}(\bar{X}) = \text{Var}(X)/n$
(c) $\text{Cov}(X_k - \bar{X}, \bar{X}) = 0 \qquad$ for any $k = 1, 2, \ldots, n$

Proof: We can write $\bar{X} = (X_1 + X_2 + \cdots + X_n)/n$, where the random variables X_i are independent and have the same probability distribution as X. Therefore,

(a) $E(\bar{X}) = E\left(\dfrac{X_1 + X_2 + \cdots + X_n}{n}\right)$

$$= \frac{1}{n} \cdot E(X_1 + X_2 + \cdots + X_n)$$

$$= \frac{1}{n} \cdot \sum_{i=1}^{n} E(X_i)$$

$$= \frac{1}{n} \cdot [nE(X)] = E(X)$$

since $E(X_i) = E(X)$ for all $i = 1, 2, \ldots, n$.

(b) $\text{Var}(\bar{X}) = \text{Var}\left(\dfrac{X_1 + X_2 + \cdots + X_n}{n}\right)$

$$= \frac{1}{n^2} \cdot \text{Var}(X_1 + X_2 + \cdots + X_n)$$

$$= \frac{1}{n^2} \cdot \sum_{i=1}^{n} \text{Var}(X_i)$$

since the $\{X_i\}$ are independent. This in turn

$$= \frac{1}{n^2} \cdot [n \, \text{Var}(X)] = \frac{\text{Var}(X)}{n}$$

since $\text{Var}(X_i) = \text{Var}(X)$ for all $i = 1, 2, \ldots, n$.

(c) We can write

$$\bar{X} = \sum_{i=1}^{n} \frac{1}{n} \cdot X_i$$

and

$$X_k - \bar{X} = \left(\frac{n-1}{n}\right) X_k - \frac{1}{n} \cdot \left(\sum_{i=1}^{n} X_i - X_k\right)$$

We then invoke Proposition 5.15(d) with $Y_1 = X_k - \bar{X}$, $Y_2 = \bar{X}$, $a_i = -1/n$ except for $a_k = (n-1)/n$, and $b_i = 1/n$ to find that

$$\text{Cov}(X_k - \bar{X}, \bar{X}) = \sum_{i=1}^{n} a_i b_i \cdot \text{Var}(X_i)$$

$$= \left[\frac{n-1}{n} + \underbrace{(-1/n) + (-1/n) + \cdots + (-1/n)}_{n-1 \text{ times}}\right] \frac{1}{n} \text{Var}(X)$$

$$= \left[\frac{n-1}{n} + (n-1)\left(\frac{-1}{n}\right)\right] \cdot \frac{1}{n} \cdot \text{Var}(X)$$

$$= \left(\frac{n-1}{n} - \frac{n-1}{n}\right) \cdot \frac{1}{n} \cdot \text{Var}(X) = 0$$

As a result of Proposition 5.20, we find that the average of the averages $[E(\bar{X})]$ of *all* samples of size n is exactly equal to the average of the population $[E(X)]$ from which the samples were taken. The variance of these sample averages is directly proportional to the population variance, and in fact decreases as the sample size gets larger. Thus, as the sample size gets larger and larger, the values of the sample mean $\bar{X}$, that is, the values of the sample averages, tend to become more and more concentrated about the population mean (their mean), even more so than the population itself. This is the justification for using the average of a large sample as a substitute for the average of the population. Although the sample average may not exactly equal the population average, it will not miss by much. It is for this reason that the term $\sigma_{\bar{X}}$ is referred to as the *standard error of the mean*.

Example 1 (continued) Notice that 80 percent of the population is located at either $+1$ or -1, at least 0.80 units away from the population mean. On the other hand, only 34 percent of the averages of samples of size 2 are located this far from the mean.

Along these same lines, we have the following important result:

Theorem 5.21 The Weak Law of Large Numbers

If $\bar{X}$ is the sample mean for samples of size n and $\varepsilon > 0$, then

$$\lim_{n \to \infty} P(|\bar{X} - \mu_X| > \varepsilon) = 0$$

or equivalently,

$$\lim_{n \to \infty} P(|\bar{X} - \mu_X| \le \varepsilon) = 1$$

Proof: The proof is a simple application of Chebyshev's theorem, with an assist from Proposition 5.20. Applying Chebyshev's inequality to $\bar{X}$, we find

$$P[|\bar{X} - E(\bar{X})| > \varepsilon] < \frac{\text{Var}(\bar{X})}{\varepsilon^2}$$

By Proposition 5.20, $E(\bar{X}) = E(X)$ and $\text{Var}(\bar{X}) = \text{Var}(X)/n$, and so we have

$$P(|\bar{X} - \mu_X| > \varepsilon) < \frac{\text{Var}(X)}{n\varepsilon^2} = \left(\frac{1}{n}\right)\left[\frac{\text{Var}(X)}{\varepsilon^2}\right]$$

Since $\text{Var}(X)$ and ε are constants, it follows that

$$\lim_{n \to \infty} \left(\frac{1}{n}\right)\left(\frac{\text{Var}(X)}{\varepsilon^2}\right) = \frac{\text{Var}(X)}{\varepsilon^2} \lim_{n \to \infty} \frac{1}{n} = 0$$

Since probabilities are nonnegative, the result follows.

Example 2 The weak law of large numbers can be used to justify the frequency theory of probability. Suppose that p is the true probability that an event E occurs. Let us conduct a compound experiment consisting of n independent trials of the sample space S, and define indicator random variables $X_1, X_2, \ldots, X_n$ as follows:

$$X_k = \begin{cases} 1 & \text{if event } E \text{ occurs on the } k\text{th trial} \\ 0 & \text{otherwise} \end{cases}$$

This is equivalent to drawing n balls from an urn containing N balls that are numbered either 0 or 1, with pN of the balls being numbered 1. The random variable X_k then gives the number on the kth ball drawn. Now $E(X_k) = P(E) = p$ and, since $X_k^2 = X_k$, $E(X_k^2) = p$, and therefore,

$$\text{Var}(X_k) = E(X_k^2) - E(X_k)^2 = p - p^2 = p(1 - p) = pq$$

(where we define $q = 1 - p$) for $k = 1, 2, \ldots, n$. If k_n denotes the number of occurrences of E over the n trials, then

$$k_n = X_1 + X_2 + \cdots + X_n$$

and if $\bar{X}$ denotes the sample mean for samples of size n, then $\bar{X} = k_n/n$. Since $E(\bar{X}) = E(X) = p$, the weak law of large numbers tells us that

$$\lim_{n \to \infty} P\left(\left|\frac{k_n}{n} - p\right| > \varepsilon\right) = 0$$

Thus, the proportion k_n/n of occurrences in trials is very nearly equal to the probability $P(E)$ if n is large enough.

Example 3 Suppose that we wished to estimate the true mileage produced by a certain brand of gasoline, so we filled the gas tanks of 50 cars and drove them until their tanks were empty. Suppose that the average mileage for these 50 cars turned out to be 15 miles per gallon. How likely is it that we have come within one mile per gallon of the true figure for this brand of gasoline? According to Chebyshev's theorem and Proposition 5.20,

$$P(|\bar{X} - \mu_X| \leq 1) \geq 1 - \frac{\text{Var}(X)}{n \cdot 1^2} = 1 - \frac{\text{Var}(X)}{n}$$

In order to proceed further, we need some figure for the variance in miles per gallon for this brand of gasoline. Assume that this variance is 3 mile^2 per gallon. Therefore,

$$P(|15 - \mu_X| \leq 1) \geq 1 - \tfrac{3}{50} = \tfrac{47}{50} = 0.85$$

so there is at least an 85 percent chance that our sample average of 15 miles per gallon is within 1 mile per gallon of the true mileage for this brand of gasoline.

Example 4 Suppose that we wish to estimate the average income of the people living in a certain city to within \$500 of the actual average. Our problem is to determine how many people we must include in our sample. As we have seen throughout this section, this problem must be stated and solved in terms of probabilities, and not certainties. We can never be certain that an arbitrarily chosen sample will produce an average within prescribed bounds of the true population average. Let us agree that we wish to be 90 percent certain that a sample of the size that we will determine will produce an average within \$500 of the true average for the city. If we let X denote the salary of a city resident chosen at random, and $\bar{X}$ the sample mean for samples of size n, with n still to be determined, then we wish to have

$$P(|\bar{X} - \mu_X| \leq 500) \geq 0.90$$

or equivalently,

$$P(|\bar{X} - \mu_X| > 500) < 0.10$$

We have also seen throughout this section that the solution of such problems requires a population variance, so we will assume that it is reasonable to believe that $\sigma_X = \$2100$. By Chebyshev's inequality, we find that

$$P(|\bar{X} - \mu_X| > 500) < \frac{\mathrm{Var}(X)}{n(500)^2} = \frac{2100^2}{n(500)^2}$$

$$= \frac{17.64}{n}$$

It is therefore sufficient to make $17.64/n < 0.10$, which can be done by choosing $n > 176.4$. Thus, a sample of size 177 (or higher) would suffice. We can be sure that the average of a sample of size 177 (or higher) will be within \$500 of the true population average 90 percent of the time.

Returning to the concept of sample mean, suppose now that the sample had been drawn *without* replacement. The random variables $X_1, X_2, \ldots, X_n$ would no longer be independent. However, although one would no longer expect them to be identically distributed either, it can be shown that they still, in fact, all have distributions identical to that of X. As a result, the proof that $E(\bar{X}) = E(X)$ follows as it did before. However, the proof that $\mathrm{Var}(\bar{X}) = \mathrm{Var}(X)/n$ is complicated by the fact that the $\{X_i\}$ are not independent. We do get the following result, however. The details of the proof are similar to the calculations involved in determining the variance of the hypergeometric distribution in Section 6.2 and can be found in another text.*

Proposition 5.22

If $\bar{X}$ is the sample mean for samples of size n taken without replacement from a population specified by a random variable X, then

(a) $E(\bar{X}) = E(X)$

(b) $\text{Var}(\bar{X}) = \dfrac{\text{Var}(X)}{n} \cdot \dfrac{N-n}{N-1}$

where N is the size of the population.

Notice that

$$\frac{N-n}{N-1} = \frac{1-n/N}{1-1/N}$$

so as $N \to \infty$,

$$\frac{N-n}{N-1} \to 1$$

Thus, for large populations, $E(\bar{X})$ and $\text{Var}(\bar{X})$ are almost identical, whether the sampling is done with or without replacement.

EXERCISES

1. If $\bar{X}$ is the sample mean for samples of size n taken with replacement from a population specified by a random variable X, prove that $M_{\bar{X}}(t) = [M_X(t/n)]^n$
2. Determine the distribution of the sample mean for samples of size 3 for the population defined in Example 1.
3. Suppose that a population is specified by the following probability distribution

1	2	3	4
$\frac{1}{5}$	$\frac{2}{5}$	$\frac{1}{5}$	$\frac{1}{5}$

Determine the distribution of the sample mean for samples of size 2. Find $E(\bar{X})$ and $\text{Var}(\bar{X})$.
4. If the average height of 100 men is 70 in., what is the probability that the average height for the population from which this sample was taken is between 69 in. and 71 in.? Assume a population standard deviation of 3 in.
5. What is the probability that the average of a sample of n items will be within one standard deviation of the true population mean?
6. How large a sample must be used so that the sample mean will be within one-half of a standard deviation of the true population mean 95 percent of the time?

* J. E. Freund, *Mathematical Statistics* (Englewood Cliffs, New Jersey: Prentice-Hall, 1971), sect. 6.2.3.

ADDITIONAL PROBLEMS

1. What is the expected number of hearts in a bridge hand?
2. If three persons are seated at random at a lunch counter with six seats, how many persons can expect to have no one sitting next to them? (*Hint*: Let I_k indicate whether seat k is occupied by an isolated person.)
3. In a game of chance, a player can either win \$5 or lose \$1, \$2, or \$3, each with probability $\frac{1}{4}$. After 500 plays, estimate the probability that the player's average gain is positive.
4. If 50 of the 120 members of a political club are Republicans, calculate the expected number of Republicans on a committee of 10 of the members.
5. A card is chosen from a standard deck, a fair coin is tossed, and a fair die is rolled. A success on the first part of the experiment is a heart, on the second part, a head, and on the final part, a six. What is the expected number of successes?
6. A store sells, on the average, \$2000 worth of merchandise each day, with a standard deviation of \$500. Find the probability that their total sales for a 320 day year are somewhere between \$600,000 and \$680,000.
7. How many light bulbs must be included in a sample if we wish to estimate the light bulbs' average life span to within 20 hours with 90 percent accuracy. Assume a standard deviation of 25 hours.

Special Discrete
Distributions

6.1 The Binomial Distribution

The simplest experiment is one which results in only two outcomes. Actually, any experiment can be looked at in this light if we use as a sample space the set $\{E, C(E)\}$, where E is some event of special interest. Looking at the experiment in this way, we can refer to the occurrence or nonoccurrence of E as a success or a failure. For this reason, we shall denote our sample space of two outcomes by $\{S, F\}$, where S stands for success and F for failure.

DEFINITION

An experiment consisting of n independent trials of the sample space $\{S, F\}$, where $P(S) = p$ and $P(F) = 1 - p = q$ for each trial, is called a *Bernoulli experiment*, and the trials are called *Bernoulli trials*.

Example 1 Suppose that we toss a fair coin and consider getting a head to be a success. If we were to toss the coin 100,000 times, we would have 100,000 independent trials of the sample space $\{S, F\} = \{H, T\}$. On each trial, we would have $P(S) = P(H) = \frac{1}{2}$ and $P(F) = P(T) = \frac{1}{2}$. This would seem to be in direct conflict with the "law of averages," which would say, for example, that if a coin showed heads on five consecutive tosses, it would be more likely to show tails on the next toss. This would mean that the probability $p = P(S)$ would vary from toss to toss, depending on the trend of the outcomes of previous tosses. To determine which theory, the Bernoulli trials or the law of averages, best described such an experiment, a computer simulation of 100,000 tosses of a fair coin was performed, with the results shown in Table 6.1. The results of this computer simulation, which made use of a uniform random number generator (we will have more to say about these in Section 8.3), seem to discredit the law of averages. Each value in the second column of the table is approximately half the preceding value. For example, five consecutive heads occurred 3070 times, and in 1543 of these cases (approximately half), a head occurred on the next toss as well. This would suggest that the probability of a head on any toss is $\frac{1}{2}$, regardless of what went before, be that a string of consecutive heads or tails, or whatever. This, of course, implies that the trials are Bernoulli.

Example 2 A fair die is rolled 10,000 times and getting a 6 is regarded as a success. As in the preceeding example, the 10,000 trials may be regarded as Bernoulli, with $P(S) = \frac{1}{6}$ for each trial.

Example 3 A keypunch operator punches thousands of cards each day. Each card may be regarded as a trial, with success being a correctly punched card. There is some doubt that the probability of success remains constant throughout the day. This probability is probably lower early in the day, when the operator first begins punching for the day, and later in the day, when the operator might be slightly fatigued. So perhaps these trials should not be considered Bernoulli.

Example 4   Player A and player B play a sequence of ten games of tennis, with player A's winning a game being regarded as a success. The ten trials (games), however, are not Bernoulli trials because $P(S)$ is dependent on whether or not player A is serving any particular game. $P(S)$ would most likely be considerably higher for those games which player A served.

Example 5   The weather of 365 days, with rain being a failure, cannot be considered a Bernoulli experiment because the probability of rain varies each day, depending on the movement of different weather systems. Although, on the average, we might expect, say, 102 days of rain per year, this does not mean that the probability of rain on any given day is $102/365 = 0.279$. On some days, the probability of rain is much higher than it is on other days.

When considering an experiment of n Bernoulli trials, with constant probability of success p on each trial, it is only natural to inquire about the number of successes that occur over the n trials. If we define a random variable X

Table 6.1
100,000 Tosses of a Fair Coin

Number of consecutive heads	Frequency of occurrence
1	50,030
2	24,920
3	12,405
4	6,203
5	3,070
6	1,543
7	771
8	394
9	197
10	104
11	59
12	36
13	21
14	15
15	10
16	6
17	4
18	3
19	2
20	1

to be the number of successes in n Bernoulli trials and denote by $b(k; n, p)$ the probability of k successes in n trials [that is, $P(X = k)$], then we have

Proposition 6.1

$b(k; n, p) = C_k^n \cdot p^k \cdot q^{n-k}$ for $k = 0, 1, 2, \ldots, n$

Proof: As a sample space, we choose the set of 2^n n-tuples $(a_1, a_2, \ldots, a_n)$ where, for each i, $a_i = S$ or $a_i = F$. Any n-tuple containing k successes must also contain $n - k$ failures and, by the independence of the trials, have probability $p^k \cdot q^{n-k}$. There are as many n-tuples containing k successes as there are ways to choose k of the n coordinates in which to place the k S's. Since there are C_k^n ways to do this, the event "k successes in n trials" corresponds to C_k^n n-tuples, each having probability $p^k \cdot q^{n-k}$, proving the stated formula.

Corollary 6.2

(a) The probability of no successes in n Bernoulli trials, $b(0; n, p)$, equals q^n.
(b) The probability of at least one success in n Bernoulli trials equals $1 - q^n$.

The random variable X defined above, the number of successes in n Bernoulli trials, therefore has probability distribution

0	1	2	$\cdots$	k	$\cdots$	n
q^n	npq^{n-1}	$C_2^n p^2 q^{n-2}$	$\cdots$	$C_k^n p^k q^{n-k}$	$\cdots$	p^n

The terms $b(k; n, p) = C_k^n \cdot p^k \cdot q^{n-k}$ are actually the terms in the binomial expansion of $(q + p)^n$, and hence, the probability distribution above is commonly referred to as the *binomial distribution* with parameters n and p. Since $q = 1 - p$, $q + p = 1$, from which it follows that the binomial distribution is indeed a probability distribution:

$$\sum_{k=0}^{n} P(X = k) = \sum_{k=0}^{n} b(k; n, p) = (q + p)^n = 1^n = 1$$

A table of binomial probabilities, for values of k as large as 25, can be found in Appendix A. Probabilities are given for values of p varying from 0.05 to 0.50 by increments of 0.05. Should the reader require a probability $b(k; n, p)$ where $p > 0.50$, he or she should reverse the roles of success and failure and calculate instead the probability $b(n - k; n, 1 - p)$. Since, in this case, $1 - p \leq 0.50$, this probability can be found in the table. The binomial distributions with parameters $n = 8$ and $p = 0.50$ and $n = 10$ and $p = 0.25$ are pictured in Figure 6.1.

Example 6   A fair die is rolled seven times, and we wish to calculate the probability that either a 5 or a 6 occurs exactly three times. We let S denote getting either a 5 or a 6, and therefore have $P(S) = \frac{2}{6}$ and $P(F) = \frac{4}{6}$ for each trial (roll). Thus, we have parameters $n = 7$, $p = \frac{1}{3}$, $q = \frac{2}{3}$, and the probability we seek is

$$b(3; 7, \tfrac{1}{3}) = C_3{}^7 \cdot (\tfrac{1}{3})^3 \cdot (\tfrac{2}{3})^4 = \frac{560}{2187}$$

By Corollary 6.2(a), the probability that neither a 5 nor a 6 occurs in the seven trials is

$$b(0; 7, \tfrac{1}{3}) = (\tfrac{2}{3})^7 = \frac{128}{2187}$$

and hence, the probability that at least one 5 or 6 occurs in the seven trials is $1 - 128/2187 = 2059/2187$.

Example 7 A family has six children, and we wish to calculate the probability that three are boys and three are girls. Each child is considered a trial, and if we let "girl" be a success, then we may take $p = \tfrac{1}{2}$. Assuming that the six trials are independent (this may not be exactly true, a woman who has already had three boys has a higher than usual probability that her fourth child will also be a boy), we have a binomial distribution with parameters $n = 6$ and $p = q = \tfrac{1}{2}$. The probability of three boys and three girls is then

$$b(3; 6, \tfrac{1}{2}) = C_3{}^6 \cdot (\tfrac{1}{2})^3 \cdot (\tfrac{1}{2})^3 = \tfrac{20}{64}$$

The probability that there are fewer girls than boys equals

$$b(0; 6, \tfrac{1}{2}) + b(1; 6, \tfrac{1}{2}) + b(2; 6, \tfrac{1}{2}) = (\tfrac{1}{2})^6 + C_1{}^6 \cdot (\tfrac{1}{2})^1 \cdot (\tfrac{1}{2})^5 + C_2{}^6 \cdot (\tfrac{1}{2})^2 \cdot (\tfrac{1}{2})^4$$
$$= \tfrac{22}{64}$$

Figure 6.1

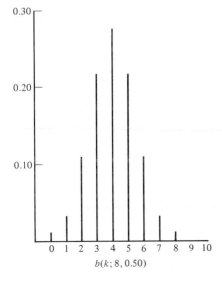

$b(k; 8, 0.50)$

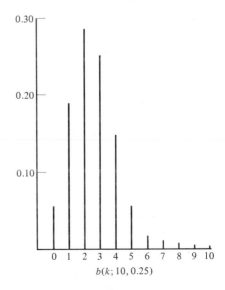

$b(k; 10, 0.25)$

Example 8 How many times must a fair die be rolled so that there is a better than even chance of obtaining at least one 6? Clearly, we have a Bernoulli experiment here, with success being getting a six, probability of success $\frac{1}{6}$, and probability of failure $\frac{5}{6}$. However, the number of trials is not specified, and in fact must be determined. By Corollary 6.2(b), the probability of at least one 6 in n trials is $1 - q^n = 1 - (\frac{5}{6})^n$. Therefore, we must choose n so that $1 - (\frac{5}{6})^n > \frac{1}{2}$, or equivalently, so that $(\frac{5}{6})^n < \frac{1}{2}$ for the first time. Since $(\frac{5}{6})^3 = 125/216 > \frac{1}{2}$ and $(\frac{5}{6})^4 = 625/1296 < \frac{1}{2}$, a total of $n = 4$ trials are necessary.

Example 9 A factory owner knows that on the average 5 percent of the items produced in his factory are defective. He ships his product in groups of 100, along with the guarantee that he will pay all expenses involved in replacing an entire shipment should there be more than m defectives in a shipment. He would like to know how large to make m so that he will have to replace no more than 5 percent of all shipments. The question that must be answered then is how likely is it that there are 0, 1, 2, ... defectives in a shipment of 100 of these items. The 100 items can be considered 100 Bernoulli trials with common probability of success—being defective—$p = 0.05$. We then calculate

$$b(0; 100, 0.05) = 0.0059$$
$$b(1; 100, 0.05) = 0.0312$$
$$b(2; 100, 0.05) = 0.0812$$
$$b(3; 100, 0.05) = 0.1396$$
$$b(4; 100, 0.05) = 0.1781$$
$$b(5; 100, 0.05) = 0.1800$$
$$b(6; 100, 0.05) = 0.1500$$
$$b(7; 100, 0.05) = 0.1060$$
$$b(8; 100, 0.05) = 0.0649$$
$$b(9; 100, 0.05) = 0.0349$$
$$\sum_{k=0}^{9} b(k; 100, 0.05) = 0.9718$$

Thus, the probability that there are anywhere between 0 and 9 defectives in a sample of size 100 equals 0.9718, or equivalently, the probability that there are more than 9 defectives in a sample of size 100 equals 0.0282. Therefore, the factory owner should guarantee that he will, at his own expense, replace any sample of 100 that contains 10 or more defectives.

Proposition 6.3

Suppose that the random variable X has binomial probability distribution with parameters n and p. Then

(a) $E(X) = np$
(b) $\text{Var}(X) = npq$
(c) $M_X(t) = [1 + p(e^t - 1)]^n$

Proof:

(a) By definition,

$$E(X) = \sum_{k=0}^{n} k \cdot b(k; n, p)$$

$$= \sum_{k=1}^{n} k \cdot \frac{n!}{k!(n-k)!} \cdot p^k \cdot q^{n-k} \qquad \text{(the } k = 0 \text{ term is 0)}$$

$$= np \sum_{k=1}^{n} \frac{(n-1)!}{(k-1)!(n-k)!} \cdot p^{k-1} \cdot q^{n-k}$$

Making the substitution $s = k - 1$ in the above expression, we find

$$E(X) = np \cdot \sum_{s=0}^{n-1} \frac{(n-1)!}{s!(n-1-s)!} \cdot p^s \cdot q^{(n-1)-s}$$

$$= np \cdot \sum_{s=0}^{n-1} b(s; n-1, p)$$

The terms $b(s; n-1, p)$ refer to a Bernoulli experiment of $n-1$ trials with constant probability of success p on each trial. Therefore,

$$\sum_{s=0}^{n-1} b(s; n-1, p) = 1$$

since the binomial distribution is a probability distribution. Hence, $E(X) = np \cdot 1 = np$.

(b) To compute $\text{Var}(X)$, we proceed in the same manner to evaluate

$$E(X^2) = \sum_{k=0}^{n} k^2 \cdot b(k; n, p)$$

$$= \sum_{k=1}^{n} k^2 \cdot \frac{n!}{k!(n-k)!} \cdot p^k \cdot q^{n-k}$$

$$= np \sum_{k=1}^{n} k \cdot \frac{(n-1)!}{(k-1)!(n-k)!} \cdot p^{k-1} \cdot q^{n-k}$$

We again let $s = k - 1$, and obtain

$$E(X^2) = np \sum_{s=0}^{n-1} (s+1) \cdot \frac{(n-1)!}{s!(n-1-s)!} \cdot p^s \cdot q^{(n-1)-s}$$

$$= np \sum_{s=0}^{n-1} (s+1) \cdot b(s; n-1, p)$$

$$= np \sum_{s=0}^{n-1} s \cdot b(s; n-1, p) + np \sum_{s=0}^{n-1} b(s; n-1, p)$$

As above,

$$\sum_{s=0}^{n-1} b(s; n-1, p) = 1$$

and by (a),

$$\sum_{s=0}^{n-1} s \cdot b(s; n-1, p) = (n-1) \cdot p$$

this summation being the mean of a binomial random variable with parameters $n - 1$ and p. Consequently,

$$
\begin{aligned}
E(X^2) &= np \cdot (n-1) \cdot p + np \cdot 1 \\
&= np \cdot (np - p + 1) \\
&= np \cdot (np + q) \\
&= (np)^2 + npq
\end{aligned}
$$

Therefore,

$$
\begin{aligned}
\mathrm{Var}(X) = E(X^2) - E(X)^2 &= \left[(np)^2 + npq\right] - (np)^2 \\
&= npq
\end{aligned}
$$

(c) If X is a Bernoulli random variable with parameters n and p, then we can write

$$X = X_1 + X_2 + \cdots + X_n$$

where the (independent) random variables X_i indicate whether or not there was a success on the ith trial. Therefore, each X_i has probability distribution

0	1
q	p

and consequently has moment generating function

$$
\begin{aligned}
M_{X_i}(t) &= e^{0t} \cdot q + e^{1t} \cdot p \\
&= q + pe^t \\
&= (1 - p) + pe^t \\
&= 1 + p(e^t - 1)
\end{aligned}
$$

Since the X_i are independent (the trials are independent) we can invoke Corollary 5.6 to prove that

$$M_X(t) = \prod_{i=1}^{n} M_{X_i}(t) = \left[1 + p(e^t - 1)\right]^n$$

Example 10 The expected number of heads in 100,000 tosses of a fair coin is $np = 100{,}000 \cdot \frac{1}{2} = 50{,}000$. The expected number of 6's in 10,000 rolls of a fair die is $np = 10{,}000 \cdot \frac{1}{6} = 1667$.

Example 11 The expected number of girls in a family of 6 is $np = 6 \cdot \frac{1}{2} = 3$.

Example 12 A toothpaste manufacturer claims that people using his brand of toothpaste will have fewer cavities than those not using his brand. It is a known fact that 50 percent of the population have fewer than two cavities per year, while a survey of 100 people using the manufacturer's toothpaste reveals that 61 of them had fewer than two cavities during the past year. Does the result of this survey lend credence to the manufacturer's claim? How likely is it that 61 (or more) people in a sample of 100 will have fewer than two cavities per year? If we consider each of these 100 people as a trial, with success being "having fewer than two cavities," then we have a binomial distribution with parameters $n = 100$ and $p = \frac{1}{2}$, since 50 percent of the population has fewer than two cavities per year. Now the probability of 61 successes in 100 trials, $b(61; 100, \frac{1}{2})$, equals 0.0071, which is quite small. For that matter, any of the probabilities $b(k; 100, \frac{1}{2})$ are small, for an n as large as 100. The important point is that the summation

$$\sum_{k=61}^{100} b(k; 100, \tfrac{1}{2}) = 0.0179$$

is also very small. Thus, it is not very likely that 61 *or more* people in a sample of 100 would have fewer than two cavities per year, *under the assumption* that 50 percent of the population from which they were chosen has fewer than two cavities per year. But we have not yet taken into consideration that the 100 people were chosen from a special subset of the total population, those people who were using this particular brand of toothpaste. At this stage, we can reach one of two conclusions. First, that the toothpaste actually has no beneficial effect, and that 50 percent of the people using this toothpaste will have fewer than two cavities per year. To reach this conclusion, we must disregard the results of the sample, which, as we have pointed out, is a rather unlikely sample anyhow. Second, we may reach the opposite conclusion, that the rarity of such a sample, together with the fact that it *did* happen, must be explained by the fact that more than 50 percent of the people using this toothpaste do have fewer than two cavities per year. That is, it is the assumption that $p = \frac{1}{2}$ that led to the low value of the probability $\sum_{k=61}^{100} b(k; 100, p)$. It is usual statistical procedure never to accept the fact that the sample we have taken could possibly be a rare, misleading one, but rather that there must be some logical reason that explains the occurrence of such a sample. Hence, we reject the assumption that $p = \frac{1}{2}$ for those using this brand of toothpaste, and accept the manufacturer's claim.

Example 13 Suppose that you toss a coin 400 times and record 148 heads and 252 tails. Should you suspect that the coin is biased so as to favor tails? Using the binomial distribution with parameters $n = 400$ and $p = \frac{1}{2}$ (assuming, initially, that the coin is not biased), we find that the expected number of heads in 400 tosses is $np = 400 \cdot \frac{1}{2} = 200$ with standard deviation $\sqrt{npq} = \sqrt{400 \cdot \frac{1}{2} \cdot \frac{1}{2}} = \sqrt{100} = 10$. Now 148 heads is more than five deviations from the mean of 200 heads, so by Chebyshev's inequality, we find that, letting

X denote the number of heads in 400 tosses,

$$P(150 \leq X \leq 250) = P(|X - 200| \leq 50)$$
$$= P(|X - 200| \leq 5\sigma_X)$$

$$\geq 1 - \frac{1}{5^2} = 0.96$$

Therefore, the probability of getting fewer than 150 heads in 400 tosses of a fair coin is less than 0.02:

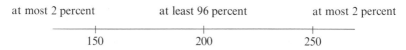

However, the coin we tossed *did* show only 148 heads in 400 tosses, so, although it is possible that an unbiased coin will show such a low number of heads (less than one chance in 50), the probability is so small that we would tend to suspect that the coin is biased rather than believe that such an unlikely sample would occur.

Example 14 A political candidate's statistician has taken a random sample of 1000 voters and found only 420 of them in favor of the candidate. On the basis of this sample, does it appear that the candidate has no reasonable hope for winning the election, or is there still some hope? More specifically, we answer the following question: If the candidate really does have the support of 50 percent of the voters, how likely is it that in a sample of 1000, only 420 would voice their support? If we assume that the opinions of the 1000 people sampled are independent of one another (which is not always a valid assumption in such a situation), and if we let "favor this candidate" be a success and X the number of those sampled who favor the candidate, then X has binomial distribution with parameters $n = 1000$ and $p = \frac{1}{2}$. Therefore, the expected number of successes is $np = 1000 \cdot \frac{1}{2} = 500$, with a standard deviation of $\sqrt{npq} = \sqrt{1000 \cdot \frac{1}{2} \cdot \frac{1}{2}} = \sqrt{250} = 15.81$. Now 420 successes is slightly more than five standard deviations below the mean. As we saw in the preceeding example, such a sample has a less than 2 percent chance of occurring. Nevertheless, this event of low probability did occur, and so we reason that it really doesn't have that low a probability. Rather, we reason, our calculation of the probability was incorrect, and the reason it was incorrect is that we used the wrong value for p. Our estimate of p was too high, and had we used a lower value, say $p = 0.40$ or $p = 0.45$, we would have found the probability of 420 successes to be more reasonable. So our conclusion is that p is somewhat less than 0.50, that it is not reasonable for the candidate to believe that he has the support of a majority of the voters.

Example 15 A machine is known to produce five percent defectives, a rate which is acceptable to its owner. However, at frequent intervals, the machine must be checked to determine if it is still producing at this acceptable rate. The owner must have some simple way to determine if the machine is still producing at this rate. He reasons as follows: If the machine is producing only five percent defectives, then in a sample of ten items, the probabilities of there being zero, one, or two defectives are

$b(0; 10, 0.05) = 0.5897$

$b(1; 10, 0.05) = 0.3151$

$b(2; 10, 0.05) = 0.0746$

and therefore,

$$\sum_{k=0}^{2} b(k; 10, 0.05) = 0.9794$$

Thus there is an approximately two percent chance that a machine producing five percent defectives will produce three or more defective items in a sample of ten. Hence, at regular intervals, the owner checks a sample of ten items produced by the machine, and if three or more are defective, he calls for a repairman. Otherwise, he has good reason to believe that the machine is still producing defectives at the five percent rate. We should point out, however, that this procedure is not foolproof. There is a two percent chance that a machine producing five percent defectives will produce three or more defectives in a sample of ten, and there is a small chance that the machine might be breaking down and still produce only zero, one, or two defectives in a given sample of ten.

Example 16 Team A and team N play a series of games, with team A having probability $p > \frac{1}{2}$ of winning any given game. We would like to know how many games the teams must play so there is at least a 90 percent probability that the better team (which is team A) is ahead in the number of games won. We have a series of n Bernoulli trials (games), success on any trial means that team A won that game, with the probability of success on any trial being $p > \frac{1}{2}$. We must determine the number of trials so that the probability that team A has won more games than it has lost exceeds 0.90 for the first time. This probability equals

$$\sum_{k=[n/2]+1}^{n} b(k; n, p)$$

where the symbol $[x]$ represents the largest integer value not exceeding x. For example, if $n = 7$, then team A has won more games than it has lost if $k = 4, 5, 6$, or 7. In this case, $[n/2] = [\frac{7}{2}] = 3$, and so the summation runs from $3 + 1 = 4$ to $n = 7$. On the other hand, if $n = 16$, then k must equal 9, 10, 11, 12, 13, 14, 15, or 16. Here, $[n/2] = [\frac{16}{2}] = 8$, and so the summation runs from $8 + 1 = 9$ to $n = 16$. These probabilities, for different values of n, are included in Table 6.2. To calculate these probabilities, we assumed that

$p = 0.60$, that is, that team A has a 60 percent chance of winning any game. As the table indicates, a series of 41 games is necessary so that the probability that team A is ahead in games won at the end of the series exceeds 90 percent.

Example 17   We refer back now to the robbery case mentioned in Section 3.4. Recall that the couple accused of the crime had been convicted because they fit the description of the robbers and because the likelihood of a couple having all the known characteristics of the robbers had been established at 1 in 12 million. The couple appealed their conviction, and used a mathematical argument of their own, which reasoned as follows. Granted that the probability p that a couple has all the characteristics of the robbers is $1/12,000,000$ (actually, they didn't grant this so readily), and granted that there is one such couple in a population of N couples, what is the probability that there is *another* such couple. We consider each of the N couples in the population as a Bernoulli trial, with success meaning that the couple has all the characteristics of the robbers. The constant probability of success on each trial is therefore $p = 1/12,000,000$. Using this binomial distribution and some

---

Table 6.2
Probabilities That Team A
Will Win an *n* Game Series,
Assuming That $p = 0.60$

n	Probability	n	Probability
3	0.648	23	0.836
4	0.475	24	0.787
5	0.683	25˙	0.846
6	0.544	26	0.801
7	0.710	27	0.855
8	0.594	28	0.813
9	0.733	29	0.864
10	0.633	30	0.825
11	0.753	31	0.872
12	0.665	32	0.835
13	0.771	33	0.879
14	0.692	34	0.845
15	0.787	35	0.886
16	0.716	36	0.854
17	0.801	37	0.892
18	0.737	38	0.862
19	0.814	39	0.898
20	0.755	40	0.870
21	0.826	41	0.903
22	0.772		

conditional probabilities, we find

P(two or more such couples | at least one such couple)

$$= \frac{P(\text{two or more such couples})}{P(\text{at least one such couple})}$$

$$= \frac{1 - b(0; N, p) - b(1; N, p)}{1 - b(0; N, p)}$$

$$= \frac{1 - (1 - p)^N - C_1^N \cdot p(1 - p)^{N-1}}{1 - (1 - p)^N}$$

Therefore, the probability of there being another couple having all the observed characteristics of the robbers is

$$\frac{1 - (1 - p)^N - Np(1 - p)^{N-1}}{1 - (1 - p)^N}$$

Table 6.3 contains these probabilities for various values of N. There were several million couples living in the Los Angeles area at the time of the crime.

**Table 6.3
Probabilities of
Another Such Couple**

Number of couples N (in millions)	Probability
1	0.0402
2	0.0786
3	0.1160
4	0.1522
5	0.1875
6	0.2216
7	0.2547
8	0.2868
9	0.3179
10	0.3479
15	0.4835
20	0.5959
25	0.6875
30	0.7610
40	0.8644
50	0.9256
75	0.9852
100	0.9973

The defense actually used a figure of N being approximately 12 million couples, and therefore came up with a probability of approximately 0.40 that there was another couple fitting the description of the robbers. Since the defendants were not clearly identified as the perpetrators of the crime, their appeal, based on the argument above, was upheld. For more details on this case, see the article already referred to.*

EXERCISES

1. Write a computer program that will calculate any (reasonable) binomial probability $b(k; n, p)$.
2. Write a computer program that will generate the table of binomial probabilities found in Appendix A.
3. Without using the table, calculate the probabilities $b(3; 7, 0.4)$, $b(2; 9, 0.2)$, $b(7; 10, 0.25)$, and $b(3; 15, 0.1)$.
4. In Example 14, how likely is the sample that occurred if the candidate really has 55 percent of the voters? If he has 60 percent?
5. What is the probability that a baseball player with a batting average of 0.300 gets two hits in five at-bats?
6. Five voters are chosen at random and asked their opinion about a certain issue. If only 40 percent of the public favors this tissue, what is the probability that a majority of the five sampled favor the issue?
7. The probability that a student will solve any given problem on an exam is 0.75. If the student must solve seven out of ten problems in order to pass the exam, what is a student's probability of passing?
8. If ten percent of a population is estimated to be honest, how many people must be interviewed so that we can be 80 percent certain that we talk to at least one honest person?
9. In Example 15, suppose that something has gone wrong with the machine, and the probability of its producing a defective has risen to 0.30. How likely is the owner's procedure to detect this?
10. The hypothesized sex ratio of the offspring of a certain species is 1:1. A sample of 25 such offspring contained 15 males. Is this sufficient reason to question the hypothesis?
11. If 60 percent of the television sets in a certain town were tuned in to the Super Bowl, what is the probability that none of the five sets on a certain block were tuned in to the Super Bowl?
12. If ten percent of the population is estimated to be allergic to ragweed, what is the probability that an entire five member family is allergic to ragweed? That an entire five member family is not allergic?
13. Suppose that the penal system reforms criminals to the extent that 75 percent of them do not commit a crime in the first three years after their release from jail. What is the probability that none in a group of five criminals is reformed?
14. If ten percent of the men in a certain city are unemployed, what is the probability that all of the 100 Smiths listed in the telephone book are employed?
15. Run a computer simulation of the roll of 10,000 dice and determine how often one, two, three, etc. consecutive heads occur. (See Example 2.)

* *Trial by Mathematics*, Time Magazine, April 26, 1968.

16. Prove the following binomial recursion formula (which should prove helpful in Exercise 2):

$$b(k + 1; n, p) = b(k; n, p) \cdot \frac{p(n - k)}{(1 - p)(k + 1)}$$

17. Suppose that X and Y are independent binomial random variables with parameters n, p and m, p, respectively. Prove that $X + Y$ has binomial distribution with parameters $n + m$ and p.

18. The following is another of Chevalier de Méré's famous problems: Which is more likely, getting one or more 6's in 4 rolls of a fair die, or getting one or more (6, 6)'s in 24 rolls of a pair of fair dice? (de Méré thought that he could prove these probabilities to be equal.)

19. Suppose that an experiment can result in any of m possible outcomes $E_1, E_2, \ldots, E_m$ with probabilities $p_1, p_2, \ldots, p_m$, respectively. Suppose that this experiment is repeated independently n times.

    (a) Prove that the probability that E_1 occurs k_1 times, E_2 occurs k_2 times, ..., and finally E_m occurs k_m times, where $k_1 + k_2 + \cdots + k_m = n$, equals

    $$\frac{n!}{k_1! k_2! \cdots k_m!} \cdot p_1^{k_1} \cdot p_2^{k_2} \cdots p_m^{k_m}$$

    This is known as the *multinomial distribution*.

    (b) Show that when $n = 2$ the multinomial distribution is nothing more than the binomial distribution.

    (c) Let X_i denote the number of occurrences of event E_i over the n trials. Identify the probability distribution of X_i.

20. A fair die is rolled ten times. What is the probability that there will be three 1's, two 2's, two 3's, and one 4, 5, and 6? What is the probability that there will be three or more 1's?

21. An urn contains five red, four blue, three black, and two white balls. If five balls are drawn with replacement from the urn, find the probability that there will be two reds, two blacks, and one white. What is the probability that there will be three or more reds in the sample?

22. The owner of a men's clothing store knows that 10 percent of the men who come into his store will purchase a suit, 20 percent will purchase at least one shirt, 30 percent will purchase at least one pair of pants, and the other 40 percent will not make any purchase. If 20 men come into the store one evening, what is the probability that 4 of them purchase a suit, 5 purchase shirts, and 6 purchase pants?

6.2 The Hypergeometric Distribution

Suppose that we have an urn containing N balls lettered either S or F (success or failure), there being Np balls lettered S and Nq balls lettered F, where $p + q = 1$. A sample of n balls is drawn from the urn, and we let X denote the number of successes in the sample.

If the sample is drawn *with* replacement, then each draw can be considered a Bernoulli trial with probability of success (drawing a ball lettered S) being p. The random variable X therefore has binomial probability distribution with parameters n and p.

On the other hand, if the sample is drawn *without* replacement, the trials cannot be Bernoulli, since the contents of the urn change with each draw, and the draws are no longer independent. In this case, X is said to have *hypergeometric* distribution, and the probability of k successes in n draws is denoted $h(k; n, p, N)$.

Example 1   If we are dealt a five card poker hand, and we let X denote the number of aces in the hand, then X is hypergeometric, and not binomial, since the cards are dealt without replacement.

Example 2   A box in a store contains 100 light bulbs, 5 of which are defective. A shopper picks 4 bulbs from the box. If we let X denote the number of defectives among the 4 bulbs chosen, then X is hypergeometric, and not binomial, because the shopper picks the bulbs without replacement, wanting to purchase 4 bulbs, not just one bulb.

We shall use the notation $M^{(k)}$ for the factorial-type product of k terms $M(M - 1)(M - 2) \cdots [M - (k - 1)]$

Proposition 6.4

$$h(k; n, p, N) = \frac{C_k^{\,n} \cdot Np^{(k)} \cdot Nq^{(n-k)}}{N^{(n)}}$$

or equivalently,

$$h(k; n, p, N) = \frac{C_k^{\,Np} \cdot C_{n-k}^{\,Nq}}{C_n^{\,N}}$$

Proof: We use the same sample space that we used for the binomial distribution:

$$S = \{(a_1, a_2, \ldots, a_n) : a_k = S \quad \text{or} \quad a_k = F\}$$

Since there are $C_k^{\,n}$ different choices of k of the n coordinates in which to place the k successes S, there are $C_k^{\,n}$ sample elements corresponding to the event $\{X = k\}$. We now prove the fact that each has the same probability. Of course, $P(a_1 = S) = Np/N = p$ and $P(a_1 = F) = Nq/N = q$, but since we are sampling without replacement, the result of the first trial will effect the probabilities on the second and subsequent trials. Thus, if $a_1 = S$, then $P(a_2 = S) = (Np - 1)/N - 1$ and $P(a_2 = F) = Nq/N - 1$, while should $a_1 = F$, then $P(a_2 = S) = Np/N - 1$ and $P(a_2 = F) = (Nq - 1)/N - 1$. In general, we have

$$P(a_k = S) = \frac{Np - \text{successes among } \{a_1, a_2, \ldots, a_{k-1}\}}{N - k + 1}$$

the quotient of the number of successes remaining in the urn by the number of balls remaining in the urn, and

$$P(a_k = F) = \frac{Nq - \text{failures among } \{a_1, a_2, \ldots, a_{k-1}\}}{N - k + 1}$$

the quotient of the number of failures remaining by the number of balls remaining. Hence, the numerator of $P(X = k)$ contains the k terms Np, $Np - 1, \ldots, Np - k + 1$, and each of these is divided by a term of the form $N - j + 1$, where j indicates on which draw the ball was drawn. The numerator also contains $n - k$ terms Nq, $Nq - 1, \ldots, Nq - (n - k) + 1$, each of which is divided by a term of the form $N - j + 1$. The denominator therefore will contain the n terms N, $N - 1, \ldots, N - n + 1$. In evaluating the probability of any n-tuple in $\{X = k\}$, we find that the terms in the numerator will not occur exactly in the order mentioned above, but rather in some permutation thereof, while the terms in the denominator will occur in the order mentioned. For example, if $n = 5$ and $k = 3$, we have

$$P[(F, S, S, F, S)] = \frac{Nq}{N} \cdot \frac{Np}{N - 1} \cdot \frac{Np - 1}{N - 2} \cdot \frac{Nq - 1}{N - 3} \cdot \frac{Np - 2}{N - 4}$$

$$= \frac{Np(Np - 1)(Np - 2)(Nq)(Nq - 1)}{N(N - 1)(N - 2)(N - 3)(N - 4)}$$

However, no matter which n-tuple in $\{X = k\}$ we choose, exactly the same terms will occur in numerator and denominator, and so each of the n-tuples in $\{X = k\}$ is equally likely. Thus,

$$P(X = k) = C_k{}^n \cdot \frac{Np(Np - 1)\cdots(Np - k + 1)Nq(Nq - 1)\cdots(Nq - (n - k) + 1)}{N(N - 1)(N - 2)\cdots(N - n + 1)}$$

$$= C_k{}^n \cdot \frac{Np^{(k)}Nq^{(n-k)}}{N^{(n)}}$$

It is then just a simple algebraic task to show that this term is equal to $C_k{}^{Np} \cdot C_{n-k}{}^{Nq}/C_n{}^N$.

Note that the term $C_k{}^{Np} \cdot C_{n-k}{}^{Nq}/C_n{}^N$ is simply the probability of taking a sample of n balls from the urn in such a way that k are successes and $n - k$ are failures. We have seen in Section 1.7 that taking a sample in this way is equivalent to sampling without replacement.

Proposition 6.5

The hypergeometric distribution is a probability distribution.

Proof: By Proposition 6.4, we have

$$\sum_{k=0}^{n} h(k; n, p, N) = \sum_{k=0}^{n} \frac{C_k{}^{Np} \cdot C_{n-k}{}^{Nq}}{C_n{}^N}$$

$$= \frac{1}{C_n{}^N} \sum_{k=0}^{n} C_k{}^{Np} \cdot C_{n-k}{}^{Nq}$$

This last summation counts $C_n{}^N$, because it counts the number of samples of size n by partitioning them according to the number of successes in the sample. Thus, every sample of size n is counted exactly once, in the term

$C_k^{Np} \cdot C_{n-k}^{Nq}$ corresponding to the number k of successes in the sample. Hence,

$$h(k; n, p, N) = \frac{1}{C_n^N} \cdot C_n^N = 1$$

Example 3 We wish to calculate the probability that a bridge hand contains exactly five hearts. Using the techniques of Chapters 1 and 2, we find this probability to be $C_5^{13} \cdot C_8^{39}/C_{13}^{52}$, which should look familiar to advocators of the hypergeometric distribution. Imagine an urn containing 52 balls, 13 of which (those corresponding to the 13 hearts) are labeled S, the other 39 being labeled F. Thus $N = 52$, $Np = 13$, and $Nq = 39$. We choose a sample of $n = 13$ without replacement and wish to find the probability of $k = 5$ successes:

$$h(5; 13, \tfrac{1}{4}, 52) = \frac{C_5^{13} \cdot C_8^{39}}{C_{13}^{52}} = 0.1247$$

Example 4 The hypergeometric distribution is useful in problems concerning quality control. Suppose that a person is interested in buying a lot containing $N = 100$ items, provided that it contains an acceptably low number m of defectives. For the sake of example, say that $m = 5$. The prospective buyer then chooses a sample of $n = 10$ items from the lot and determines that there are k defectives in his sample. Since the process of determining whether or not a sampled item is defective can sometimes be a destructive one (that is, the item must be used to the extent that it cannot be used again), we must assume that our sample is taken without replacement. The value k obtained from the sample will then allow the prospective buyer to decide if the value m is a good estimate for the number of defectives in the entire lot. If the probability

$$h\left(k; n, \frac{m}{N}, N\right) = \frac{C_k^m \cdot C_{n-k}^{N-m}}{C_n^N}$$

is very small, then it is not very likely that we could choose a sample of n items containing k defectives from a population of N items containing m defectives. But if we *did* choose such a sample, then we would be inclined to believe instead that our assumption concerning m is incorrect, either too high or too low. On the other hand, if the probability above is relatively high, then we would be inclined to believe that our estimate of m is fairly accurate. Using the estimate $m = 5$, we get the following probabilities:

k	$h(k; 10, 0.05, 100)$
0	0.5838
1	0.3394
2	0.0702
3	0.0064
4	0.0002
5	0.0000

Therefore, a sample of size 5 containing either none or one defective would tend to support the hypothesis that $m = 5$.

Proposition 6.6

If X is a random variable having hypergeometric distribution $h(k; n, p, N)$, then

(a) $E(X) = np$

(b) $\text{Var}(X) = npq[(N - n)/(N - 1)]$

Proof: For $k = 1, 2, \ldots, n$, we let I_k be the indicator of a success on the kth draw. That is,

$$I_k = \begin{cases} 1 & \text{if a success on the } k\text{th draw} \\ 0 & \text{otherwise} \end{cases}$$

We claim that $P(I_k = 1) = p$ for all $k = 1, 2, \ldots, n$, as would be the case if the draws were made with replacement. Imagine that the Np balls lettered S are distinguishable—that they are marked $S_1, S_2, \ldots, S_{Np}$, for example. The probability that a particular one of these balls, S_j for example, is drawn on the kth draw is

$$\underbrace{\frac{N-1}{N} \cdot \frac{N-2}{N-1} \cdots \cdot \frac{N-k+1}{N-k+2}}_{k-1 \text{ terms}} \cdot \frac{1}{N-k+1} = \frac{1}{N}$$

which is simply the product of the probabilities of not drawing *that* ball on the first $k - 1$ draws by the probability of then drawing *that* ball on the kth draw. Since there are a total of Np successes, and since

$$\{I_k = 1\} = S_1 \cup S_2 \cup \cdots \cup S_{Np}$$

the probability of a success on the kth draw is $Np \cdot (1/N) = p$. Hence, the $\{I_k\}$ are identically distributed (each takes the values 0 and 1 with the same probability), and so $E(I_k) = p$ for all $k = 1, 2, \ldots, n$. Since the I_k are indicators, we have $I_k^2 = I_k$, and so $\text{Var}(I_k) = p - p^2 = pq$. Therefore, since $X = I_1 + I_2 + \cdots + I_n$, we have

$$E(X) = E\left(\sum_{k=1}^{n} I_k \right) = \sum_{k=1}^{n} E(I_k) = np$$

Thus, (a) is true. We also have

$$P(I_j \cdot I_k = 1) = P(I_j = 1 \cap I_k = 1)$$
$$= P(I_j = 1) \cdot P(I_k = 1 | I_j = 1)$$
$$= \frac{Np}{N} \cdot \frac{Np - 1}{N - 1}$$

where the evaluation of the conditional probability is similar to the calculation described above. As a result,

$$E(I_j \cdot I_k) = \frac{Np(Np - 1)}{N(N - 1)} \qquad \text{for any pair } j, k$$

and so

$$\text{Cov}(I_j, I_k) = E(I_j \cdot I_k) - E(I_j) \cdot E(I_k)$$

$$= \frac{Np(Np - 1)}{N(N - 1)} - \frac{Np}{N} \cdot \frac{Np}{N}$$

$$= \frac{-Np(N - Np)}{N^2(N - 1)} \qquad \text{for any pair } j, k$$

Therefore,

$$\text{Var}(X) = \text{Var}(I_1 + I_2 + \cdots + I_n)$$

$$= \sum_{k=1}^{n} \text{Var}(I_k) + 2 \sum_{j<k} \text{Cov}(I_j, I_k) \qquad \text{(by Proposition 5.15)}$$

$$= n(pq) + 2 C_2{}^n \cdot \frac{-Np(N - Np)}{N^2(N - 1)}$$

there being $C_2{}^n$ pairs (j, k) with $j < k$ (there are as many such pairs as there are two-element subsets of $\{1, 2, \ldots, n\}$). Arithmetic manipulations then give

$$\text{Var}(X) = npq - 2 \cdot \frac{n(n - 1)}{2} \cdot \frac{Np}{p} \cdot \frac{N - Np}{N} \cdot \frac{1}{N - 1}$$

$$= npq - n(n - 1)p \cdot \frac{Nq}{N} \cdot \frac{1}{N - 1}$$

$$= npq - \frac{n(n - 1)pq}{N - 1}$$

$$= npq \frac{1 - (n - 1)}{N - 1}$$

$$= npq \frac{N - n}{N - 1}$$

Example 3 (continued)   The expected number of hearts in a bridge hand is $np = 13 \cdot \frac{1}{4} = \frac{13}{4}$.

Notice that as $N \to \infty$, the quotient

$$\frac{N - n}{N - 1} = \frac{1 - (n/N)}{1 - (1/N)} \to 1$$

so that the binomial and hypergeometric probability distributions have identical means and almost identical variances, with the difference in variances decreasing as the population size increases. Therefore, it should not come as a surprise that these two probability distributions are very close, and get closer and closer as the population size increases.

Proposition 6.7

$$\lim_{N \to \infty} h(k; n, p, N) = b(k; n, p)$$

Proof: By Proposition 6.4,

$$h(k; n, p, N) = C_k^n \cdot \frac{Np^{(k)} \cdot Nq^{(n-k)}}{N^{(n)}}$$

$$= C_k^n \cdot \frac{Np(Np-1)\cdots(Np-k+1) \cdot Nq(Nq-1)\cdots[Nq-(n-k)+1]}{N(N-1)(N-2)\cdots(N-n+1)}$$

there being n terms in both numerator and denominator. If we divide both numerator and denominator by N^n, we get

$$h(k; n, p, N)$$

$$= C_k^n \cdot \frac{p(p-1/N)\cdots[p-(k-1)/N]q(q-1/N)\cdots[q-(n-k-1)/N]}{1(1-1/N)(1-2/N)\cdots[1-(n-1)/N]}$$

Letting $N \to \infty$ in this expression, we find that

$$\lim_{N \to \infty} h(k; n, p, N) = C_k^n \cdot \frac{p^k \cdot q^{n-k}}{1^n} = b(k; n, p)$$

Therefore, for large populations, not only do the binomial and hypergeometric distributions have identical means and almost identical variances, but they are also almost identically distributed. The probability of drawing k successes from an urn containing N balls, Np of which are successes, is practically the same if we draw the balls with or without replacement.

The following table demonstrates the closeness of these two probability distributions in the case that $N = 100$ (which is irrelevant for the binomial), $p = \frac{1}{4}$, and $n = 4$:

k	$b(k; n, p)$	$h(k; n, p, N)$
0	0.316	0.310
1	0.422	0.431
2	0.211	0.212
3	0.047	0.044
4	0.004	0.003

EXERCISES

1. Show that the two equivalent expressions for $h(k; n, p, N)$ in Proposition 6.4 are, in fact, equivalent.
2. Determine the moment-generating function of the hypergeometric random variable with parameters n, p, and N.

3. Prove the hypergeometric recursion formula

$$h(k + 1, n, p, N) = \frac{(n - k)(Np - k)}{(k + 1)(Nq - n + k + 1)} \cdot h(k; n, p, N)$$

4. In Example 2, what is the expected number of defectives in the sample of 4?
5. What is the expected number of aces in a five-card poker hand?
6. A toy bank contains 5 quarters, 8 dimes, 10 nickels, and 25 pennies. What is the expected number of dimes in a sample of 5 coins taken from the bank?
7. Half the 40 members of a political organization are women. What is the probability that there will be no women on a committee of 5?
8. If 25 percent of the 100 people who apply for one of five identical jobs are black, what is the probability that none of the jobs are taken by blacks? What is the probability they take all five jobs?

6.3 The Geometric Distribution

Consider a sequence of Bernoulli trials, each of which produces a success with probability p. How many trials are required to achieve a first success? To answer this question, we define a random variable X to be the number of the trial on which the first success occurs. X is a discrete random variable with range the positive integers $\{1, 2, \ldots\}$ and $P(X = k) = p(1 - p)^{k-1}$, since the event $\{X = k\}$ means that the first $k - 1$ trials must result in failure (each with probability $1 - p$), and then finally, the last trial must result in a success.

DEFINITION

A discrete random variable X with probability function

$$P(X = k) = \begin{cases} p(1 - p)^{k-1} & \text{for } k = 1, 2, 3, \ldots \\ 0 & \text{otherwise} \end{cases}$$

is called a *geometric random variable* with parameter p.

Proposition 6.8

The geometric distribution is a probability distribution.

Proof:

$$\sum_{k=1}^{\infty} P(X = k) = \sum_{k=1}^{\infty} p(1 - p)^{k-1}$$

$$= p \sum_{k=1}^{\infty} (1 - p)^{k-1}$$

$$= p \sum_{k=0}^{\infty} (1 - p)^{k}$$

$$= p \cdot \frac{1}{[1 - (1 - p)]} = p \cdot \frac{1}{p} = 1$$

Proposition 6.9

If X is a geometric random variable with parameter p, then

(a) $E(X) = 1/p$
(b) $\text{Var}(X) = (1 - p)/p^2$

Proof: Consider the geometric series

$$\sum_{k=1}^{\infty} (1 - p)^k = \frac{1 - p}{1 - (1 - p)} = \frac{1 - p}{p}$$

Differentiating this series term by term with respect to p, we find that

$$-\sum_{k=1}^{\infty} k(1 - p)^{k-1} = \frac{p(-1) - (1 - p) \cdot 1}{p^2} = \frac{-1}{p^2}$$

Therefore,

$$(*) \qquad \sum_{k=1}^{\infty} k(1 - p)^{k-1} = \frac{1}{p^2}$$

Since

$$E(X) = \sum_{k=1}^{\infty} kp(1 - p)^{k-1} = p \sum_{k=1}^{\infty} k(1 - p)^{k-1}$$

we conclude from $(*)$ that $E(X) = p(1/p^2) = 1/p$. Multiplying both sides of $(*)$ by $1 - p$, we obtain

$$\sum_{k=1}^{\infty} k(1 - p)^k = \frac{1 - p}{p^2}$$

and differentiating both sides of this equation with respect to p, we obtain

$$-\sum_{k=1}^{\infty} k^2(1 - p)^{k-1} = \frac{p^2(-1) - (1 - p) \cdot 2p}{p^4} = \frac{p - 2}{p^3}$$

Therefore,

$$\sum_{k=1}^{\infty} k^2(1 - p)^{k-1} = \frac{2 - p}{p^3}$$

and so

$$E(X^2) = \sum_{k=1}^{\infty} k^2 p(1 - p)^{k-1}$$

$$= p \sum_{k=1}^{\infty} k^2(1 - p)^{k-1}$$

$$= p\left(\frac{2 - p}{p^3}\right) = \frac{2 - p}{p^2}$$

Therefore,

$$\text{Var}(X) = \frac{2-p}{p^2} - \left(\frac{1}{p}\right)^2 = \frac{1-p}{p^2}$$

Example 1 Suppose that a fair coin is tossed until a head occurs. Let X denote the number of the toss on which the first head occurs. Then X is geometric with parameter $p = \frac{1}{2}$, and so the average number of tosses required before a head occurs is $E(X) = 1/p = 1/\frac{1}{2} = 2$.

Example 2 A baseball player with a batting average of 0.250 must wait an average of $E(X) = 1/p = 1/0.250 = 4$ at-bats before getting his first hit. The probability that a hitter with a 0.250 average will go hitless in ten at-bats before getting his first hit in a Sunday doubleheader is $P(X = 11) = p(1-p)^{10} = 0.250 \cdot (0.750)^{10} = 0.014$

Example 3 An urn contains five red and three white balls. Let X denote the number of draws with replacement until a red ball is drawn. Then X is geometric (this would *not* be the case if the draws were made *without* replacement) with parameter $p = \frac{5}{8}$. The average number of draws required before a red ball is drawn is $E(X) = 1/p = 1/\frac{5}{8} = \frac{8}{5} = 1.6$. If we wish to determine the probability that fewer than ten draws will be needed before the first red ball is drawn, then we calculate

$$\sum_{k=1}^{9} P(X = k) = \sum_{k=1}^{9} \left(\tfrac{3}{8}\right)^{k-1}\left(\tfrac{5}{8}\right) \qquad (k - 1 \text{ whites, then a red})$$

$$= \tfrac{5}{8} \cdot \sum_{k=1}^{9} \left(\tfrac{3}{8}\right)^{k-1}$$

$$= \tfrac{5}{8} \cdot \sum_{k=0}^{8} \left(\tfrac{3}{8}\right)^{k}$$

$$= \tfrac{5}{8} \cdot \frac{1 - \left(\tfrac{3}{8}\right)^{9}}{1 - \left(\tfrac{3}{8}\right)}$$

$$= 1 - \left(\tfrac{3}{8}\right)^{9} = 0.99985$$

Example 4 (Full Occupancy) Suppose that we have an unlimited supply of balls which are to be distributed among m urns. We let X denote the number of balls required before each of the m urns contains at least one ball (this is called full occupancy). We define a sequence of random variables as follows:

$X_1 =$ the number of balls required until one urn is occupied. (Obviously, $X_1 = 1$.)

$X_2 =$ the number of additional balls required until a second urn is occupied.

X_3 = the number of additional balls required until a third urn is occupied.
$\vdots$

X_m = the number of additional balls required until the last urn is occupied.

Clearly, $X = X_1 + X_2 + \cdots + X_m$, and therefore, $E(X) = E(X_1) + E(X_2) + \cdots + E(X_m)$. As mentioned above $X_1 = 1$, and so $E(X_1) = 1$. We claim that X_2 is geometric with parameter $p = (m-1)/m$. After the first ball is placed in some urn, each succeeding ball can be regarded as a Bernoulli trial, with success meaning that the ball is placed in an unoccupied urn. At this stage, there are $m-1$ unoccupied urns, so the probability of success p equals $(m-1)/m$. X_2 is then counting the number of balls until the first success, and therefore, X_2 is geometric with parameter $p = (m-1)/m$ and mean $E(X_2) = 1/p = m/(m-1)$. Similarly, X_3 is geometric with parameter $p = (m-2)/m$, since there are now just $m-2$ unoccupied urns. In general, X_k is geometric with parameter $p = [m - (k-1)]/m$, since at this stage, $k-1$ urns are unoccupied. Therefore, $E(X_k) = 1/p = m/[m - (k-1)]$. Consequently,

$$E(X) = \sum_{k=1}^{m} E(X_k) = \sum_{k=1}^{m} \frac{m}{m-k+1} = m \cdot \sum_{k=1}^{m} \frac{1}{m-k+1}$$

so that

$$E(X) = m \cdot \left(\frac{1}{m} + \frac{1}{m-1} + \cdots + \frac{1}{2} + 1 \right)$$

$$= m \cdot \left(1 + \frac{1}{2} + \cdots + \frac{1}{m} \right)$$

EXERCISES

1. Prove that the moment-generating function of a geometric random variable with parameter p is $pe^t/[1 - (1-p)e^t]$.

2. What is the expected number of throws before a pair of coins both show heads?

3. The drunk with five keys on his keychain is once again trying to open his front door. This time he is so drunk that he can't tell from one moment to the next which keys he has already tried. What now is the expected number of keys that he must try before staggering into his wife's loving arms?

4. A radio newscaster is interviewing people on the street. If only one of every 10 people are willing to be interviewed, what is the expected number of people who must be asked before finding one willing to be interviewed? What is the probability that such a person will not be found among the first 20 people to be asked?

5. Two parents decide to continue having children until they have a boy. What is the expected number of children in their family? How likely is it that they will still be trying for a boy after having five children?

6. The average thoroughbred race horse wins one race in every nine tries. What is the expected number of attempts before the average horse wins his first race of the year? How likely is it that an average horse will still be looking for its first win of the year after 20 starts?

7. Under the same assumptions made for a geometric random variable, let X denote the number of trials required until the kth success. Prove that

$$P(X = n) = C_{k-1}^{n-1} \cdot p^k \cdot (1 - p)^{n-k}$$

(X is said to have a negative binomial distribution. Instead of calculating the probability of k successes in n trials, we are calculating the probability that it will take n trials for k successes to occur.)

8. What is the probability that it will take 20 rolls for a fair die to come up 6 three times? That it will take 20 or more rolls?

9. If 40 percent of the fish in a lake are of a certain species, what is the probability it will take a sample of ten (caught and thrown back) to find five of the given species?

10. If it takes catching (and throwing back) 50 fish before identifying ten of the given species, is it very likely that 25 percent of the fish in the lake are of the given species?

6.4 The Poisson Distribution

Calculations involving binomial probabilities can often become very tedious. Approximations to binomial probabilities are therefore necessary. In Chapter 8, we shall study the normal distribution and its use as an approximation to the binomial. Unfortunately, the normal distribution does not provide a very accurate approximation to the binomial when the value of p is close to 0.

Let us suppose that X is a binomial random variable with parameters n and p, where n is large and p is close to 0. We define λ to be np, the expected number of successes, and find that

$$b(0; n, p) = (1 - p)^n = \left(1 - \frac{\lambda}{n}\right)^n$$

Therefore,

$$\log_e b(0; n, p) = n \log_e \left(1 - \frac{\lambda}{n}\right)$$

Substituting $-\lambda/n$ into the Maclaurin series

$$\log_e(1 + x) = x - \frac{x^2}{2} + \frac{x^3}{3} - \frac{x^4}{4} + \cdots$$

$$= \sum_{n=1}^{\infty} (-1)^{n+1} \frac{x^n}{n}$$

we have

$$\log_e b(0; n, p) = n\left(-\frac{\lambda}{n} - \frac{\lambda^2}{2n^2} - \frac{\lambda^3}{3n^3} - \cdots\right)$$

$$= -\lambda - \frac{\lambda^2}{2n} - \frac{\lambda^3}{3n^2} - \cdots$$

Since n is assumed large, we have

$\log_e b(0; n, p) \cong -\lambda$

and so

$b(0; n, p) \cong e^{-\lambda}$

Furthermore, since

$$\frac{b(k; n, p)}{b(k-1; n, p)} = \frac{C_k^n \cdot p^k \cdot q^{n-k}}{C_{k-1}^n \cdot p^{k-1} \cdot q^{n-k+1}}$$

$$= \frac{(n-k+1)p}{kq}$$

$$= \frac{\lambda - (k+1)p}{kq}$$

which, since p is approximately 0 and q is approximately 1, is approximately λ/k. Therefore,

$$\frac{b(k; n, p)}{b(k-1; n, p)} \cong \frac{\lambda}{k}$$

that is,

$$b(k; n, p) \cong \frac{\lambda}{k} \cdot b(k-1; n, p)$$

Using the approximation $b(0; n, p) \cong e^{-\lambda}$ as a starting point, we find iteratively that

$$b(1; n, p) \cong \frac{\lambda}{1} \cdot b(0; n, p) \cong \lambda e^{-\lambda}$$

$$b(2; n, p) \cong \frac{\lambda}{2} \cdot b(1; n, p) \cong \left(\frac{\lambda^2}{2}\right)e^{-\lambda}$$

$$b(3; n, p) \cong \frac{\lambda}{3} \cdot b(2; n, p) \cong \left(\frac{\lambda^3}{3!}\right)e^{-\lambda}$$

and inductively, that if

$$b(k; n, p) \cong \left(\frac{\lambda^k}{k!}\right)e^{-\lambda}$$

then

$$b(k+1; n, p) \cong \frac{\lambda}{k+1} \cdot b(k; n, p)$$

$$\cong \frac{\lambda^{k+1}}{(k+1)!}e^{-\lambda}$$

In general, therefore, we have the approximation

$$b(k; n, p) \cong e^{-\lambda} \frac{\lambda^k}{k!}$$

We are now ready to make the following definition:

DEFINITION

A random variable X with probability function

$$P(X = k) = \begin{cases} e^{-\lambda}\lambda^k/k! & \text{for } k = 0, 1, 2, 3, \ldots \\ 0 & \text{otherwise} \end{cases}$$

is called a Poisson random variable with parameter $\lambda > 0$. We will use the notation $p(k; \lambda)$ for the probability $P(X = k)$.

Proposition 6.10

The Poisson distribution is a probability distribution.

Proof: The proof follows from the Maclaurin series expansion

$$e^x = \sum_{k=0}^{\infty} \frac{x^k}{k!}$$

and

$$\sum_{k=0}^{\infty} p(k; \lambda) = \sum_{k=0}^{\infty} \frac{\lambda^k e^{-\lambda}}{k!}$$

$$= e^{-\lambda} \sum_{k=0}^{\infty} \frac{\lambda^k}{k!} = e^{-\lambda}e^{\lambda} = e^0 = 1$$

A table of Poisson probabilities, for values of λ ranging from 0.1 to 5.0 by increments of 0.1, and from 5.0 to 25.0 by increments of 1.0, can be found in Appendix B. Only those values of k which result in significantly positive probabilities have been included. The Poisson distributions with parameter $\lambda = 1$ and $\lambda = 5$ are pictured in Figure 6.2.

The derivation above is the proof of the following:

Proposition 6.11

If p is small and n is large, the binomial probability distribution with parameters n and p can be closely approximated by the Poisson probability distribution with parameter $\lambda = np$.

With this in mind, we can think of the Poisson distribution as calculating the probability of k successes in a large number n trials in terms of the average number of successes $\lambda = np$, where the event denoted "success" is assumed to be rare.

Proposition 6.12

If X is a Poisson random variable with parameter λ, then

(a) $E(X) = \lambda$
(b) $\mathrm{Var}(X) = \lambda$

Proof:

(a) By definition

$$E(X) = \sum_{k=0}^{\infty} k \cdot p(k; \lambda) = \sum_{k=1}^{\infty} k \cdot \frac{\lambda^k e^{-\lambda}}{k!}$$

$$= \lambda \sum_{k=1}^{\infty} \frac{\lambda^{k-1} e^{-\lambda}}{(k-1)!}$$

$$= \lambda \sum_{k=0}^{\infty} \frac{\lambda^k e^{-\lambda}}{k!}$$

$$= \lambda \sum_{k=0}^{\infty} p(k; \lambda)$$

$$= \lambda \cdot 1 = \lambda$$

Figure 6.2

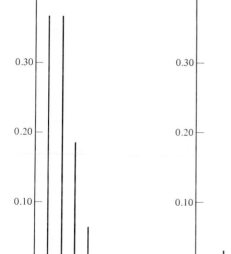

$p(k; 1)$

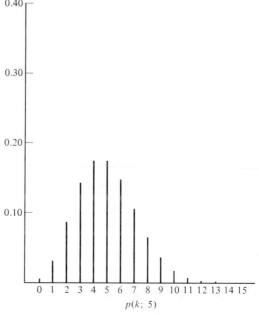

$p(k; 5)$

(b) In similar fashion,

$$E(X^2) = \sum_{k=0}^{\infty} k^2 \cdot p(k; \lambda)$$

$$= \sum_{k=1}^{\infty} k^2 \cdot \frac{\lambda^k e^{-\lambda}}{k!}$$

$$= \lambda \sum_{k=1}^{\infty} k \cdot \frac{\lambda^{k-1} e^{-\lambda}}{(k-1)!}$$

Letting $s = k - 1$, we obtain

$$E(X^2) = \lambda \sum_{s=0}^{\infty} (s + 1)\left(\frac{\lambda^s e^{-\lambda}}{s!}\right)$$

$$= \lambda \sum_{s=0}^{\infty} s \cdot \frac{\lambda^s e^{-\lambda}}{s!} + \lambda \sum_{s=0}^{\infty} \frac{\lambda^s e^{-\lambda}}{s!}$$

$$= \lambda \cdot \lambda + \lambda \cdot 1 = \lambda^2 + \lambda$$

by (a) and Proposition 6.10. Thus, $E(X^2) = \lambda^2 + \lambda$, and hence, $\text{Var}(X) = (\lambda^2 + \lambda) - \lambda^2 = \lambda$.

The mean and variance we obtained for the Poisson distribution are to be expected in light of the close approximation to the binomial distribution. The latter has mean np and variance npq. Since $\lambda = np$, and p is assumed to be small, then q is close to 1. Therefore, $npq = (np)q$ is close to $(np)1 = np$, the same as the mean.

Example 1 Suppose that 300 misprints are randomly distributed throughout a book containing 500 pages. We wish to calculate the probability that a given page contains exactly 2 misprints. Each of the 300 misprints has probability 1/500 of being on any particular page. If we look at each of the 300 misprints as a trial, with success meaning that the misprint occurs on the given page, then we have a binomial probability situation and the probability we seek is

$$b\left(2; 300, \frac{1}{500}\right) = C_2^{300} \cdot \left(\frac{1}{500}\right)^2 \cdot \left(\frac{499}{500}\right)^{298}$$

which equals 0.09879. Without a computer or hand calculator, this binomial would be rather tedious to calculate. To avoid these calculations, we could use the Poisson approximation $p(2; \lambda)$, where $\lambda = np = 300 \cdot (1/500) = 0.6$. We can expect our approximation to be highly accurate since $p = 1/500$ *is* quite small. From our table of Poisson probabilities, we find that $p(2; 0.6) = 0.0988$, almost identical to the correct answer. If we wish to calculate the probability that at least two misprints occurred on the given page, we calculate

$$1 - [p(0; 0.6) + p(1; 0.6)] = 1 - (0.549 + 0.329) = 0.122$$

to approximate

$$1 - \left[b\left(0; 300, \frac{1}{500}\right) + b\left(1; 300, \frac{1}{500}\right) \right]$$

$$= 1 - (0.548 + 0.330) = 0.12176$$

Example 2   From previous experience, an airline knows that approximately three percent of those who reserve plane tickets fail to make the flight. As a result, the airline regularly sells 100 tickets for a flight which seats 98. Is this a safe business practice? How likely is it that the airline will oversell a flight? If we consider each person who buys a ticket as a trial, with success being "failing to make the flight," then we are given that $n = 100$ and $p = 0.03$. If we have either $k = 0$ or $k = 1$ successes in the 100 trials, then more than 98 people show up for the flight, causing the airline problems. The probability of this happening is

$$b(0; 100, 0.03) + b(1; 100, 0.03)$$

which, of course, is somewhat tedious to calculate (in fact, it equals 0.1946). Using the Poisson approximation ($p = 0.03$ is very small), we have $\lambda = np = 100 \cdot 0.03 = 3$, and

$$p(0; 3) + p(1; 3) = 0.0498 + 0.1494 = 0.1992.$$

Example 3   An insurance company insures 10,000 people (for one year each) against a particular kind of accident, knowing that one person in 500 can be expected to have such an accident during a calendar year. We wish to calculate the probability that 25 or more people (attempt to) collect on their policy in a given year. If we consider each of the 10,000 people as a trial, with success being "having the given accident," then $p = 1/500 = 0.002$ by the insurance company's estimate. The probability we desire is

$$1 - \sum_{k=0}^{24} b(k; 10{,}000, 0.002)$$

Using the Poisson approximation ($p = 0.002$ is very small), with parameter $\lambda = 10{,}000 \cdot 0.002 = 20$, we have

$$1 - \sum_{k=0}^{24} p(k; 20) = 1 - 0.8431 = 0.1569$$

(This probability was obtained by adding the first 25 probabilities in the $\lambda = 20$ column of the Poisson probabilities table in Appendix B.)

Notice that in each of these examples, our experiment (Bernoulli trial) could result in one of two possible outcomes. The misprint either is or is not on the given page, the person either does or does not show up for the flight, or the person either does or does not have the specified accident. In each case, we chose as a success that event, of the two possible, which had the lower probability. In each case, this probability was extremely small, allowing us to use the Poisson approximation to the binomial probabilities.

Example 4   The telephone company has found that during peak hours an average of five lines are in use between two towns and wishes to determine how many lines are needed so that there is a less than five percent chance that a customer will find all lines in use. Notice that this problem is worded differently from the preceeding problems. The major use of the Poisson distribution is as an approximation to the binomial distribution, and there does not seem to be a binomial distribution present. Actually a binomial distribution is implicit in the wording of the problem. At any given moment, we may consider each of the telephone company's many customers as a trial, with success being picking up the telephone to make a call. We assume, of course, that n is large and p is small. But we do not know how large n is, nor how small p is, only that their product $np = \lambda = 5$. The value $p(k; 5)$ is the probability of k successes (having k telephone lines in use) at any given moment, given that there is an average of five lines in use at any given moment. Since

$$\sum_{k=0}^{8} p(k; 5) = 0.9319$$

while

$$\sum_{k=0}^{9} p(k; 5) = 0.9682$$

nine telephone lines are needed. If there are nine lines, the probability of an overload (ten or more lines needed) is 0.0328. If there are eight lines, the probability of an overload (nine or more lines needed) is 0.0681, which exceeds 0.05.

The Poisson distribution is quite capable of standing on its own two feet. It has uses other than as an approximation to the binomial distribution. The Poisson distribution arises in a number of cases where it is necessary to count the number of occurrences of a rare and random event in a temporal or spatial setting. Some examples are the following:

1. the number of calls received at a telephone switchboard during a given interval of time;
2. the number of traffic accidents per hour on a given highway, or at a given intersection;
3. the number of parasites on a given host;
4. the number of moss plants in a sampling quadrat on a hillside;
5. the number of bacteria in a specified volume of liquid.

All of these examples share a certain characteristic: Although the number of times the event occurs can be counted easily, it makes no sense to count the number of times the event does not occur. For example, it makes no sense to talk about the number of telephone calls that were *not* made, the number of traffic accidents that did *not* occur, or the number of bacteria that are *not* in a specified volume of liquid. Consequently, we do not have a binomial experiment, in the strict sense of the term, where each trial will result in one of two outcomes, success or failure. Even the concept of trials is vague in this situation.

We will now show that under certain assumptions about the time interval (these assumptions can be generalized to intervals of space, such as area and volume) a Poisson distribution will arise. We assume that our time interval has been divided into a large number of subintervals of equal length Δt such that

1. the probability that the given event will occur more than once during a time interval of length Δt is 0;
2. the probability that the given event will occur exactly once during a time interval of length Δt is directly proportional to the length of the interval that is, this probability equals $\alpha \Delta t$, where $\alpha > 0$);
3. the occurrence of the event in one time interval of length Δt is independent of the occurrence of the event in some other disjoint time interval of length Δt.

Condition 3. implies that the number of times the event occurs in a time interval of length t_1 is independent of the number of times the event occurs in a time interval of length t_2.

Imitating our notation for Poisson probabilities, we define

$p(k; t) =$ probability that the event will occur
$\quad\quad\quad k$ times during an interval of length t

From the partition theorem, it follows that $p(k; t + \Delta t)$ equals

$P([\{k \text{ occurrences in time } t\} \cap \{0 \text{ occurrences in time } \Delta t\}]$
$\quad \cup [\{k - 1 \text{ occurrences in time } t\} \cap \{1 \text{ occurrence in time } \Delta t\}])$

Since these two events are disjoint, this probability equals

$P(\{k \text{ occurrences in time } t\} \cap \{0 \text{ occurrences in time } \Delta t\})$
$\quad + P(\{k - 1 \text{ occurrences in time } t\} \cap \{1 \text{ occurrence in time } \Delta t\})$

By the independence (3) of the time intervals of lengths t and Δt, this probability equals

$P(k \text{ occurrences in time } t) \cdot P(\text{no occurrences in time } \Delta t)$
$\quad + P(k - 1 \text{ occurrences in time } t) \cdot P(1 \text{ occurrence in time } \Delta t)$

which, by definition, equals

$p(k; t) \cdot p(0; \Delta t) + p(k - 1; t) \cdot p(1; \Delta t)$

However, $p(1; \Delta t) = \alpha \Delta t$ by (2) and $p(0; \Delta t) = 1 - \alpha \Delta t$ by (1) and (2).

Therefore,

$$p(k; t + \Delta t) = p(k; t)(1 - \alpha \Delta t) + p(k - 1; t)(\alpha \Delta t)$$
$$= p(k; t) - \alpha \Delta t \, p(k; t) + \alpha \Delta t \, p(k - 1; t)$$

Subtracting $p(k; t)$ from both sides of this equation, and then dividing by Δt yields

$$\frac{p(k; t + \Delta t) - p(k; t)}{\Delta t} = \frac{\alpha \Delta t [p(k - 1; t) - p(k, t)]}{\Delta t}$$

Taking the limit as $\Delta t \to 0$ gives

$$\frac{d[p(k; t)]}{dt} = \alpha[p(k - 1; t) - p(k; t)]$$

We leave it as an exercise for the reader to supply the inductive proof that this system of differential equations, with the initial condition $p(k; 0) = 0$ for all $k > 0$, has solution

$$p(k; t) = \frac{e^{-\alpha t}(\alpha t)^k}{k!}$$

for any $k = 0, 1, 2, \ldots$ and for any t. Therefore, if we choose $\lambda = \alpha t$, the expected number of occurrences of the event over a time interval of length t, we have $p(k; t) = p(k; \lambda)$, the Poisson probability with parameter λ.

The fact that we did arrive at a Poisson probability is not surprising. If we divide our time interval of length t into n equal subintervals, each of length Δt, then each subinterval can be considered a Bernoulli trial with constant probability of success, the occurrence of the event, equaling $\alpha \Delta t$. We therefore have a binomial experiment with parameters n and $\alpha \Delta t$. If n is large and $\alpha \Delta t$ close to 0, both of which appear reasonable under the assumption that the event is rare, then our binomial probabilities can be approximated by Poisson probabilities with parameter

$$\lambda = n \cdot \alpha \Delta t = \frac{t}{\Delta t} \alpha \Delta t = \alpha t$$

Example 5 As indicated above, it is reasonable to assume that the number of incoming telephone calls received per minute at a switchboard is Poisson distributed, with the parameter λ equaling the average number of calls received per minute. If this average is, in fact, six calls per minute, then the probability of an incoming call during a Δt second time interval is $\Delta t/10$. The constant α is $\frac{1}{10}$ in this case, since there are, on the average, six calls for every 60 seconds. If we wish to calculate the probability that there will be between five and ten calls for any given minute, we are faced with the simple Poisson calculation

$$\sum_{k=5}^{10} p(k; 6) = 0.1606 + 0.1606 + 0.1377 + 0.1033 + 0.0688 + 0.0413 = 0.6723$$

Example 6 If there are, on the average, eight accidents per hour on the Long Island Expressway, then the probability that there are fewer than eight accidents during the evening rush hour (4:30–6:00) equals

$$\sum_{k=0}^{7} p(k; 12) = 0.0000 + 0.0001 + 0.0004 + 0.0018 + 0.0053 + 0.0127$$

$$+ 0.0255 + 0.0437 = 0.0895$$

Notice that the parameter $\lambda = 12$ was used because 12 is the average number of accidents per $1\frac{1}{2}$ hours.

Example 7 A flask contains 500 cubic centimeters of vaccine drawn from a vat which contains, on the average, five live viruses per 1000 cubic centimeters. How likely is it that the flask contains no live viruses? Since 500 cubic centimeters contain, on the average, $\lambda = 2.5$ live viruses, a simple Poisson calculation $p(0; 2.5) = 0.0821$ gives us our answer.

EXERCISES

1. Prove that the moment-generating function of a Poisson random variable with parameter λ is $e^{\lambda(e^t - 1)}$.
2. If five percent of all theater reservations are cancelled and if a theater takes 102 reservations for its 100 seats, how often will too many people appear for a show?
3. It is estimated that the probability that a golfer shoots a hole-in-one during a hole-in-one contest is 1/5000. What is the probability that the contest will have no winner if 3000 golfers enter.
4. A car rental agency has three Firebirds and estimates the probability that a customer will request a Firebird to be $\frac{4}{100}$. If 50 people request cars on a given day, find the probability that the agency will have enough Firebirds.
5. On the average there are three rainy days during July (a thirty-day month). Find the probability that there are at most two rainy days during a given July.
6. If a fisherman catches, on the average, three fish per hour, what is the probability that he will catch no fish during a given hour?
7. At a supermarket checkout, customers arrive at a rate of one every two minutes. What is the probability that fewer than five customers will arrive during a fifteen-minute period?
8. The average number of suicides per week in Washington D.C. is 8. In the week after the suicide of a famous politician, there were 12 suicides. Is this unusual?
9. The average person suffers four colds during a calendar year. What is the probability of escaping a cold for an entire year?
10. A computer suffers, on the average, three breakdowns over any twenty-four–hour period. What is the probability the computer will get through an entire day without a breakdown? What is the probability that the computer will have five or more breakdowns during a twenty-four–hour period?
11. New York City has, on the average, four labor strikes per year. What is the probability that New York City will have no labor strikes during a given year? What is the probability during a given three-month period?
12. The mortality rate for a certain disease is five percent. What is the probability that more than 10 people in 100 suffering from this disease will die?

ADDITIONAL PROBLEMS

1. Eight graduate students share the same office. Each is just as likely to study at home as in the office. How many desks are needed so that each student has a desk at least 90 percent of the time he is in the office?

2. One hundred cards marked $1, 2, 3, \ldots, 100$ are randomly arranged in a line. Find the probability that there are exactly two evens among the first 20 positions.

3. A die is tossed until either a 1 or a 6 appears. You win if the 6 appears before the 1 appears. Find the probability that you win.

4. If you have seven spades in your bridge hand, what is the probability that your partner has none?

5. An urn contains m black and n white balls. Then k balls are withdrawn from the urn, color unnoticed. What is the probability the next ball drawn is white?

6. Two dice are rolled repeatedly. What is the probability that sum 5 occurs before sum 7?

7. Eight dice are rolled simultaneously. What is the probability that each face shows at least once?

8. Two people alternately roll a pair of dice, with the first to roll sum 10 the winner. Find their respective probabilities of winning.

9. Three people alternately toss a coin, with the first to toss a head the winner. Find their respective probabilities of winning.

10. Two people with an equal chance of winning play a game until one of them wins twice in succession. Find the probability that it takes an odd number of games before one of them emerges victorious. What is this same probability if one of the players has a two-thirds chance of winning any particular game?

11. If two people both toss a coin n times, find the probability that they both get the same number of heads.

12. An urn contains nine balls, three of each of the colors red, white, and blue. What is the probability that a sample of four balls from the urn contains balls of just two colors? What is the expected number of colors in such a sample?

13. Three children are matching pennies, with the odd man the winner. Let X denote the number of times a particular child wins in six plays. Find $E(X)$.

Markov Chains

7.1 Introduction

In this chapter, we shall study the theory of Markov chain processes, a branch of probability theory whose origins trace back to the Russian mathematician A. A. Markov in the first decade of this century. Markov chain processes are but one of a more general type of species known as stochastic processes. A stochastic process can most easily be described as a process (or experiment, or sequence of experiments) whose development in time is governed by the laws of probability. The time parameter may or may not be discrete, although in this book we shall consider only the discrete case.

The theory of stochastic processes, and the theory of Markov chain processes in particular, has been growing in importance in recent years. The reason for this growth is that many important applications have been found, especially in the sciences. Examples include the Mendelian theory of inheritance, the growth of populations, the diffusion of gases (Brownian motion), the movement of particles from a radioactive source, and the theory of queues (waiting lines).

DEFINITION

A *Markov chain process* is a discrete sequence of experiments (stochastic process) with the following properties:

1. Each experiment results in one of the outcomes $\{a_1, a_2, \ldots, a_k\}$. These outcomes are called the *states* of the process, and may be either qualitatively or quantitatively defined.
2. The probability that a particular outcome occurs on any given experiment depends only on the outcome of the preceding experiment.

Example 1   A person's political preference (Democratic, Republican, Liberal, Conservative, or independent) can be looked upon as a Markov chain process. Political preference is measured each year on election day, so our time parameter is quite discrete, with constant increment of one year. In order to have a Markov chain process, however, we must assume that a person's political preference this year depends only on his preference last year (and the performance of those he voted for last year).

Example 2   Suppose, in a binomial experiment, we define as states the possible number of consecutive successes after each trial. Our states are therefore $a_1 = 0$, $a_2 = 1$, $a_3 = 2$, and so forth. If the process is presently in state a_k, it can then move, after the next trial, to state a_{k+1} (as a result of a success) or to state a_1 (as a result of a failure) in accordance with the binomial probabilities p and q.

Example 3   The number of mice in a biologist's laboratory can be looked upon as a Markov chain process. The number of mice present at a given time depends only on the number present the last time a count was taken. The states are the different possible population sizes.

Example 4   The Mendelian theory of inheritance is rich in applications of Markov chains. According to Mendel, each individual possesses two genes relating to each inheritable characteristic (such as eye color, hair color, etc.). These form one of the three genotypes AA (pure dominant), Aa (hybrid), or aa (pure recessive). The genotype of a child is determined solely by the genotypes of its parents. Each parent donates one of its two genes, each with probability $\frac{1}{2}$, to the child. Consequently, the following experiment is an example of a Markov chain process. We choose a person from a population consisting of equal numbers of pure dominants, hybrids, and pure recessives. We mate this person to a pure dominant, then mate the child to a pure dominant, the subsequent child to another pure dominant, and so forth. Our states will be the genotypes of the child: $a_1 = $ AA, $a_2 = $ Aa, and $a_3 = $ aa. Since the genetic makeup of the parents alone determines the genetic makeup of the child, this process is a Markov process.

Associated with a Markov chain process is a sequence of random variables $X_1, X_2, X_3, \ldots$ which indicate the outcomes of the individual experiments. These random variables are sometimes called the *Markov chain* and satisfy the following two properties, which follow from 1 and 2 above:

1. $X_i(S) = \{a_1, a_2, \ldots, a_k\},$ for each i

2. $P\left(X_m = a_{j_m} \middle| \bigcap_{i=1}^{m-1} X_i = a_{j_i}\right) = P(X_m = a_{j_m} | X_{m-1} = a_{j_{m-1}})$

Condition 2 is simply saying that the future X_m is independent of the past $X_1, X_2, \ldots, X_{m-2}$ and depends only on the present X_{m-1}.

We will make the additional restriction that

$$P(X_m = a_j | X_{m-1} = a_i)$$

is independent of the value m. That is, the probability of moving from one state to another does not depend on the individual trial, but rather is the same for all trials. We can then make the following definition:

DEFINITION

The probability that a Markov chain process which is presently in state a_i will move next to state a_j is called the *transition probability* from state a_i to state a_j and is denoted p_{ij}. The $k \times k$ matrix $P = (p_{ij})$ is called the *transition matrix* of the Markov chain.

Example 4 (continued)   We wish to calculate the transition probabilities for the Mendelian Markov process described above. To do this, we use the following table of probabilities that indicate the likelihood of the child's possible genotypes in terms of the genotype of the "other" parent. Recall that one parent is always required to be a pure dominant AA.

		Child's genotype		
		AA	*Aa*	*aa*
Other	*AA*	1	0	0
parent's	*Aa*	$\frac{1}{2}$	$\frac{1}{2}$	0
genotype	*aa*	0	1	0

For example, the mating AA × AA must produce another AA, since each parent must donate an A gene. On the other hand, the mating AA × Aa can produce either an AA or an Aa, since the pure dominant parent must donate an A gene, while the hybrid parent may donate either an A or an a gene, each with probability $\frac{1}{2}$. Therefore, our transition matrix is

$$P = \begin{pmatrix} 1 & 0 & 0 \\ \frac{1}{2} & \frac{1}{2} & 0 \\ 0 & 1 & 0 \end{pmatrix}$$

Example 5 A New York City commuter buys either the New York Times or the Daily News each day before boarding the train. If he buys the Times one day, the chances that he will buy the Times again the next day are 75 percent. On the other hand, if he buys the News one day, the chances he will buy the News again the next day are only 40 percent. We therefore have a Markov chain process with states a_1 = New York Times and a_2 = Daily News, and transition probability matrix

$$P = \begin{pmatrix} 0.75 & 0.25 \\ 0.60 & 0.40 \end{pmatrix}$$

The probability that the Markov chain will be in prescribed states at each stage of the experiment can be calculated by the following rule:

Proposition 7.1

$$P\left(\bigcap_{i=1}^{\infty} X_i = a_{j_i} \right) = P(X_1 = a_{j_1}) \cdot \prod_{i=1}^{\infty} p_{j_i j_{i+1}}$$

Proof: By the Multiplication Theorem (3.1), we have

$$P(X_1 = a_{j_1} \cap X_2 = a_{j_2} \cap X_3 = a_{j_3} \cap \cdots)$$

$$= P(X_1 = a_{j_1})P(X_2 = a_{j_2} | X_1 = a_{j_1})P\left(X_3 = a_{j_3} \middle| \bigcap_{i=1}^{2} X_i = a_{j_i} \right) \cdots$$

$$= P(X_1 = a_{j_1})P(X_2 = a_{j_2} | X_1 = a_{j_1})P(X_3 = a_{j_3} | X_2 = a_{j_2}) \cdots$$

the last step following from the definition of a Markov chain. In general, the definition allows us to conclude that

$$P\left(X_m = a_{j_m} \middle| \bigcap_{i=1}^{m-1} X_i = a_{j_i}\right) = P(X_m = a_{j_m} | X_{m-1} = a_{j_{m-1}})$$

and so we have

$$P\left(\bigcap_{i=1}^{\infty} X_i = a_{j_i}\right) = P(X_1 = a_{j_1}) \cdot \prod_{i=2}^{\infty} P(X_i = a_{j_i} | X_{i-1} = a_{j_{i-1}})$$

$$= P(X_1 = a_{j_1}) \cdot \prod_{i=2}^{\infty} p_{j_{i-1}j_i}$$

$$= P(X_1 = a_{j_1}) \cdot \prod_{i=1}^{\infty} p_{j_{i-1}j_i}$$

Proposition 7.2

For each $i = 1, 2, \ldots, k$, we have $p_{i1} + p_{i2} + \cdots + p_{ik} = 1$.

Proof: The elements of row i in the transition matrix are the probabilities of moving from state a_i to each of the states $a_1, a_2, \ldots, a_k$. Since a Markov process must result in one of these states, the result follows.

Given the initial state of the experiment and the transition matrix P, we have sufficient information to calculate (with the help of tree diagrams) the probability that the chain will be in any particular state at any given stage of the process. Along these lines, we make the following definition:

DEFINITION

Given that a Markov chain is presently in state a_i, the probability that the chain will be in state a_j after n steps will be denoted $p_{ij}^{(n)}$. We shall let $P^{(n)}$ denote the matrix $[p_{ij}^{(n)}]$ of n-step probabilities.

Example 4 (continued) Suppose that we wish to calculate the three-step transition matrix $P^{(3)}$. The first row of $P^{(3)}$ corresponds to the process starting in state a_1; that is, the first "other" parent is a pure dominant AA. Since the process demands that one parent always be pure dominant, we have a situation in this case where both parents are pure dominants, and therefore, all subsequent offspring must be pure dominant. Therefore, $p_{11}^{(3)} = 1$ and $p_{12}^{(3)} = p_{13}^{(3)} = 0$. To calculate the probabilities in the second and third rows of $P^{(3)}$, we use the tree diagrams found in Figure 7.1. Therefore, we have

$$p_{21}^{(3)} = \tfrac{1}{2} \cdot 1 \cdot 1 + \tfrac{1}{2} \cdot \tfrac{1}{2} \cdot 1 + \tfrac{1}{2} \cdot \tfrac{1}{2} \cdot \tfrac{1}{2} = \tfrac{7}{8}$$
$$p_{22}^{(3)} = \tfrac{1}{2} \cdot \tfrac{1}{2} \cdot \tfrac{1}{2} = \tfrac{1}{8}$$
$$p_{23}^{(3)} = 0$$

Figure 7.1

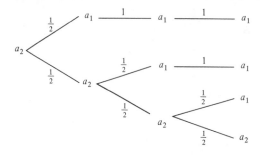

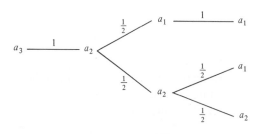

Figure 7.2

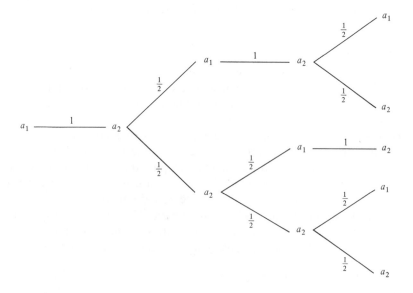

| Monday | Tuesday | Wednesday | Thursday | Friday |

and

$$p_{31}{}^{(3)} = 1 \cdot \tfrac{1}{2} \cdot 1 + 1 \cdot \tfrac{1}{2} \cdot \tfrac{1}{2} = \tfrac{3}{4}$$

$$p_{32}{}^{(3)} = 1 \cdot \tfrac{1}{2} \cdot \tfrac{1}{2} = \tfrac{1}{4}$$

$$p_{33}{}^{(3)} = 0$$

We therefore have the following three-step transition matrix:

$$P^{(3)} = \begin{pmatrix} 1 & 0 & 0 \\ \tfrac{7}{8} & \tfrac{1}{8} & 0 \\ \tfrac{3}{4} & \tfrac{1}{4} & 0 \end{pmatrix}$$

Notice that after three steps it is impossible for the process to be in state $a_3 = $ aa. This is indicated by the three zeros in column 3 of $P^{(3)}$.

Example 6   A man either drives his car or takes the train to work each day. He never takes the train two consecutive days, but if he drives one day, he is just as likely to drive again the next day. We have a Markov chain process with the two states

$a_1 = $ "takes the train"

$a_2 = $ "drives his car"

and transition matrix

$$P = \begin{pmatrix} 0 & 1 \\ \tfrac{1}{2} & \tfrac{1}{2} \end{pmatrix}$$

Suppose that the man takes the train to work on Monday. We wish to calculate the probability that he also takes the train to work on Friday of that week. Since Friday is the fourth day after Monday, we must calculate the four-step probability $p_{11}{}^{(4)}$. Using the tree diagram in Figure 7.2, we find that there are two branches leading to state a_1 on Friday, and so

$$p_{11}{}^{(4)} = 1 \cdot \tfrac{1}{2} \cdot 1 \cdot \tfrac{1}{2} + 1 \cdot \tfrac{1}{2} \cdot \tfrac{1}{2} \cdot \tfrac{1}{2} = \tfrac{3}{8}$$

Therefore, the probability that the man will take the train on Friday, given that he took the train on Monday, is $\tfrac{3}{8}$.

Example 7   Suppose that urn A contains two white balls, while urn B contains three red balls. A ball is chosen at random from each urn and placed in the other urn. This process is repeated many times, defining a Markov chain. We define the states of the process to be the number of red balls in urn A. Although there are three red balls, urn A must always contain exactly two balls, and so we have three states:

$a_1 = $ "no red balls in urn A"

$a_2 = $ "one red ball in urn A"

$a_3 = $ "two red balls in urn A"

To determine the probabilities p_{ij}, we must consider separately the three possible states for urn A:

1. If the process is in state a_1 (Figure 7.3), then a white ball must be drawn from urn A, and a red ball must be drawn from urn B. When these two balls are interchanged, state a_2 results. Therefore, $p_{11} = p_{13} = 0$ and $p_{12} = 1$.
2. If the process is in state a_2 (Figure 7.4), then, in order to move to state a_1,

Figure 7.3

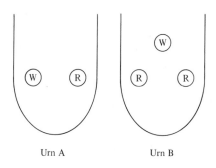

Urn A   Urn B

Figure 7.4

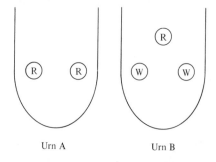

Urn A   Urn B

Figure 7.5

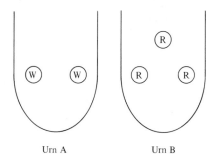

Urn A   Urn B

the red ball must be chosen from urn A and the white ball must be chosen from urn B. Therefore, by the multiplication theorem, we have $p_{21} = \frac{1}{2} \cdot \frac{1}{3} = \frac{1}{6}$. The process can move to state a_3 provided that the white ball is chosen from urn A and a red ball is chosen from urn B. Therefore, $p_{23} = \frac{1}{2} \cdot \frac{2}{3} = \frac{1}{3}$. Consequently, $p_{22} = 1 - \frac{1}{6} - \frac{1}{3} = \frac{1}{2}$. (The process remains in state a_2 if balls of the same color are drawn from the two urns. Therefore, $p_{22} = \frac{1}{2} \cdot \frac{1}{3} + \frac{1}{2} \cdot \frac{2}{3} = \frac{1}{2}$.)

3. Finally, if the process is in state a_3 (Figure 7.5), then a red ball must be drawn from urn A. If a red ball is also drawn from urn B, then the process remains in state a_3. Hence, $p_{33} = 1 \cdot \frac{1}{3} = \frac{1}{3}$. If a white ball is drawn from urn B, then the process reverts back to state a_2. Therefore, $p_{32} = 1 \cdot \frac{2}{3} = \frac{2}{3}$. The process can never move from state a_3 to state a_1 in one step, and so $p_{31} = 0$.

Therefore, the transition matrix for this Markov process is

$$P = \begin{pmatrix} 0 & 1 & 0 \\ \frac{1}{6} & \frac{1}{2} & \frac{1}{3} \\ 0 & \frac{2}{3} & \frac{1}{3} \end{pmatrix}$$

Example 8 The following generalization of Example 7 is used by physicists as a model for the diffusion of gases. Suppose that urn A contains n black balls and urn B contains n white balls. The process is the same as above: Balls are drawn at random from the two urns and interchanged. We define $n + 1$ states indicating the number of black balls in urn A; that is, state

$a_j = "j - 1$ black balls in urn A"

for $j = 1, 2, \ldots, n - 1$. Suppose that the process is presently in state a_j (Figure 7.6). In one step, the process can only move to one of the states a_{j-1}, a_j, a_{j+1}, as follows:

1. The process can move to state a_{j-1} if a black ball is drawn from urn A and a white ball is drawn from urn B.
2. The process will remain in state a_j if balls of the same color are drawn from the two urns.

Figure 7.6

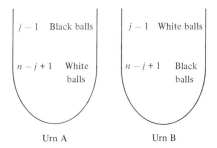

Urn A

Urn B

3. The process can move to state a_{j+1} if a white ball is drawn from urn A and a black ball is drawn from urn B.

Therefore, we have the following transition probabilities:

$$p_{j,j-1} = \frac{j-1}{n} \cdot \frac{j-1}{n} = \left(\frac{j-1}{n}\right)^2$$

$$p_{j,j} = \frac{j-1}{n} \cdot \frac{n-j+1}{n} + \frac{n-j+1}{n} \cdot \frac{j-1}{n}$$

$$p_{j,j+1} = \frac{n-j+1}{n} \cdot \frac{n-j+1}{n} = \left(\frac{n-j+1}{n}\right)^2$$

This gives

$$\begin{cases} p_{j,j-1} = \left(\dfrac{j-1}{n}\right)^2 \\[2mm] p_{j,j} = \dfrac{2(j-1)(n-j+1)}{n^2} \\[2mm] p_{j,j+1} = \left(\dfrac{n-j+1}{n}\right)^2 \\[2mm] p_{j,k} = 0 \qquad \text{if } |j-k| > 1 \end{cases}$$

A physicist would be interested in the composition of the urns after a certain number of exchanges have taken place. It might be expected that the effect of the initial distribution would wear off after a large number of exchanges.

EXERCISES

1. Five points are marked on a circle. A process moves clockwise with probability $\frac{2}{3}$ and counterclockwise with probability $\frac{1}{3}$. Find the transition matrix of this Markov process. If the process starts in state a_3 (any state, for that matter), what is the probability that it returns there after two steps?

2. Suppose that today's weather depends only on yesterday's weather. If it rained yesterday, the chances of rain today are 70 percent. If it did not rain yesterday, the chances of rain today are 40 percent. What is the transition matrix for this Markov process? If it rains today, what is the probability that it will rain four days from today?

3. Suppose that the probability that coin 1 comes up heads is 0.70, while the probability that coin 2 comes up heads is 0.60. If today's coin comes up heads, we flip coin 1 tomorrow, but if today's coin comes up tails, we flip coin 2 tomorrow. Assuming that the initial choice of coins is random, find the probability that the coin flipped on the third day is coin 1.

4. An urn contains two white balls. At each stage of the experiment, a ball is drawn at random from the urn. If the ball is white, a fair coin is tossed, and if it comes up heads, the ball is painted red, but if it comes up tails, the ball is painted blue. If the ball drawn from the urn is either red or blue, it is painted the opposite color. We define three states to indicate the number of white balls in the urn. Find the transition

matrix for this process. What is the probability that the process will be in state $a_1 = 0$ after four steps?

5. In Example 1, suppose that the five states a_1, a_2, a_3, a_4, and a_5 are defined in the order listed in the text. If the transition matrix for this process is

$$\begin{pmatrix} 0.70 & 0.10 & 0.10 & 0.05 & 0.05 \\ 0.15 & 0.60 & 0.05 & 0.15 & 0.05 \\ 0.25 & 0.05 & 0.60 & 0.00 & 0.10 \\ 0.05 & 0.20 & 0.00 & 0.70 & 0.05 \\ 0.20 & 0.20 & 0.10 & 0.10 & 0.40 \end{pmatrix}$$

calculate the percentage of independent voters four years from now if the percentage today is ten percent.

6. In Example 5, if the New York City commuter buys the Daily News on Monday, what is the probability that he also buys the Daily News on Friday?

7. Suppose that a voter is allowed to change his party affiliation for the purpose of voting in a primary if he abstains from voting for one primary. Statistics indicate that a Republican will do this 15 percent of the time, while a Democrat will do this 25 percent of the time. A voter who has abstained for one primary stands a 50-50 chance of registering for either major party. Each voter therefore can be classified into one of three categories at primary time: Democrat, Republican, or abstaining. What is the transition matrix for this process? If a voter is a registered Democrat this year, what is the probability he will be a registered Democrat in five years? What is the probability he will be a registered Republican in five years?

7.2 Random Walks

In this section, we will study a particular type of Markov chain called a random walk, and we will consider one of its interesting applications to the theory of gambling.

We will be working with the following general situation. We assume that a person is standing at an integral point on a life-size version of the x-axis, somewhere between the origin and the point $x = N$, where $N > 0$. Each minute, he takes a step of length 1, either to his right (towards N) or to his left (towards the origin). We assume that at each stage of this process, he will take a step to the right with probability p or a step to the left with probability $q = 1 - p$.

The Markov chain process just described is known as a *random walk*. There are $N + 1$ different states, corresponding to the $N + 1$ different points at which the person might be standing. Hence, we have state

$$a_k = \text{"person is standing at } x = k - 1\text{"}$$

for $k = 1, 2, 3, \ldots, N + 1$.

We have not yet discussed what happens when the person finds himself at either the origin or the point $x = N$. In the first case, he can no longer take a step to his left, while in the latter case, he cannot take a step to his right. Two possibilities seem to be of particular interest and significance:

1. If the person is at the origin, he will always take a step to his right to the point $x = 1$, while if the person is at the point $x = N$, he will always take a step to his left to the point $x = N - 1$. Random walks of this type are called *random walks with reflecting barriers*, and have a transition matrix of the form

$$P = \begin{pmatrix} 0 & 1 & 0 & 0 & \cdots & 0 & 0 & 0 \\ q & 0 & p & 0 & \cdots & 0 & 0 & 0 \\ 0 & q & 0 & p & \cdots & 0 & 0 & 0 \\ \vdots & & & & & & & \vdots \\ 0 & 0 & 0 & 0 & \cdots & q & 0 & p \\ 0 & 0 & 0 & 0 & \cdots & 0 & 1 & 0 \end{pmatrix}$$

Example 1 In a psychological experiment, a rat is placed in a narrow "room" which has several doors along one wall (Figure 7.7). Whenever the rat touches one of these doors, it is given an electrical shock, which causes the frightened animal to rush to an adjacent door in an attempt to escape from the "room." Assuming that the shocked rat has no memory, the process just described is a random walk with reflecting barriers, with $p = q = \frac{1}{2}$.

2. If, on the other hand, the person always remains at the origin or the point $x = N$ once he arrives there, we have what is called a *random walk with absorbing barriers*. In this case, the transition matrix is

$$P = \begin{pmatrix} 1 & 0 & 0 & 0 & \cdots & 0 & 0 & 0 \\ q & 0 & p & 0 & \cdots & 0 & 0 & 0 \\ 0 & q & 0 & p & \cdots & 0 & 0 & 0 \\ \vdots & & & & & & & \vdots \\ 0 & 0 & 0 & 0 & \cdots & q & 0 & p \\ 0 & 0 & 0 & 0 & \cdots & 0 & 0 & 1 \end{pmatrix}$$

Figure 7.7

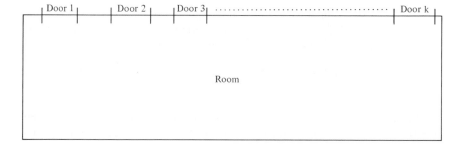

Example 2   Player A, possessing $A, plays a game in which he has probability p of winning against player B, who has $B. The game is over when either player A or player B has all of his opponent's money. We have just described a random walk with absorbing barriers, where $N = A + B$.

For the remainder of this section, we shall discuss a random walk with absorbing barriers known as *gambler's ruin*.

Suppose that you are gambling against a professional gambler (such as the "house" in Las Vegas), playing a game for which you have probability p of winning. Of course, the nature of the game is such that $p < \frac{1}{2}$, although the value of p is probably close enough to $\frac{1}{2}$ to make the game enticing. An example of such a game is roulette, where $p = \frac{18}{38}$ for reds.

At the start of the game, you have $A and the gambler has $B. Betting $1 a game, you play until one of you is ruined (has no money left). We wish to calculate the probability that it is you who will be ruined, and we will derive a formula for this probability in terms of the parameters p, A, and B upon which this probability would seem to depend.

We set up the problem as a random walk. Let $N = A + B$, the total amount of money in the game. We define $N + 1$ states as follows:

$a_k = $ "you possess $k - 1$ dollars"

for $k = 1, 2, 3, \ldots, N + 1$. The random walk moves right (when you win a game) with probability p, and moves left (when you lose a game) with probability $q = 1 - p$. If the random walk reaches either state a_1 or state a_{N+1}, then either you or the gambler has been ruined and can no longer play. Hence, we have a random walk with absorbing barriers. We are interested in the probability that the random walk reaches the origin—that is, that you are ruined.

If we let R_k denote the probability that you will be ruined, given that you now have $k, then by the partition theorem, we have

$$R_k = p \cdot R_{k+1} + q \cdot R_{k-1}$$

since the right-hand side of this equality can be interpreted as

({probability of winning the first game}

   $\times$ {probability of ruin, starting with $k + 1$ dollars})

   $+$ ({probability of losing the first game}

   $\times$ {probability of ruin, starting with $k - 1$ dollars})

Since $p + q = 1$, we have

$$(p + q) \cdot R_k = p \cdot R_{k+1} + q \cdot R_{k-1}$$

or equivalently,

$$p \cdot (R_k - R_{k+1}) = q \cdot (R_{k-1} - R_k)$$

Therefore,

$$R_{k-1} - R_k = r \cdot (R_k - R_{k+1})$$

where $r = p/q < 1$ since $p < \frac{1}{2}$.

This leads us to the set of equations

$$(*) \quad \begin{cases} R_0 - R_1 = r \cdot (R_1 - R_2) \\ R_1 - R_2 = r \cdot (R_2 - R_3) \\ R_2 - R_3 = r \cdot (R_3 - R_4) \\ \vdots \\ R_{N-2} - R_{N-1} = r(R_{N-1} - R_N) \end{cases}$$

However, $R_0 = 1$ (if you start with no money, you already are ruined!) and $R_N = 0$ (if you start with all the money, you can't lose!). From the last equation of $(*)$, we find that

$$R_{N-2} - R_{N-1} = r \cdot R_{N-1}$$

which may then be substituted in the preceding equation, and so forth, to give the following simplified set of equations (we have reversed the order of equations $(*)$ and added the first equation, an obvious identity):

$$(**) \quad \begin{cases} R_{N-1} = 1 \cdot R_{N-1} \\ R_{N-2} - R_{N-1} = r \cdot R_{N-1} \\ R_{N-3} - R_{N-2} = r^2 \cdot R_{N-1} \\ \vdots \\ R_2 - R_3 = r^{N-3} \cdot R_{N-1} \\ R_1 - R_2 = r^{N-2} \cdot R_{N-1} \\ R_0 - R_1 = r^{N-1} \cdot R_{N-1} \end{cases}$$

Adding both sides of the equations $(**)$, we find that

$$R_0 = (1 + r + r^2 + \cdots + r^{N-1}) \cdot R_{N-1}$$

and using the identity

$$1 - r^N = (1 - r)(1 + r + r^2 + \cdots + r^{N-1})$$

we solve for R_{N-1} and find

$$R_{N-1} = \frac{1 - r}{1 - r^N}$$

From $(**)$ we see that $R_{N-2} = (1 + r) \cdot R_{N-1}$, and so we obtain

$$R_{N-2} = \frac{1 - r^2}{1 - r^N}$$

Likewise, we obtain from (**) that

$$R_{N-3} = R_{N-2} + r^2 \cdot R_{N-1}$$
$$= \frac{1 - r^2}{1 - r^N} + \frac{r^2(1 - r)}{1 - r^N}$$
$$= \frac{1 - r^3}{1 - r^N}$$

Inductively, it can be shown that

$$R_{N-j} = \frac{1 - r^j}{1 - r^N}$$

or equivalently, that

$$R_k = \frac{1 - r^{n-k}}{1 - r^N}$$

Hence, if you start with A dollars against the gambler's B dollars, the probability of your ruin equals

$$R_A = \frac{1 - r^B}{1 - r^N}$$

Since $r < 1$, we have $1 - r^N < 1$, and so $R_A > 1 - r^B$.

This is a rather remarkable result. It says that the probability of your ruin does *not* depend on A, the amount of money you begin with, but only on your probability p of winning and the amount of money B that the gambler has. Since $r < 1$, we can make r^B as close to 0 as we please, and consequently, $1 - r^B$ as close to 1 as we please by making B sufficiently large. As a result, R_A, the probability of your ruin, is very close to 1. In other words, the more money the gambler has, the more certain it is that you will be ruined, regardless of how much money you have!

For example, if the gambler wishes to be 99.9 percent certain of ruining an opponent, he must have enough money B to guarantee that $r^B < 0.001$. If his opponent has probability $p = 0.495$ of winning (almost an even chance), then $r = 495/505$, and to guarantee that $(495/505)^B < 0.001$, we must have $B \geq 346$. So in the case $p = 0.495$, the gambler needs only $346 to be 99.9 percent sure of ruining you, even if you have one million dollars available! If $p = 0.48$, he needs only $87.

Example 3 A ball is drawn with replacement from an urn which contains 9 white and 11 red balls. If the ball is white, you win $1, but if the ball is red, you lose $1. You begin with $20 and your opponent begins with $10, and you play until one is ruined. We wish to find the probability that you are the one who is ruined. In this case, we have $A = 20$, $B = 10$, $N = A + B = 30$,

$p = 0.45$, $q = 0.55$, and $r = \frac{9}{11}$. Hence, the probability of your ruin is

$$R_{20} = \frac{1 - \left(\frac{9}{11}\right)^{10}}{1 - \left(\frac{9}{11}\right)^{30}} = 0.868$$

Example 4   Mr. Jones lives on a street that is 100 steps long. At one end of the street is his home, and at the other end is a lake. A bar is situated exactly in the middle of the block. One night, Mr. Jones staggers out of the bar and attempts to walk home. Assume that his probability of taking a step in the direction of his home is $p = 0.49$. We wish to find the probability that he will sober up in the lake. Assume that the lake is located at the origin, the bar at $x = 50$, and his home at $x = 100$

$$\leftarrow q = 0.51 \qquad\qquad p = 0.49 \rightarrow$$

Lake	Bar	Home
—+———————————+———————————+—		
0	50	100

We therefore have a random walk with $A = B = 50$, $N = 100$, $p = 0.49$, $q = 0.51$, and $r = \frac{49}{51}$. Mr. Jones' probability of ruin (ending up in the lake) is

$$R_{50} = \frac{1 - \left(\frac{49}{51}\right)^{50}}{1 - \left(\frac{49}{51}\right)^{100}} = 0.881$$

EXERCISES

1. Player A has $50 and player B has $25. They play a game involving two dice, where player A wins $1 if the sum of the dice is 6, 7, or 8, and player B wins $1 otherwise. What is the probability of player A being ruined?
2. A coin is biased so as to come up heads only 49 percent of the time. As the coin is tossed repeatedly, we keep track of the difference H − T between the number of heads that have occurred and the number of tails that have occurred. What is the probability that H − T reaches as low as −20 before it reaches as high as +20?
3. A point moves 10 degrees clockwise on the circumference of a circle with probability 0.48, or 10 degrees counterclockwise with probability 0.52. What is the probability the point makes a full circle in a clockwise direction before it makes a full circle in a counterclockwise direction?

7.3 Stability in Markov Processes

In order to get a Markov process started, we must specify how the initial state is to be chosen. This may be done by means of a chance device (such as a random number table, a flip of a coin, a roll of a die, to name three) which selects state a_j with probability $p_j^{(0)}$. We can express these initial probabilities by means of the *initial vector*

$$\mathbf{p}^{(0)} = \left[p_1^{(0)}, p_2^{(0)}, \ldots, p_k^{(0)}\right]$$

Alternatively, we may decide to start the process in a particular state, in which case one of the initial probabilities $p_j^{(0)}$ equals 1, while all the other initial probabilities equal 0.

By means of a tree diagram, we can then calculate the probability that the process will be in state a_j after n steps, for any state a_j and any number of steps n. We denote this probability by $p_j^{(n)}$, and express these n-step probabilities by means of the *n-step vector*

$$\mathbf{p}^{(n)} = [p_1^{(n)}, p_2^{(n)}, \ldots, p_k^{(n)}]$$

DEFINITION

A vector $\mathbf{p} = (p_1, p_2, \ldots, p_k)$ is called a *probability vector* if

1. $p_i \geq 0$ for all $i = 1, 2, \ldots, k$
2. $p_1 + p_2 + \cdots + p_k = 1$

Clearly, both $\mathbf{p}^{(0)}$ and each $\mathbf{p}^{(n)}$ are probability vectors, because the Markov process must be in one of the k states at each step of the process.

Example 1   Using the tree diagram in Figure 7.8, we calculate the n-step probability vectors for the Markov process described in Example 6 of Section 7.1, assuming the initial probability vector $\mathbf{p}^{(0)} = (\frac{1}{2}, \frac{1}{2})$ to be

$$\mathbf{p}^{(1)} = (\tfrac{1}{4}, \tfrac{3}{4})$$
$$\mathbf{p}^{(2)} = (\tfrac{3}{8}, \tfrac{5}{8})$$
$$\mathbf{p}^{(3)} = (\tfrac{5}{16}, \tfrac{11}{16})$$
$$\mathbf{p}^{(4)} = (\tfrac{11}{32}, \tfrac{21}{32})$$
$$\vdots$$

Figure 7.8

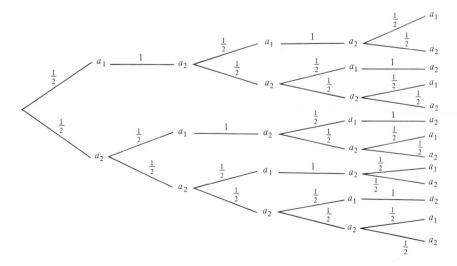

As an example of these calculations, we have

$$p_1{}^{(2)} = \tfrac{1}{2} \cdot 1 \cdot \tfrac{1}{2} + \tfrac{1}{2} \cdot \tfrac{1}{2} \cdot \tfrac{1}{2} = \tfrac{1}{4} + \tfrac{1}{8} = \tfrac{3}{8}$$
$$p_2{}^{(2)} = \tfrac{1}{2} \cdot 1 \cdot \tfrac{1}{2} + \tfrac{1}{2} \cdot \tfrac{1}{2} \cdot 1 + \tfrac{1}{2} \cdot \tfrac{1}{2} \cdot \tfrac{1}{2} = \tfrac{5}{8}$$

The process of constructing a tree diagram to calculate the coordinates of $\mathbf{p}^{(n)}$ becomes very tedious as n increases. We now derive an alternate method that will reduce the problem to one of matrix multiplication, which can be handled by means of a computer program.

Proposition 7.3

$$\mathbf{p}^{(n)} = \mathbf{p}^{(0)} \cdot P^n$$

Proof: The following set of equations are an immediate consequence of the partition theorem:

$$p_1{}^{(n)} = p_1{}^{(n-1)} \cdot p_{11} + p_2{}^{(n-1)} \cdot p_{21} + \cdots + p_k{}^{(n-1)} \cdot p_{k1}$$
$$p_2{}^{(n)} = p_1{}^{(n-1)} \cdot p_{12} + p_2{}^{(n-1)} \cdot p_{22} + \cdots + p_k{}^{(n-1)} \cdot p_{k2}$$
$$\vdots$$
$$p_k{}^{(n)} = p_1{}^{(n-1)} \cdot p_{1k} + p_2{}^{(n-1)} \cdot p_{2k} + \cdots + p_k{}^{(n-1)} \cdot p_{kk}$$

Each of these equations calculates the probability that the process is in a given state after n steps, in terms of the possible states the process might have been in after $n - 1$ steps and the proper transition probabilities. These equations can be expressed more simply by the matrix equation

$$\mathbf{p}^{(n)} = \mathbf{p}^{(n-1)} \cdot P$$

Using iteration on this result, we find that

$$\begin{aligned}
\mathbf{p}^{(n)} = \mathbf{p}^{(n-1)} \cdot P &= \left[\mathbf{p}^{(n-2)} \cdot P\right] \cdot P \\
&= \mathbf{p}^{(n-2)} \cdot P^2 \\
&= \left[\mathbf{p}^{(n-3)} \cdot P\right] \cdot P^2 \\
&= \mathbf{p}^{(n-3)} \cdot P^3 \\
&\ \ \vdots \\
&= \mathbf{p}^{(0)} \cdot P^n
\end{aligned}$$

Example 1 (continued) We calculate the same four n-step probability vectors, using Proposition 7.3 instead of the tree diagram:

$$\mathbf{p}^{(1)} = (\tfrac{1}{2}, \tfrac{1}{2}) \begin{pmatrix} 0 & 1 \\ \tfrac{1}{2} & \tfrac{1}{2} \end{pmatrix} = (\tfrac{1}{4}, \tfrac{3}{4})$$

$$\mathbf{p}^{(2)} = (\tfrac{1}{2}, \tfrac{1}{2}) \begin{pmatrix} 0 & 1 \\ \tfrac{1}{2} & \tfrac{1}{2} \end{pmatrix}^2$$

$$= (\tfrac{1}{4}, \tfrac{3}{4}) \begin{pmatrix} 0 & 1 \\ \tfrac{1}{2} & \tfrac{1}{2} \end{pmatrix} = (\tfrac{3}{8}, \tfrac{5}{8})$$

$$\mathbf{p}^{(3)} = (\tfrac{1}{2}, \tfrac{1}{2}) \begin{pmatrix} 0 & 1 \\ \tfrac{1}{2} & \tfrac{1}{2} \end{pmatrix}^3$$

$$= (\tfrac{3}{8}, \tfrac{5}{8}) \begin{pmatrix} 0 & 1 \\ \tfrac{1}{2} & \tfrac{1}{2} \end{pmatrix} = (\tfrac{5}{16}, \tfrac{11}{16})$$

$$\mathbf{p}^{(4)} = (\tfrac{1}{2}, \tfrac{1}{2}) \begin{pmatrix} 0 & 1 \\ \tfrac{1}{2} & \tfrac{1}{2} \end{pmatrix}^4$$

$$= (\tfrac{5}{16}, \tfrac{11}{16}) \begin{pmatrix} 0 & 1 \\ \tfrac{1}{2} & \tfrac{1}{2} \end{pmatrix} = (\tfrac{11}{32}, \tfrac{21}{32})$$

Proposition 7.3 can be used to derive the following result, which relates the n-step transition matrix $P^{(n)}$ to the powers of the transition matrix P.

Proposition 7.4

$$P^{(n)} = P^n$$

Proof: We first choose our initial vector $\mathbf{p}^{(0)}$ to be $(1, 0, 0, \ldots, 0)$, which is equivalent to assuming that the process starts in state a_1. We apply Proposition 7.3 to find

$$\mathbf{p}^{(0)} \cdot P^n = \mathbf{p}^{(n)}$$

Since we are assuming the process starts in state a_1, we have

$$\mathbf{p}^{(n)} = [p_{11}^{(n)}, p_{12}^{(n)}, \ldots, p_{1k}^{(n)}]$$

On the other hand,

$$\mathbf{p}^{(0)} \cdot P^n = (1, 0, 0, \ldots, 0) \cdot P^n$$

gives the first row of P^n. Therefore, the first row of the matrix P^n is identical to the first row of the matrix $P^{(n)}$. By a similar process, using the $k - 1$ probability vectors $(0, 1, 0, \ldots, 0), \ldots, (0, 0, 0, \ldots, 1)$ that indicate that the process starts in each of the $k - 1$ remaining states $a_2, a_3, \ldots, a_k$, we find that each row of P^n is identical to the corresponding row of $P^{(n)}$. Hence, these two matrices are identical.

DEFINITION

A vector $\mathbf{v}$ is called a *fixed point* for the matrix Q if $\mathbf{v} \cdot Q = \mathbf{v}$.

Example 1 (continued) In order to find a fixed point probability vector for the transition matrix

$$P = \begin{pmatrix} 0 & 1 \\ \tfrac{1}{2} & \tfrac{1}{2} \end{pmatrix}$$

we must find a vector $(p, 1 - p)$ such that

$$(p, 1 - p) \begin{pmatrix} 0 & 1 \\ \tfrac{1}{2} & \tfrac{1}{2} \end{pmatrix} = (p, 1 - p)$$

This matrix equation gives rise to the simultaneous equations

$$\tfrac{1}{2} \cdot (1 - p) = p$$
$$1 \cdot p + \tfrac{1}{2} \cdot (1 - p) = 1 - p$$

which have the unique solution $p = \tfrac{1}{3}$, $q = 1 - p = \tfrac{2}{3}$. Thus, $(\tfrac{1}{3}, \tfrac{2}{3})$ is the unique fixed point probability vector for P. Note that

$$P^2 = \begin{pmatrix} \tfrac{1}{2} & \tfrac{1}{2} \\ \tfrac{1}{4} & \tfrac{3}{4} \end{pmatrix} \qquad P^3 = \begin{pmatrix} \tfrac{1}{4} & \tfrac{3}{4} \\ \tfrac{3}{8} & \tfrac{5}{8} \end{pmatrix}$$

$$P^4 = \begin{pmatrix} \tfrac{3}{8} & \tfrac{5}{8} \\ \tfrac{5}{16} & \tfrac{11}{16} \end{pmatrix} \qquad P^5 = \begin{pmatrix} \tfrac{5}{16} & \tfrac{11}{16} \\ \tfrac{11}{32} & \tfrac{21}{32} \end{pmatrix}$$

so it would appear that the sequence of matrices $\{P^n\}$ is going to converge to the matrix

$$W = \begin{pmatrix} \tfrac{1}{3} & \tfrac{2}{3} \\ \tfrac{1}{3} & \tfrac{2}{3} \end{pmatrix}$$

both of whose rows are the fixed-point vector $(\tfrac{1}{3}, \tfrac{2}{3})$.

We note that the number of fixed points varies, depending on the transition matrix. There may be none, one, or an infinite number of fixed points.

Example 2 The transition matrix for the Mendelian Markov process described in Example 4 of Section 7.1 has a unique fixed point. For if (p, q, r) is a fixed point, then from

$$(p, q, r) \begin{pmatrix} 1 & 0 & 0 \\ \tfrac{1}{2} & \tfrac{1}{2} & 0 \\ 0 & 1 & 0 \end{pmatrix} = (p, q, r)$$

we get the two simultaneous equations

$$p + \tfrac{1}{2} \cdot q + = p$$
$$\tfrac{1}{2} \cdot q + r = q$$

together with the equation $p + q + r = 1$ that holds for all probability vectors. The first of these equations yields $q = 0$, which forces $r = 0$ in the second equation. The third equation then gives $p = 1$, and so there is the unique fixed-point vector $(1, 0, 0)$.

Example 3 The transition matrix described in Example 7 of Section 7.1 also has a unique fixed-point vector. If

$$(p, q, r) \begin{pmatrix} 0 & 1 & 0 \\ \tfrac{1}{6} & \tfrac{1}{2} & \tfrac{1}{3} \\ 0 & \tfrac{2}{3} & \tfrac{1}{3} \end{pmatrix} = (p, q, r)$$

we have the simultaneous equations

$$\left(\tfrac{1}{6}\right) \cdot q \qquad = p$$
$$p + \left(\tfrac{1}{2}\right) \cdot q + \left(\tfrac{2}{3}\right) \cdot r = q$$
$$\left(\tfrac{1}{3}\right) \cdot q + \left(\tfrac{1}{3}\right) \cdot r = r$$

which have solution $(p, 6p, 3p)$. Since this vector must be a probability vector, we have $p + 6p + 3p = 1$, which implies that $p = \tfrac{1}{10}$. We therefore have the unique fixed-point probability vector $(\tfrac{1}{10}, \tfrac{6}{10}, \tfrac{3}{10})$.

DEFINITION

A transition matrix P is said to be *regular* if some power P^m has only strictly positive entries.

Note that, according to Proposition 7.4, the matrices P^m are m-step transition matrices, so the individual entries in these matrices are probabilities. (As a matter of fact, the sum of each row in these matrices is 1.) Consequently, the entries in P^m are nonnegative to start with, and our definition of regular is adding the stipulation that none of the entries be zero.

Regular transition matrices are the transition matrices for those Markov processes in which it is possible to reach any state after a certain number of steps m, regardless of the state in which the process started. The significance of regular transition matrices is contained in the following theorem, the proof of which is beyond the scope of this book. For more details, the reader is referred to another text.*

Theorem 7.5

If P is a regular transition matrix, then

1. P has a unique fixed-point probability vector $\mathbf{w}$;
2. the components of $\mathbf{w}$ are all strictly positive;
3. the sequence of matrices $\{P^n\}$ converges to the matrix W, each of whose rows is the vector $\mathbf{w}$;
4. if $\mathbf{p}$ is any probability vector, the sequence of vectors $\{\mathbf{p}P^n\}$ converges to $\mathbf{w}$.

If, in 4, we take $\mathbf{p}$ to be the vector $\mathbf{p}^{(0)}$ of initial probabilities, then according to Proposition 7.3, the sequence of vectors $\{\mathbf{p}^{(n)}\}$ converges to $\mathbf{w}$. Therefore, no matter what the initial probabilities of the various states are, if the transition matrix P is regular, then after a sufficient number of steps, the probability that the process is in any particular state depends only on the coordinates of the vector $\mathbf{w}$. Such a process is said to be in *equilibrium*.

* J. G. Kemeny and J. L. Snell, *Finite Markov Chains* (New York, New York: Van Nostrand 1960).

Example 1 (continued) Since

$$P^2 = \begin{pmatrix} \frac{1}{2} & \frac{1}{2} \\ \frac{1}{4} & \frac{3}{4} \end{pmatrix}$$

the transition matrix P is regular. We have seen that P has a unique fixed point vector $\mathbf{w} = (\frac{1}{3}, \frac{2}{3})$ whose entries are strictly positive. We have also seen that the matrices $\{P^m\}$ appear to be converging to the matrix

$$W = \begin{pmatrix} \frac{1}{3} & \frac{2}{3} \\ \frac{1}{3} & \frac{2}{3} \end{pmatrix}$$

If we choose $\mathbf{p}^{(0)} = (\frac{5}{6}, \frac{1}{6})$ to be our initial vector (that is, we assume the probability of taking the train the first day to be $\frac{5}{6}$), we find that

$$\mathbf{p}^{(4)} = \mathbf{p}^{(0)} \cdot P^4 = (0.365, 0.635)$$

and

$$\mathbf{p}^{(5)} = \mathbf{p}^{(0)} \cdot P^5 = (0.318, 0.682)$$

which are already fairly close to the fixed point $(\frac{1}{3}, \frac{2}{3})$. Since $\mathbf{p}^{(n)}$ converges to $\mathbf{w}$ as $n \to \infty$, we see that, in the long run, the man drives to work twice as often as he takes the train, regardless of what he did the first day.

Example 2 (continued) The transition matrix P is *not* regular since the first row in any power P^m will contain 0's in the second and third columns. However, we have seen that this matrix has a unique fixed-point probability vector $(1, 0, 0)$. If we choose $(\frac{1}{3}, \frac{1}{3}, \frac{1}{3})$ to be our initial probability vector, then

$$(\tfrac{1}{3}, \tfrac{1}{3}, \tfrac{1}{3}) \cdot P^1 = (\tfrac{1}{2}, \tfrac{1}{2}, 0)$$
$$(\tfrac{1}{3}, \tfrac{1}{3}, \tfrac{1}{3}) \cdot P^2 = (\tfrac{3}{4}, \tfrac{1}{4}, 0)$$
$$(\tfrac{1}{3}, \tfrac{1}{3}, \tfrac{1}{3}) \cdot P^3 = (\tfrac{7}{8}, \tfrac{1}{8}, 0)$$
$$(\tfrac{1}{3}, \tfrac{1}{3}, \tfrac{1}{3}) \cdot P^4 = (\tfrac{15}{16}, \tfrac{1}{16}, 0)$$

which appear to be converging to $(1, 0, 0)$. Notice how state $a_3 = $ aa was eliminated in the first step, and how state $a_2 = $ Aa is being eliminated as the process moves along. The pure dominant state $a_1 = $ AA is acting as its name suggests.

Example 3 (continued) The transition matrix P is regular, since the matrix P^2 has no zero entries, and P has a unique fixed-point vector $(0.1, 0.6, 0.3)$. This Markov process began in state a_1, and so $\mathbf{p}^{(0)} = (1, 0, 0)$. Consequently,

$$\mathbf{p}^{(1)} = \mathbf{p}^{(0)} \cdot P = (0, 1, 0)$$
$$\mathbf{p}^{(2)} = \mathbf{p}^{(0)} \cdot P^2 = (\tfrac{1}{6}, \tfrac{1}{2}, \tfrac{1}{3})$$
$$\mathbf{p}^{(3)} = \mathbf{p}^{(0)} \cdot P^3 = (\tfrac{1}{12}, \tfrac{23}{36}, \tfrac{5}{18})$$
$$\vdots$$

Hence, the probabilities that there are none, one, and two red balls in urn A after 3 steps are $\frac{1}{12}$, $\frac{23}{36}$, and $\frac{5}{18}$, respectively. In the long run, according to the coordinates of the fixed-point vector $(0.1, 0.6, 0.3)$, there will be none, one, and two red balls in urn A approximately 10, 60, and 30 percent of the time, respectively. A computer simulation of 1000 steps of this Markov process found the three states occurring with the following frequencies:

State	Frequency
$a_1 = 0$	95
$a_2 = 1$	602
$a_3 = 2$	303

These frequencies compare very favorably with the probabilities given by the fixed-point vector.

We conclude this chapter with a brief mention of two important concepts similar in nature to the concept of regular.

DEFINITION

A Markov process is called *ergodic* if, given any pair of states a_i and a_j, there is an integer m such that $p_{ij}^{(m)} \neq 0$. That is, it is always possible for the Markov process to move from a given state a_i to another given state a_j in a finite number of steps m.

Obviously, if the transition matrix P of the Markov process is regular, the process is ergodic. A two-state Markov process with nonregular transition matrix

$$P = \begin{pmatrix} 0 & 1 \\ 1 & 0 \end{pmatrix}$$

is ergodic, since it is possible to move from a_1 to a_2, or from a_2 to a_1 in one step, and from a_1 to itself, or from a_2 to itself, in two steps.

DEFINITION

A state in a Markov process is called *absorbing* if, once the process enters the state, it cannot leave the state. That is, a_i is absorbing if $p_{ii}^{(1)} = 1$ and $p_{ij}^{(1)} = 0$ for any $j \neq i$.

The concept of absorbing states is directly opposite to the concepts of regular transition matrices and ergodic Markov processes. The purely dominant state $a_1 = AA$ in the Mendelian Markov process in Example 2 above is an example of an absorbing state.

EXERCISES

1. Referring to Example 5 in Section 7.1, find the unique fixed-point vector for the Markov process described. Over the long run, what percentage of the time will the commuter buy the New York Times?

2. Referring to Exercise 2 in Section 7.1, what is the long-range probability of a rainy day?
3. Referring to Exercise 3 in Section 7.1, what percentage of the time, over the long run, will coin 1 be tossed?
4. Referring to Exercise 4 in Section 7.1, is the transition matrix P regular? Does it have a unique fixed-point vector? Is this Markov process ergodic? What are the long-range probabilities of the three states?
5. Referring to Exercise 5 in Section 7.1, is this transition matrix regular? Does it have a unique fixed-point vector? What is the long-range percentage of independent voters?
6. Referring to Exercise 7 in Section 7.1, is this transition matrix regular? Does it have a unique fixed-point vector? What is the long-range percentage of registered Republican voters?
7. Referring to Example 3, carry out the computer simulation mentioned.

ADDITIONAL PROBLEMS

1. Suppose that the occupation of a son depends on the occupation of his father (and on nothing else) according to the following transition matrix:

upper class $\begin{pmatrix} 0.45 & 0.48 & 0.07 \\ 0.05 & 0.70 & 0.25 \\ 0.01 & 0.50 & 0.49 \end{pmatrix}$
middle class
lower class

What is the long-range percentage of upper-, middle-, and lower-class occupations?
2. Suppose that you have two coins, one of which is fair and the other biased so as to come up heads only 25 percent of the time. You are to play a game in which you will toss a coin n times and get $1.00 for each head. Unfortunately, you have forgotten how to distinguish between the two coins, and so decide upon the following strategy. After choosing the first coin at random, you will use it again if it came up heads; otherwise, you will next toss the other coin. Find the long-range probabilities of tossing a head and of tossing the fair coin.

8

Continuous Random
Variables

8.1 Introduction

In Chapter 4 we studied discrete random variables—those random variables, defined on arbitrary sample spaces, whose range was restricted to being at most a countably infinite set. We shall now study the one remaining case—random variables whose range is a continuum of real numbers. By a continuum, we shall generally mean an interval, either a finite interval

$$[a, b] = \{x \in \mathbb{R} : a \leq x \leq b\}$$

or an infinite interval of the form

$$[a, \infty) = \{x \in \mathbb{R} : a \leq x\}$$

or of the form

$$(-\infty, a] = \{x \in \mathbb{R} : x \leq a\}$$

Such random variables occur frequently, particularly in experiments in which measurements are taken. In order to obtain a continuum of real numbers as possible measurements, we must assume that the measurement could be taken with any degree of accuracy desired, even though the measurement probably will be rounded off to just a few decimal places.

Consider the following example: Suppose that each of 200 people measures the same object, whose length is approximately 25 inches. What would the sample space of all possible measurements look like? Usually, when measurements of this sort are taken, the answer is rounded off to a certain number of decimal places. If, in our example, the measurements are rounded off to the nearest inch, then a possible sample space might be the finite set

$$\{22, 23, 24, 25, 26, 27, 28\}$$

If the measurements are rounded off to the nearest one-hundredth of an inch, then a possible sample space might be the finite set

$$\{22.00, 22.01, 22.02, \ldots, 27.99, 28.00\}$$

On the other hand, it is conceivable, before roundoff, that any value in the interval $[22, 28]$ could be a possible value for this measurement. Hence, the interval $[22, 28]$ can be used as a sample space for the range of these measurements, with the understanding that any one of the infinite number of elements in this set is a possible outcome of the experiment.

One problem immediately arises here that is quite similar to a problem that arose in the study of nondiscrete sample spaces. Just as we were then not able to assign a probability to a simple event $\{x\}$, where $x \in S$, we now will not be able to assign probabilities to events of the form $\{X = a\}$, where $a \in X(S)$, X being a random variable whose range is a continuum of real numbers. We therefore will not be able to talk about probability functions for this type of random variable. In order to motivate what we shall use to replace this concept, we consider the following example.

A company surveys each of its 10,000 employees regarding their heights. The responses, when rounded to the nearest inch, form the histogram in Figure 8.1, in which the height of each bar is accurate to the nearest 50 people, and in which each bar is eight units wide. The area of each bar represents the probability that a person employed by the company has height as indicated below the bar. Should we wish to determine the probability that a person employed by the company has height between 70 and 72 in., say, we would simply add the areas of the bars centered over 70, 71, and 72 in., and compare to the total area of all the bars. If we consider the histogram as being a step-function, then the probability we have just calculated is represented by the area under this function (curve) between $x = 69.5$ and $x = 72.5$ in., relative to the total area under the curve.

If the results of this same survey had been rounded, instead, to the nearest half-inch, the histogram in Figure 8.2 would have resulted. Again, the heights are accurate to the nearest 50 people, and in this case, each bar is four units

Figure 8.1

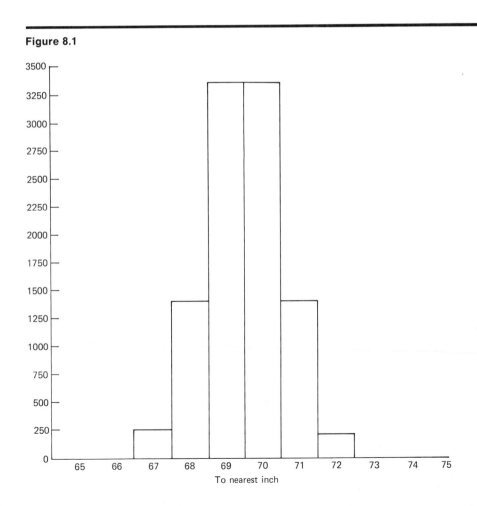

To nearest inch

Figure 8.2

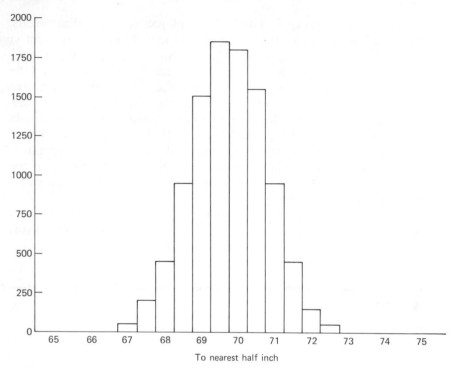

To nearest half inch

Figure 8.3

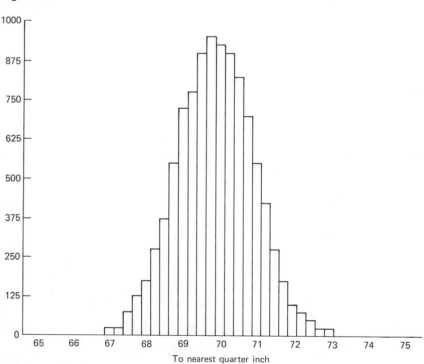

To nearest quarter inch

wide. The area of a bar again represents the probability that an employee of the company has the height indicated below the bar. To calculate the probability that an employee of the company has height between 70 and 72 in., we would now add the area of the bars centered over 70, 70.5, 71, 71.5, and 72 in., which is the area under the histogram-defining step function between $x =$ 69.75 and $x = 72.25$ in.

The results of the same survey, rounded to the nearest quarter inch and eighth inch, can be seen in Figure 8.3 and 8.4, respectively. In Figure 8.3, the heights of the bars are accurate to the nearest 25 people, and each bar is two units wide. In Figure 8.4, the heights are accurate to the nearest 10 people, and each bar is one unit wide. Notice that as the round-off error (in the heights) is reduced, the number of bars in the histogram is increased, and the step function looks more and more like a smooth curve. If the actual heights in inches of the employees of the company can be any value in the interval [65, 75] (assuming accuracy of measurement to any degree desired), then the

Figure 8.4

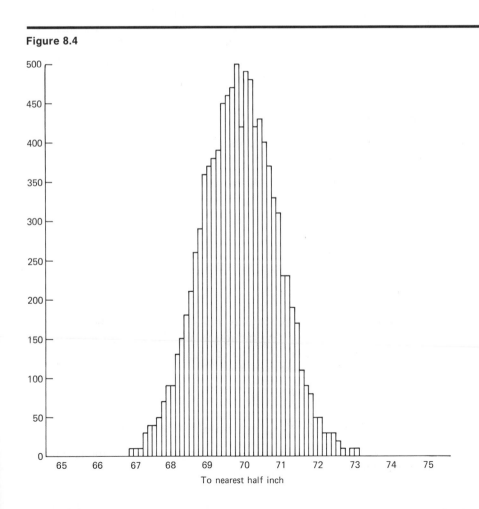

To nearest half inch

distribution of the heights can no longer be represented by a histogram with a finite number of bars. The distribution can, however, be represented by the area under a smooth curve, and this is the basis for the following definition:

DEFINITION

If X is a random variable defined on a nondiscrete sample space S such that $X(S)$ is a continuum of real numbers, and if there exists a piecewise continuous function $f : \mathbb{R} \to \mathbb{R}$ such that

$$P(a \leq X \leq b) = \int_a^b f(t)\, dt$$

for any choice of real numbers a, b, then X is called a *continuous random variable*, and the function f is called the *density function* of X.

Note that a function is piecewise continuous if it is continuous except for a finite number of jumps.

Therefore, the probability that a continuous random variable assumes a value between the limits a and b can be calculated by means of an integral— the area under the density function of the random variable between the given limits, just as was suggested in the example above. The density function therefore is similar to the step function that forms the top of the histogram.

Proposition 8.1

The density function f of a continuous random variable X must satisfy the following two properties:

(a) $f(x) \geq 0$ "almost everywhere"

(b) $\int_{-\infty}^{\infty} f(t)\, dx = 1$

Proof:

(a) If there is some interval $[a, b]$ such that $f(x) < 0$ for all $x \in [a, b]$, then

$$\int_a^b f(t)\, dt < 0$$

But this integral is the probability $P(a \leq X \leq b)$, and therefore must be at least 0. Therefore, $f(x)$ cannot be negative over an interval, and so can be negative only for isolated values of x, which is essentially what we mean by "almost everywhere."

(b) $P(-\infty < X < \infty) = \int_{-\infty}^{\infty} f(t)\, dt$

and since the left-hand side of this expression is 1, (b) follows.

Extending the definition of mean as it was stated for countable random variables, we have

DEFINITION

If X is a continuous random variable with density function f, then we define

$$E(X) = \int_{-\infty}^{\infty} t \cdot f(t)\, dt$$

whenever the integral $\int_{-\infty}^{\infty} |t| \cdot f(t)\, dt$ exists.

The requirement that the latter integral exist is equivalent to the requirement in the discrete case that the infinite series defining the mean be absolutely convergent.

The following result, which we state without proof, is the extension of Proposition 4.3 to the continuous case. The proof is similar to that of Proposition 4.3, although somewhat more complicated.

Proposition 8.2

If X is a continuous random variable, if $g : \mathbb{R} \to \mathbb{R}$ is a function, and if $Y = g(X)$, then

$$E(Y) = \int_{-\infty}^{\infty} g(t) \cdot f(t)\, dt$$

whenever this integral is absolutely convergent. (That is, whenever

$$\int_{-\infty}^{\infty} |g(t) \cdot f(t)|\, dt$$

exists.)

Since the variance of a random variable X is defined to equal $E[(X - \mu_X)^2]$, it follows from Proposition 8.2 that

$$\mathrm{Var}(X) = \int_{-\infty}^{\infty} (t - \mu_X)^2 \cdot f(t)\, dt$$

whenever this integral exists. As in the discrete case, the existence of $E(X)$ and $E(X^2)$ guarantee the existence of $\mathrm{Var}(X)$ and the validity of the formula

$$\mathrm{Var}(X) = E(X^2) - E(X)^2$$

Chebyshev's inequality, and the weak law of large numbers. All proofs are analogous to those already presented in the discrete case.

The concept of moments and moment-generating functions also carry over very easily to the continuous case.

DEFINITION

If X is a continuous random variable with density function f, then the moments μ_k' of X *about the origin* are defined by

$$\mu_k' = \int_{-\infty}^{\infty} t^k \cdot f(t)\, dt$$

and the *moments* μ_k of X *about its mean* μ_X are defined by

$$\mu_k = \int_{-\infty}^{\infty} (t - \mu_X)^k \cdot f(t)\, dt$$

whenever these integrals exist. The *moment-generating function* $M_X(t)$ of X is defined by

$$M_X(t) = E(e^{tX}) = \int_{-\infty}^{\infty} e^{tx} \cdot f(x)\, dx$$

whenever this integral exists.

As in the discrete case, it is easy to verify that

$$M_X(t) = \sum_{k=0}^{\infty} \mu_k' \cdot t^k/k!$$

and that the results contained in Proposition 4.17 hold.

It therefore would seem that the density function is quite important, occurring, as it does, in all of these formulas, and one might wonder how it is determined. On many occasions, the density function will be evident from the description of the problem, while on other occasions, a density function that approximates the given data will become evident after expressing the data in the form of a histogram.

Still another method for determining the density function is by means of the (cumulative) distribution function

$$F(x) = P(X \le x) = \int_{-\infty}^{x} f(t)\, dt$$

Notice that although the probability function does *not* exist for continuous random variables, the (cumulative) distribution function does exist. This is because the latter is defined in terms of the interval $[-\infty, x]$, while the former is defined in terms of individual points.

We have seen in Proposition 4.2 that the (cumulative) distribution function of a finite random variable is a nondecreasing step function. In the continuous case, it is no longer possible that $F(x)$ be a step function, but it will still be nondecreasing, and, like the density function, $F(x)$ will be piecewise continuous. Furthermore, one of the basic results of elementary calculus tells us that

$$F'(x) = f(x)$$

which gives a simple method for determining $f(x)$ provided that $F(x)$ is known. On many occasions, it will be much easier to calculate $F(x)$, and then $f(x)$, rather than calculate $f(x)$ directly.

Example 1 Let X be a continuous random variable with density function

$$f(x) = \begin{cases} x/2 & \text{if } 0 \le x \le 2 \\ 0 & \text{otherwise} \end{cases}$$

The graph of $f(x)$, in Figure 8.5, suggests that X takes values only in the interval $[0, 2]$ and that the right-hand side of this interval is favored. This is verified by the following calculations:

$$P(0 \le X \le 2) = \int_0^2 f(x)\,dx = \int_0^2 \left(\frac{x}{2}\right) dx = \left.\frac{x^2}{4}\right|_0^2 = 1$$

and

$$P(1 \le X \le 2) = \int_1^2 f(x)\,dx = \int_1^2 \left(\frac{x}{2}\right) dx = \left.\frac{x^2}{4}\right|_1^2 = \frac{3}{4}$$

Because of the nature of this distribution, one would expect the mean to lie somewhere between 1 and 2. In fact,

$$E(X) = \int_{-\infty}^{\infty} x \cdot f(x)\,dx = \int_0^2 x\left(\frac{x}{2}\right) dx = \int_0^2 \left(\frac{x^2}{2}\right) dx = \left.\frac{x^3}{6}\right|_0^2 = \frac{4}{3}$$

Likewise,

$$E(X^2) = \int_{-\infty}^{\infty} x^2 \cdot f(x)\,dx = \int_0^2 \left(\frac{x^3}{2}\right) dx = \left.\frac{x^4}{8}\right|_0^2 = 2$$

so that $\mathrm{Var}(X) = E(X^2) - E(X)^2 = 2 - (\frac{4}{3})^2 = \frac{2}{9}$. The (cumulative) distribution function for X is

$$F(x) = \int_{-\infty}^x f(t)\,dt = \begin{cases} 0 & \text{if } x \le 0 \\ \int_0^x f(t)\,dt & \text{if } 0 \le x \le 2 \\ 1 & \text{if } x \ge 2 \end{cases}$$

and since

$$\int_0^x f(t)\,dt = \int_0^x \left(\frac{t}{2}\right) dt = \left.\frac{t^2}{4}\right|_0^x = \frac{x^2}{4}$$

Figure 8.5

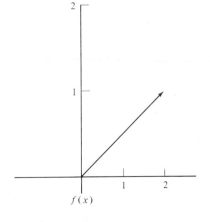

$f(x)$

we have

$$F(x) = \begin{cases} 0 & \text{if } x \leq 0 \\ x^2/4 & \text{if } 0 \leq x \leq 2 \\ 1 & \text{if } x \geq 2 \end{cases}$$

Example 2 Suppose that the life X in hours of an electronic tube is a continuous random variable with density function

$$f(t) = \begin{cases} e^{-t/1000}/1000 & \text{if } t \geq 0 \\ 0 & \text{otherwise} \end{cases}$$

Since

$$\int_{-\infty}^{\infty} f(t)\, dt = \int_{0}^{\infty} \frac{e^{-t/1000}}{1000}\, dt$$

$$= -e^{-t/1000} \Big|_{0}^{\infty}$$

$$= 1$$

$f(t)$ is a possible density function for a continuous random variable. We calculate the average lifespan for these electronic tubes as follows:

$$E(X) = \int_{0}^{\infty} \frac{t e^{-t/1000}}{1000}\, dt$$

$$= -t e^{-t/1000} \Big|_{0}^{\infty} + \int_{0}^{\infty} e^{-t/1000}\, dt \qquad \text{(integration by parts)}$$

$$= \qquad 0 \qquad + \qquad 1000$$

$$= 1000 \text{ hours}$$

Example 3 Suppose that X denotes the square of a value chosen at random in the unit interval $[0, 1]$. X is a continuous random variable, since it can take any value in the interval $[0, 1]$. Since

$$\{X \leq a\} = \{x \in [0, 1] : x^2 \leq a\} = \{x \in [0, 1] : x \leq \sqrt{a}\}$$

we find that $P(X) \leq a = \sqrt{a}$, the quotient of the length of the subinterval $[0, \sqrt{a}] = \{X \leq a\}$ by the length of the sample interval $[0, 1]$. We have thus identified the (cumulative) distribution function of X (Figure 8.6) to be

$$F(x) = \begin{cases} 0 & \text{if } x \leq 0 \\ \sqrt{x} & \text{if } 0 \leq x \leq 1 \\ 1 & \text{if } x \geq 1 \end{cases}$$

Differentiating the distribution function, we find the density function f of X to be

$$f(x) = \begin{cases} 1/2\sqrt{x} & \text{if } 0 \leq x \leq 1 \\ 0 & \text{otherwise} \end{cases}$$

We can then evaluate

$$E(X) = \int_{-\infty}^{\infty} t \cdot f(t)\, dt = \int_0^1 \left(\frac{t}{2\sqrt{t}} \right) dt$$

$$= \frac{1}{2} \int_0^1 \sqrt{t}\, dt$$

$$= \frac{1}{2} \left(\frac{t^{3/2}}{\frac{3}{2}} \right) \Big|_0^1$$

$$= \frac{t^{3/2}}{3} \Big|_0^1 = \frac{1}{3}$$

So the average value of the square of a number in the interval $[0, 1]$ is $\frac{1}{3}$.

Example 4 A point is chosen at random within a circle of radius R, and we let X denote the distance of the point from the center of the circle. We have seen in Example 1 of Section 2.5 that for any $0 \leq a \leq R$

$$P(X \leq a) = \frac{\pi a^2}{\pi R^2} = \left(\frac{a}{R} \right)^2$$

Therefore, the (cumulative) distribution function of X is

$$F(x) = \begin{cases} 0 & \text{if } x \leq 0 \\ (x/R)^2 & \text{if } 0 \leq x \leq R \\ 1 & \text{if } x \geq R \end{cases}$$

$F(R/2)$ is the probability that a point chosen at random in the circle lies closer to the center than to the circumference. Since

$$F\left(\frac{R}{2} \right) = \left(\frac{R/2}{R} \right)^2 = \frac{1}{4}$$

Figure 8.6

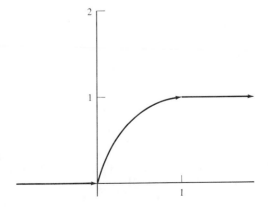

we are in agreement with the result obtained in Section 2.5. Differentiating $F(x)$, we obtain

$$f(x) = \begin{cases} 2x/R^2 & \text{if } 0 \le x \le R \\ 0 & \text{otherwise} \end{cases}$$

the density function of X. [Actually, at $x = R$, $F(x)$ is not differentiable, so to overcome this problem, we simply define $f(R) = 0$.] The probability

$$P(a \le X \le b) = \int_a^b f(t)\, dt = \int_a^b \left(\frac{2t}{R^2}\right) dt$$

$$= \frac{1}{R^2} \int_a^b 2t\, dt$$

$$= \frac{t^2}{R^2} \Big|_a^b$$

$$= \frac{b^2 - a^2}{R^2}$$

provided that $0 \le a < b \le R$. Finally, the average distance from the center of the circle is

$$E(X) = \int_{-\infty}^{\infty} t \left(\frac{2t}{R^2}\right) dt = \frac{2}{R^2} \int_0^R t^2\, dt$$

$$= \left(\frac{2}{R^2}\right)\left(\frac{t^3}{3}\right)\Big|_0^R$$

$$= \frac{2R}{3}$$

Let us now suppose that we have a random variable Y which is a function of another random variable X, say $Y = u(X)$. We have already seen that it is a fairly simple process to determine the distribution of Y from the distribution of X in the case that both are discrete. But how are we to determine the density function of Y from the density function of X, should both random variables be continuous? The following example will give some idea of how this is to be done.

Example 5 Suppose that X is a continuous random variable with density function

$$f(x) = \begin{cases} 4x^3 & \text{if } 0 \le x \le 1 \\ 0 & \text{otherwise} \end{cases}$$

Suppose further that $Y = u(X)$, where u is the function $u(x) = 3x + 5$. We wish to determine the density function g of Y, and we do so by means of the (cumulative) distribution function G of Y. Now

$$G(y_0) = P(Y \leq y_0) = P(3X + 5 \leq y_0)$$

$$= P\left(X \leq \frac{y_0 - 5}{3}\right)$$

$$= \int_{-\infty}^{(y_0 - 5)/3} 4t^3 \, dt$$

$$= \left(\frac{y_0 - 5}{3}\right)^4$$

provided that $0 \leq (y_0 - 5)/3 \leq 1$, or equivalently, provided that $5 \leq y_0 \leq 8$. Therefore,

$$G(y_0) = \begin{cases} 0 & \text{if } y_0 \leq 5 \\ [(y_0 - 5)/3]^4 & \text{if } 5 \leq y_0 \leq 8 \\ 1 & \text{if } y_0 \geq 8 \end{cases}$$

Differentiating this distribution function, we obtain the density function

$$g(y_0) = \begin{cases} \frac{4}{3}[(y_0 - 5)/3]^3 & \text{if } 5 \leq y_0 \leq 8 \\ 0 & \text{otherwise} \end{cases}$$

The following Proposition gives us a quick method for determining the density function of Y, provided that Y is a certain special type of function of X:

Proposition 8.3

Suppose that X and Y are continuous random variables with density functions f and g, and distribution functions F and G, respectively. If $Y = u(X)$, where u is a differentiable function that is either increasing or decreasing over the range of X, then

$$g(y) = f(x) \cdot \left|\frac{dx}{dy}\right|$$

where $dy/dx \neq 0$.

Proof: We shall assume that u is an increasing function, and leave the other case, which is quite similar, as an exercise. By definition, we have

$$F(a) = P(X \leq a) = \int_{-\infty}^{a} f(x) \, dx$$

and, since u is an increasing function,

$$P(X \leq a) = P[Y \leq u(a)]$$

and therefore,

$$G(u(a)) = \int_{-\infty}^{a} f(x) \, dx$$

If we let v be the inverse function of u (which exists since u is increasing, and is differentiable since u is differentiable), then under the change of variable $y = u(x)$, we have $x = v(y)$ and $dx = v'(y)\,dy$. Consequently,

$$G(u(a)) = P[Y \le u(a)] = \int_{-\infty}^{u(a)} f(v(y)) \cdot v'(y)\,dy$$

for any value $y = u(a)$ within the range of Y. As a result, we have found the density function of Y to be

$$g(y) = f(v(y)) \cdot v'(y)$$

which, when rewritten in terms of x, gives

$$g(y) = f(x) \cdot \frac{dx}{dy}$$

(Note that the absolute values will be needed only in the case where u is a decreasing function.)

Example 6   If $u(x) = 2x$ (that is, $Y = 2X$), then $dy = 2\,dx$, so that $dx/dy = \frac{1}{2}$. Consequently, the density function of Y is simply

$$g(y) = f(x) \cdot |\tfrac{1}{2}| = \frac{f(x)}{2}$$

where $f(x)$ is the density function of X.

Example 5 (*continued*)   The function $y = 3x + 5$ is certainly a differentiable and increasing function for all values of x. Since $y = 3x + 5$ is equivalent to $x = (y - 5)/3$, we have $dx/dy = \frac{1}{3}$, and so

$$g(y) = f(x) \cdot |\tfrac{1}{3}| = \tfrac{1}{3} \cdot f\left(\frac{y-5}{3}\right)$$

giving

$$g(y) = \begin{cases} \frac{4}{3}[(y-5)/3]^3 & \text{if } 0 \le (y-5)/3 \le 1 \\ 0 & \text{otherwise} \end{cases}$$

Example 7   Suppose that X is a continuous random variable with density function

$$f(x) = \begin{cases} 3x^2 & \text{if } 0 \le x \le 1 \\ 0 & \text{otherwise} \end{cases}$$

and suppose that $Y = X^3$. We wish to find the density function g of Y. Since the function $y = x^3$ is differentiable and increasing for all values of x, and since $x = \sqrt[3]{y}$, we have

$$\frac{dx}{dy} = \tfrac{1}{3} \cdot y^{-2/3} = \tfrac{1}{3} \cdot (\sqrt[3]{y})^2$$

Consequently,

$$g(y) = f(x) \cdot \left| \tfrac{1}{3} (\sqrt[3]{y})^2 \right|$$
$$= f(\sqrt[3]{y}) \left[\tfrac{1}{3} (\sqrt[3]{y})^2 \right]$$
$$= \frac{3(\sqrt[3]{y})^2}{3(\sqrt[3]{y})^2} = 1$$

Therefore,

$$g(y) = \begin{cases} 1 & \text{if } 0 \le y \le 1 \\ 0 & \text{otherwise} \end{cases}$$

EXERCISES

1. Under the assumption that $E(X)$ and $E(X^2)$ exist, verify that $\text{Var}(X)$ exists, that $\text{Var}(X) = E(X^2) - E(X)^2$, and that Chebyshev's inequality and the law of large numbers hold for continuous random variables.
2. Prove that Proposition 4.17 is true for continuous random variables and that $M_X(t) = \sum_{k=0}^{\infty} \mu_k' \cdot t^k / k!$ is true for continuous random variables.
3. Verify Proposition 8.3 in the case that u is a decreasing function.
4. Suppose that X is a continuous random variable. Prove
   (a) $E(aX + b) = aE(X) + b$
   (b) $\text{Var}(aX + b) = a^2 \cdot \text{Var}(X)$
5. If X is a continuous random variable with density function $f(x) = (1/\pi)[1/(1 + x^2)]$, find $E(X)$.
6. A point r is chosen at random in the interval $(0, 1)$, and a circle having radius r and center at the origin is formed. Find the probability that the area of this circle is less than 1.
7. Suppose that the life span (in hours) of a light bulb is a continuous random variable X with density function

   $$f(x) = \begin{cases} A/x^3 & \text{if } 1500 \le x \le 2500 \\ 0 & \text{otherwise} \end{cases}$$

   What value must A take so that $f(x)$ will, in fact, be a density function? Find $P(X > 2000)$, and calculate $E(X)$.
8. Suppose that the number of hours X between malfunctions of an electronic computer part is a continuous random variable with density function

   $$f(x) = \begin{cases} e^{-x} & \text{if } x > 0 \\ 0 & \text{otherwise} \end{cases}$$

   where x is in units of hundreds of hours. Find $P(X < 9)$ and calculate $E(X)$.

8.2 Joint Distributions of Continuous Random Variables

In this section, we will extend many of the concepts and results of Chapter 5 to continuous random variables. This section will be almost entirely theoretical, but the results obtained will find numerous applications in later sections.

We will assume throughout this section that X and Y are continuous random variables with density functions f_1 and f_2, and (cumulative) distribution functions F_1 and F_2, respectively.

As was the case with a single continuous random variable, we are *not* able to define a *joint* continuous probability function

$$f(x_0, y_0) = P(X = x_0 \cap Y = y_0)$$

However, we proceed as we did in the previous section, and define a joint density function:

DEFINITION

A piecewise continuous function $f : \mathbb{R} \times \mathbb{R} \to \mathbb{R}$ is called a *joint probability density function* of the random variables X and Y if

$$P(a \leq X \leq b \cap c \leq Y \leq d) = \int_c^d \int_a^b f(x, y)\, dx\, dy$$

The following are simple consequences of the fact that the double integral above represents a probability (see Proposition 8.1):

Proposition 8.4

(a) $f(x, y) \geq 0$ for almost all pairs (x, y)

(b) $\int_{-\infty}^{\infty} \int_{-\infty}^{\infty} f(x, y)\, dx\, dy = 1$

Example 1 We wish to determine the value of the constant k so that the following function of two variables might qualify as a possible joint density of two continuous random variables:

$$f(x, y) = \begin{cases} kxy & \text{if } 0 \leq x, y \leq 1 \\ 0 & \text{otherwise} \end{cases}$$

This function is continuous and nonnegative for all values (x, y), provided that $k \geq 0$, and so we need only use (b) of Proposition 8.4 to determine the value of k:

$$\int_{-\infty}^{\infty} \int_{-\infty}^{\infty} f(x, y)\, dx\, dy = \int_0^1 \int_0^1 kxy\, dx\, dy$$

$$= \int_0^1 ky \left\{ \int_0^1 x\, dx \right\} dy$$

$$= \int_0^1 ky \left\{ \frac{x^2}{2} \Big|_0^1 \right\} dy$$

$$= \int_0^1 \frac{k}{2} \cdot y\, dy$$

$$= \left(\frac{k}{2} \right) \left(\frac{y^2}{2} \right) \Big|_0^1 = \frac{k}{4}$$

By (b), we have $k/4 = 1$, so that $k = 4$.

Since the (cumulative) distribution function was of such great importance in the case of one continuous random variable, we now extend that definition to the case of two continuous random variables.

DEFINITION

The *joint distribution function* of two continuous random variables X and Y is defined, for any choice of $a, b \in \mathbb{R}$, to equal

$$F(a, b) = P(X \leq a \cap Y \leq b) = \int_{-\infty}^{b} \int_{-\infty}^{a} f(x, y) \, dx \, dy$$

The joint distribution function F turns out to be monotonic nondecreasing in each variable.

Example 1 (continued) We have already derived the joint density

$$f(x, y) = \begin{cases} 4xy & \text{if } 0 \leq x, y \leq 1 \\ 0 & \text{otherwise} \end{cases}$$

Consequently,

$$F(a, b) = \int_{-\infty}^{b} \int_{-\infty}^{a} f(x, y) \, dx \, dy = \int_{0}^{b} \int_{0}^{a} 4xy \, dx \, dy$$

provided that $0 \leq a, b \leq 1$. In this case,

$$F(a, b) = \int_{0}^{b} 4y \left\{ \int_{0}^{a} x \, dx \right\} dy$$

$$= \int_{0}^{b} 4y \left\{ \frac{x^2}{2} \Big|_{0}^{a} \right\} dy$$

$$= \int_{0}^{b} 2a^2 \cdot y \, dy$$

$$= 2a^2 \left\{ \frac{y^2}{2} \Big|_{0}^{b} \right\} = a^2 b^2$$

Therefore,

$$F(a, b) = \begin{cases} a^2 b^2 & \text{if } 0 \leq a, b \leq 1 \\ 0 & \text{if } a < 0 \text{ or } b < 0 \\ 1 & \text{if } a \geq 1 \text{ and } b \geq 1 \end{cases}$$

Notice that we have not yet defined $F(a, b)$ in the two cases $0 \leq a \leq 1$, $b \geq 1$ and $0 \leq b \leq 1$, $a \geq 1$. In the first case, for example, we would have to calculate

$$P(X \leq a \cap Y \leq b) = P(X \leq a)$$

since the event $\{Y \leq b\} = S$. In order to carry out this calculation, we must be able to determine the density and distribution functions of X from the joint density function of X and Y. We shall discuss how this is to be done momentarily. But first, we prove

Proposition 8.5

For any choice of $x_1 \leq x_2$ and $y_1 \leq y_2$, we have

$$F(x_2, y_2) - F(x_1, y_2) - F(x_2, y_1) + F(x_1, y_1) \geq 0$$

Proof: The fact that the term

$$F(x_2, y_2) - F(x_1, y_2) - F(x_2, y_1) + F(x_1, y_1)$$

is nonnegative follows from the fact that this term is precisely the probability

$$P(x_1 \leq X \leq x_2 \cap y_1 \leq Y \leq y_2)$$

To verify this, we use the following set-theoretic argument:

$$\{x_1 \leq X \leq x_2\} \cap \{y_1 \leq Y \leq y_2\}$$
$$= \{x_1 \leq X \leq x_2\} \cap \{\{Y \leq y_2\} - \{Y \leq y_1\}\}$$
$$= \{x_1 \leq X \leq x_2\} \cap \{Y \leq y_2\} - \{x_1 \leq X \leq x_2\} \cap \{Y \leq y_1\}$$

But

$$\{x_1 \leq X \leq x_2\} = \{X \leq x_2\} - \{X \leq x_1\}$$

and so, for $i = 1, 2$,

$$\{x_1 \leq X \leq x_2\} \cap \{Y \leq y_i\}$$
$$= \{X \leq x_2\} \cap \{Y \leq y_i\} - \{X \leq x_1\} \cap \{Y \leq y_i\}$$

Therefore,

$$\{x_1 \leq X \leq x_2\} \cap \{y_1 \leq Y \leq y_2\}$$
$$= \{\{X \leq x_2\} \cap \{Y \leq y_2\} - \{X \leq x_1\} \cap \{Y \leq y_2\}\}$$
$$- \{\{X \leq x_2\} \cap \{Y \leq y_1\} - \{X \leq x_1\} \cap \{Y \leq y_1\}\}$$

The desired result now follows from the elementary laws of probability functions.

An immediate consequence of this result is the following generalization of the fact that the derivative of the distribution function of a continuous random variable is equal to the density function:

Proposition 8.6

$$\frac{\delta^2 F(x, y)}{\delta x \, \delta y} = f(x, y)$$

Proof:

$$\frac{\delta^2 F(x, y)}{\delta x \, \delta y} = \frac{\delta}{\delta x} \left\{ \lim_{\Delta y \to 0} \frac{F(x, y + \Delta y) - F(x, y)}{\Delta y} \right\}$$

$$= \lim_{\substack{\Delta x \to 0 \\ \Delta y \to 0}} \frac{F(x + \Delta x, y + \Delta y) - F(x, y + \Delta y) - F(x + \Delta x, y) + F(x, y)}{\Delta x \, \Delta y}$$

which, by Proposition 8.5, equals

$$= \lim_{\substack{\Delta x \to 0 \\ \Delta y \to 0}} \frac{P(x \le X \le x + \Delta x \cap y \le Y \le y + \Delta y)}{\Delta x \, \Delta y}$$

$$= \lim_{\substack{\Delta x \to 0 \\ \Delta y \to 0}} \frac{\int_y^{y + \Delta y} \int_x^{x + \Delta x} f(s, t) \, ds \, dt}{\Delta x \, \Delta y}$$

$$= f(x, y)$$

Marginal densities can be defined in much the same manner as in the discrete case. However, rather than sum (over k) terms of the form $f(x, y_k)$ or $f(x_k, y)$, we instead integrate out the "other" variable.

DEFINITION

If X and Y are continuous random variables with joint density $f(x, y)$, then the marginal densities f_X and f_Y of X and Y, respectively, are defined to equal

$$f_X(x) = \int_{-\infty}^{\infty} f(x, y) \, dy$$

and

$$f_Y(y) = \int_{-\infty}^{\infty} f(x, y) \, dx$$

The following result indicates how the (cumulative) distribution functions of X and Y can be obtained from the marginal densities:

Proposition 8.7

Suppose that X and Y are continuous random variables with joint density function $f(x, y)$. Then

(a) $F_1(a) = P(X \le a) = \int_{-\infty}^a f_X(x) \, dx$

(b) $F_2(b) = P(Y \le b) = \int_{-\infty}^b f_Y(y) \, dy$

Proof: We give the proof of (a); the proof of (b) is identical. For any real number a,

$$F_1(a) = P(X \le a) = P(X \le a \cap -\infty \le Y \le \infty)$$

$$= \int_{-\infty}^{\infty} \int_{-\infty}^a f(x, y) \, dx \, dy$$

$$= \int_{-\infty}^a \left\{ \int_{-\infty}^{\infty} f(x, y) \, dy \right\} dx$$

$$= \int_{-\infty}^a f_X(x) \, dx$$

Example 1 (continued)   From the density function

$$f(x, y) = \begin{cases} 4xy & \text{if } 0 \leq x, y \leq 1 \\ 0 & \text{otherwise} \end{cases}$$

we determine the marginal density

$$f_X(x) = \int_{-\infty}^{\infty} f(x, y) \, dy = \int_0^1 4xy \, dy$$

$$= 4x \left\{ \frac{y^2}{2} \Big|_0^1 \right\}$$

$$= 2x$$

Therefore,

$$f_X(x) = \begin{cases} 2x & \text{if } 0 \leq x \leq 1 \\ 0 & \text{otherwise} \end{cases}$$

Likewise,

$$f_Y(y) = \begin{cases} 2y & \text{if } 0 \leq y \leq 1 \\ 0 & \text{otherwise} \end{cases}$$

From the marginal density function, we can determine the distribution function

$$F_1(a) = \int_{-\infty}^a f_X(x) \, dx = \int_0^a 2x \, dx = a^2$$

provided that $0 \leq a \leq 1$, so that

$$F_1(a) = \begin{cases} 0 & \text{if } a < 0 \\ a^2 & \text{if } 0 \leq a \leq 1 \\ 1 & \text{if } a \geq 1 \end{cases}$$

Likewise,

$$F_2(b) = \begin{cases} 0 & \text{if } b < 0 \\ b^2 & \text{if } 0 \leq b \leq 1 \\ 1 & \text{if } b \geq 1 \end{cases}$$

We can now completely define the joint distribution function of X and Y:

$$F(a, b) = \begin{cases} a^2 b^2 & \text{if } 0 \leq a, b \leq 1 \\ a^2 & \text{if } 0 \leq a \leq 1, b \geq 1 \\ b^2 & \text{if } 0 \leq b \leq 1, a \geq 1 \\ 0 & \text{if } a < 0 \text{ or } b < 0 \\ 1 & \text{if } a \geq 1 \text{ and } b \geq 1 \end{cases}$$

Notice that, since $F_1(a) = \int_{-\infty}^a f_X(x) \, dx$, the marginal density f_X is, in fact, the density function f_1 of X. Likewise, the marginal f_Y is the density f_2 of Y.

Example 2 Suppose that X and Y represent the percentage of responses to two different types of mail order solicitation, and suppose that X and Y have the following joint density function:

$$f(x, y) = \begin{cases} 2(x + 4y)/5 & \text{if } 0 \le x, y \le 1 \\ 0 & \text{otherwise} \end{cases}$$

Then

$$F(a, b) = P(X \le a \cap Y \le b) = \tfrac{2}{5} \int_0^b \int_0^a (x + 4y) \, dx \, dy$$

$$= \frac{a^2b + 4ab^2}{5}$$

provided that $0 \le a, b \le 1$. Also, the marginal density of X equals

$$f_X(x) = \int_{-\infty}^{\infty} f(x, y) \, dx \, dy = \tfrac{2}{5} \int_0^1 (x + 4y) \, dy = \frac{2(x + 2)}{5}$$

so that

$$f_X(x) = \begin{cases} 2(x + 2)/5 & \text{if } 0 \le x \le 1 \\ 0 & \text{otherwise} \end{cases}$$

Likewise,

$$f_Y(y) = \begin{cases} (1 + 8y)/5 & \text{if } 0 \le y \le 1 \\ 0 & \text{otherwise} \end{cases}$$

Consequently, if $0 \le a \le 1$, we have

$$F_1(a) = P(X \le a) = \int_0^a \frac{2(x + 2)}{5} \, dx = \frac{a^2 + 4a}{5}$$

and, if $0 \le b \le 1$,

$$F_2(b) = P(Y \le b) = \int_0^b \frac{1 + 8y}{5} \, dy = \frac{b + 4b^2}{5}$$

So the probability that there will be a less than 40 percent response to the first type of mail order solicitation is

$$F_1(0.40) = \frac{(0.40)^2 + 4(0.40)}{5} = 0.352$$

and the probability that there will be a less than 20 percent response to the second type of mail order solicitation is

$$F_2(0.20) = \frac{0.20 + 4(0.20)^2}{5} = 0.072$$

On the other hand, the probability that the first type of solicitation will produce less than 40 percent response while at the same time the second type produces less than 20 percent response is

$$F(0.40, 0.20) = \frac{(0.40)^2(0.20) + 4(0.40)(0.20)^2}{5}$$

$$= 0.0192$$

Example 3 Suppose that a point is chosen at random in the unit square $\{(x, y) : 0 \leq x, y \leq 1\}$, and we let X represent its x-coordinate and Y its y-coordinate. We wish to determine the joint density function of X and Y. To do so we first determine their joint distribution function. If $0 \leq x, y \leq 1$, then the probability $P(X \leq x \cap Y \leq y)$ is represented by the area of the rectangle that is cross-hatched in Figure 8.7. Therefore, $F(x, y) = xy$ whenever $0 \leq x, y \leq 1$ (as is always the case in this example). Since $\delta^2 F(x, y)/\delta x \, \delta y = 1$, we have, by Proposition 8.6, that

$$f(x, y) = \begin{cases} 1 & \text{if } 0 \leq x, y \leq 1 \\ 0 & \text{otherwise} \end{cases}$$

Consequently, the marginal distribution of X is

$$f_X(x) = \int_{-\infty}^{\infty} f(x, y) \, dy = \int_0^1 1 \, dy = 1$$

giving

$$f_X(x) = \begin{cases} 1 & \text{if } 0 \leq x \leq 1 \\ 0 & \text{otherwise} \end{cases}$$

Likewise,

$$f_Y(y) = \begin{cases} 1 & \text{if } 0 \leq y \leq 1 \\ 0 & \text{otherwise} \end{cases}$$

Figure 8.7

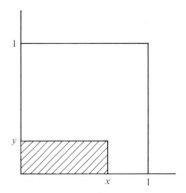

Then if $0 \leq x \leq 1$, we have

$$F_1(x) = \int_{-\infty}^{x} f_X(t) \, dt = \int_{0}^{x} 1 \, dt = x$$

so that

$$F_1(x) = \begin{cases} 0 & \text{if } x \leq 0 \\ x & \text{if } 0 \leq x \leq 1 \\ 1 & \text{if } x \geq 1 \end{cases}$$

Likewise,

$$F_2(y) = \begin{cases} 0 & \text{if } y \leq 0 \\ y & \text{if } 0 \leq y \leq 1 \\ 1 & \text{if } y \geq 1 \end{cases}$$

Note that these two distribution functions are precisely what we would have expected for sampling a point at random in the unit interval $[0, 1]$.

Next, exactly as we did in the discrete case, we may define the two conditional densities as follows:

DEFINITION

If X and Y are continuous random variables with joint density $f(x, y)$ and marginal densities $f_X(x)$ and $f_Y(y)$, then we define the *conditional density* of X, given Y, to equal

$$h_1(x \mid y) = \frac{f(x, y)}{f_Y(y)}$$

and the conditional density of Y, given X, to equal

$$h_2(y \mid x) = \frac{f(x, y)}{f_X(x)}$$

The following two results are immediate consequences of the definition:

1. $\displaystyle \int_{-\infty}^{\infty} h_1(x \mid y) \, dx = \int_{-\infty}^{\infty} h_2(y \mid x) \, dy = 1$

2. $\displaystyle P(a \leq X \leq b \mid y) = \int_{a}^{b} h_1(x \mid y) \, dx$ and

   $\displaystyle P(c \leq Y \leq d \mid x) = \int_{c}^{d} h_2(y \mid x) \, dy$

The concept of independent random variables can also be extended to the continuous case. Our definition will be stated in terms of the distribution functions, although, as we shall see in Proposition 8.8, the definition can just as easily be stated in terms of the (marginal) density functions, which is how the definition is stated in the discrete case.

DEFINITION

Let X and Y be continuous random variables with joint distribution function F. Then X and Y are said to be independent if and only if $F(x, y) = F_1(x) \cdot F_2(y)$ for all pairs (x, y).

Proposition 8.8

Let X and Y be continuous random variables with joint density function f. Then X and Y are independent if and only if $f(x, y) = f_X(x) \cdot f_Y(y)$ for all pairs (x, y).

Proof: We let F be the joint distribution function of X and Y.
($\Rightarrow$) If X and Y are independent, then $F(x, y) = F_1(x) \cdot F_2(y)$, and so

$$f(x, y) = \frac{\delta^2 F(x, y)}{\delta x \, \delta y}$$

$$= \frac{\delta^2 F_X(x) \cdot F_Y(y)}{\delta x \, \delta y}$$

$$= \frac{dF_X(x)}{dx} \cdot \frac{dF_Y(y)}{dy}$$

$$= f_X(x) \cdot f_Y(y)$$

($\Leftarrow$) On the other hand, if $f(x, y) = f_X(x) \cdot f_Y(y)$, then

$$F(x, y) = \int_{-\infty}^{y} \int_{-\infty}^{x} f(s, t) \, ds \, dt$$

$$= \int_{-\infty}^{y} \int_{-\infty}^{x} f_X(s) \cdot f_Y(t) \, ds \, dt$$

$$= \int_{-\infty}^{y} f_Y(t) \left\{ \int_{-\infty}^{x} f_X(s) \, ds \right\} dt$$

$$= \int_{-\infty}^{y} f_Y(t) \cdot F_X(x) \, dt$$

$$= F_X(x) \int_{-\infty}^{y} f_Y(t) \, dt$$

$$= F_X(x) \cdot F_Y(y)$$

Notice that the random variables X and Y in Examples 1 and 3 above are independent, while those in Example 2 are not independent.

As in the discrete case, the independence of X and Y has the following implications concerning the conditional density functions:

1. $h_1(x \mid y) = f_X(x)$
2. $h_2(y \mid x) = f_Y(y)$

The result of Proposition 5.1 also extends to the continuous case.

All of the concepts discussed thus far in this section can be extended, in a straightforward manner, to the case of n continuous random variables. We, however, will continue to discuss just the case $n = 2$ and will now extend the results of the preceding section concerning functions of a random variable.

Our goal is to obtain a result analogous to Proposition 8.3 for a function of two continuous random variables. We will achieve this goal indirectly, by first considering the case of two functions of two continuous random variables.

Let us assume that X and Y are continuous random variables with joint density function $f(x, y)$. Also, let us assume that $u, v : \mathbb{R} \times \mathbb{R} \to \mathbb{R}$ are real-valued functions of two real variables such that the function

$$(x, y) \longrightarrow [u(x, y), v(x, y)]$$

is one-to-one, and both it and its inverse have continuous partial derivatives. If we denote this inverse function by

$$(u, v) \longrightarrow [w(u, v), z(u, v)]$$

then we have $w(u, v) = x$ and $z(u, v) = y$

Proposition 8.9

Under the assumptions above, the random variables $U = u(X, Y)$ and $V = v(X, Y)$ have joint density $g(u, v)$ given by

$$g(u, v) = f[w(u, v), z(u, v)] \cdot |J|$$

where

$$J = \left(\frac{\delta x}{\delta u}\right)\left(\frac{\delta y}{\delta v}\right) - \left(\frac{\delta x}{\delta v}\right)\left(\frac{\delta y}{\delta u}\right)$$

is known as the *Jacobian*. Equivalently, $g(u, v) = f(x, y) \cdot |J|$.

Proof: The proof of this result is somewhat above the level of this text, and therefore is omitted. For details, see another text.*

We can use this result to obtain a formula for the density function of a random variable of the form $u(X, Y)$, where X and Y are continuous random variables. To do this, we must introduce a second function of X and Y, which we define to be $v(X, Y) = Y$. Consequently, we have a mapping

$$(x, y) \longrightarrow [u(x, y), y]$$

which, should the assumptions above hold, would assist us in calculating the joint density of $u(X, Y)$ and $v(X, Y) = Y$, and, in turn, the density function of $u(X, Y)$.

---

* C. P. Tsokos, *Probability Distributions: An Introduction to Probability Theory with Applications* (Belmont, California: Duxbury Press 1972).

Proposition 8.10

If the density function of the random variable $U = u(X, Y)$ is denoted $h(u)$, then

$$h(u) = \int_{-\infty}^{\infty} f(x, y) \cdot \left|\frac{\delta x}{\delta u}\right| dv$$

where $f(x, y)$ is the joint density of X and Y.

We have $u = u(x, y)$ and $v = v(x, y) = y$, so that $\delta y/\delta u = 0$ and $\delta y/\delta v = 1$. Consequently, the Jacobian simplifies to just $\delta x/\delta u$. We then apply Proposition 8.9 to find that the joint density of U and $V = Y$ is

$$g(u, v) = f(x, y) \cdot \left|\frac{\delta x}{\delta u}\right|$$

Integrating out the dummy variable v, we obtain the (marginal) density function of U as stated above.

Corollary 8.11

Suppose that X and Y are continuous random variables with joint density function $f(x, y)$. Suppose that U is a function of X and Y having density function $h(u)$. If

(a) $U = X + Y$, then $h(u) = \int_{-\infty}^{\infty} f(u - v, v) \, dv$;

(b) $U = XY$, then $h(u) = \int_{-\infty}^{\infty} f(u/v, v) \cdot |1/v| \, dv$;

(c) $U = X/Y$, then $h(u) = \int_{-\infty}^{\infty} f(uv, v) \cdot |v| \, dv$.

Proof:

(a) We have $u = x + y$ and $v = y$. Consequently, $x = u - v$ and $y = v$, so that $\delta x/\delta u = 1$.

(b) We have $u = xy$ and $v = y$, so that $x = u/v$ and $y = v$, so that $\delta x/\delta u = 1/v$.

(c) We have $u = x/y$ and $v = y$, so that $x = uv$ and $y = v$, so that $\delta x/\delta u = v$.

We point out that the definition of the joint density function $f(x, y)$ will usually place some restrictions on the domains of the two arguments x and y. As the following example illustrates, care must be taken that these restrictions be carried over to the new arguments $w(u, v)$ and $z(u, v)$.

Example 1 (continued) We let $U = X + Y$. According to Proposition 8.10, the joint density of U and $V = Y$ is

$$f(u - v, v) = 4(u - v)v$$

provided that $0 \leq u - v \leq 1$. This condition implies that

$0 \leq v \leq u$ when $0 \leq u \leq 1$

and

$u - 1 \leq v \leq 1$ when $1 \leq u \leq 2$

(Recall that $U = X + Y$ will exist in the interval $[0, 2]$.) When $0 \leq u \leq 1$, we have

$$h(u) = \int_0^u f(u - v, v) \, dv = \int_0^u 4(u - v)v \, dv$$

$$= \left(2uv^2 - \frac{4v^3}{3} \right) \bigg|_0^u$$

$$= \frac{2u^3}{3}$$

and when $1 \leq u \leq 2$, we have

$$h(u) = \int_{u-1}^1 f(u - v, v) \, dv = \int_{u-1}^1 4(u - v)v \, dv$$

$$= 4u - \frac{2u^3}{3} - \frac{8}{3}$$

Thus, the density function h of the random variable $U = X + Y$ is

$$h(u) = \begin{cases} 2u^3/3 & \text{if } 0 \leq u \leq 1 \\ 4u - 2u^3/3 - \frac{8}{3} & \text{if } 1 \leq u \leq 2 \end{cases}$$

It is then straightforward to check that the integral of $h(u)$ over the interval $[0, 2]$ is, in fact, equal to 1, as it should be, $h(u)$ being the density function of a continuous random variable.

Most of the definitions and results in Chapter 5 can be extended to the case of functions of two (or more) continuous random variables. In particular, we can define

$$E(XY) = \int_{-\infty}^\infty \int_{-\infty}^\infty xy \cdot f(x, y) \, dx \, dy$$

and prove that the equivalent of Proposition 5.3

$$E[g(X, Y)] = \int_{-\infty}^\infty \int_{-\infty}^\infty g(x, y) \cdot f(x, y) \, dx \, dy$$

remains true. From this, the equivalent of Corollary 5.4 follows easily. Also, under the additional assumption that X and Y are independent, the results of Propositions 5.8 and 5.10 can be verified. These, and the result of Corollary 5.6, all extend to the case of n independent continuous random variables.

In addition, the definitions of covariance and the coefficient of correlation can be extended, and all of the results in Section 5.3 can be extended.

EXERCISES

1. For what value of k is the function

$$f(x, y) = \begin{cases} kx^2y & \text{if } 0 \leq x, y \leq 1 \\ 0 & \text{otherwise} \end{cases}$$

a possible joint density function for two continuous random variables?

2. For what value of k is the function

$$f(x, y) = \begin{cases} ke^{x+y} & \text{if } 0 \leq x, y \leq 1 \\ 0 & \text{otherwise} \end{cases}$$

a possible joint density function for two continuous random variables?

3. Suppose that the continuous random variables X and Y have joint density function

$$f(x, y) = \begin{cases} xe^{-xy} & \text{if } x \geq 0 \text{ and } y \geq 1 \\ 0 & \text{otherwise} \end{cases}$$

Find the joint distribution function of X and Y and the marginal density of X.

4. Suppose that the continuous random variables X and Y have joint density function

$$f(x, y) = \begin{cases} e^{-(x+y)} & \text{if } x, y \geq 0 \\ 0 & \text{otherwise} \end{cases}$$

Find the joint distribution function of X and Y, and find the marginal densities of both X and Y.

5. Find the density function of the random variable $X + Y$, where X and Y are continuous random variables having the joint density function defined in Example 2.

6. Find the density function of the random variable $X + Y$, where X and Y are continuous random variables having the joint density function defined in Exercise 1.

7. Suppose that X and Y are independent continuous random variables with common density function

$$f(x) \quad \begin{cases} e^{-x} & \text{if } x \geq 0 \\ 0 & \text{otherwise} \end{cases}$$

Prove that $X + Y$ has density xe^{-x} and X/Y has density $1/(1 + x)^2$.

8.3 Special Continuous Distributions

In this section, we shall study the gamma and beta distributions and special derivatives thereof, the exponential, chi-square, and uniform distributions.

The gamma distribution is important in its own right, but one particular gamma distribution, the exponential distribution, is perhaps of even greater significance. And we shall study in Chapter 10 the importance of another gamma distribution, the chi-square distribution, in the field of statistical inference.

DEFINITION

A function of the form

$$f(x) = \begin{cases} kx^{\alpha-1}e^{-x/\beta} & \text{if } x > 0 \\ 0 & \text{otherwise} \end{cases}$$

where $\alpha > 0$, $\beta > 0$ (and the constant k is to be determined) is called a *gamma density* with parameters α and β.

The value of k must be chosen so that the integral

$$\int_0^\infty kx^{\alpha-1}e^{-x/\beta} \, dx = 1$$

If we make the change of variables

$$y = \frac{x}{\beta}$$

$$\beta \, dy = dx$$

we obtain

$$\int_0^\infty kx^{\alpha-1}e^{-x/\beta} \, dx = k\beta^\alpha \int_0^\infty y^{\alpha-1}e^{-y} \, dy$$

This last integral is called the *gamma function*

$$\Gamma(\alpha) = \int_0^\infty y^{\alpha-1}e^{-y} \, dy \qquad (\alpha > 0)$$

and is usually studied in second-year calculus. Integration by parts gives

$$\Gamma(\alpha) = (\alpha - 1)\Gamma(\alpha - 1)$$

which is equivalent to $\Gamma(\alpha) = (\alpha - 1)!$ whenever α is a positive integer. We therefore have found that

$$1 = \int_0^\infty kx^{\alpha-1}e^{-x/\beta} \, dx = k\beta^\alpha\Gamma(\alpha)$$

and consequently, that

$$k = \frac{1}{\beta^\alpha\Gamma(\alpha)}$$

Two special cases of the gamma distribution are important enough to be given special names. First, there is the case $\alpha = 1$ which gives rise to the *exponential distribution*, which has density function

$$f(x) = \begin{cases} (1/\beta)e^{-x/\beta} & \text{if } x > 0 \\ 0 & \text{otherwise} \end{cases}$$

The graphs of exponential densities with parameters $\beta = 1, 2, 3$, and 4 are in Figure 8.8.

A second interesting family of gamma distributions occurs when $\alpha = k/2$ and $\beta = 2$. These are the *chi-square* (χ^2) *distributions*, and they have densities

$$f(x) = \begin{cases} \dfrac{1}{2^{k/2}\Gamma(k/2)} x^{(k-2)/2} e^{-x/2} & \text{if } x > 0 \\ 0 & \text{otherwise} \end{cases}$$

The value k is called the number of *degrees of freedom* of the χ^2 random variable.

Proposition 8.12

If X has gamma distribution with parameters α and β, then

(a) $\mu_k' = \beta^k \Gamma(\alpha + k)/\Gamma(\alpha)$
(b) $E(X) = \alpha\beta$
(c) $\text{Var}(X) = \alpha\beta^2$
(d) $M_X(t) = (1 - \beta t)^{-\alpha}$

Proof:

(a) We evaluate the moments about the origin directly:

$$\mu_k' = \int_0^\infty x^k \frac{1}{\beta^\alpha \Gamma(\alpha)} x^{\alpha-1} e^{-x/\beta} \, dx$$

Figure 8.8

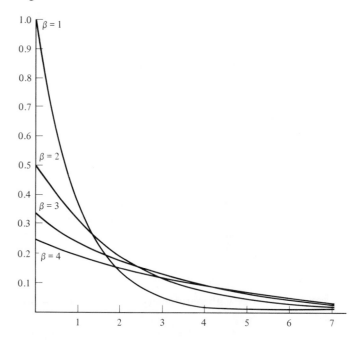

Using the same change of variables as above, $y = x/\beta$, we obtain

$$\mu'_k = \frac{\beta^k}{\Gamma(\alpha)} \int_0^\infty y^{\alpha+k-1} e^{-y} \, dy$$

$$= \frac{\beta^k \Gamma(\alpha + k)}{\Gamma(\alpha)}$$

(b) Since $E(X) = \mu'_1$, we have

$$E(X) = \frac{\beta \Gamma(\alpha + 1)}{\Gamma(\alpha)}$$

$$= \frac{\beta \alpha \Gamma(\alpha)}{\Gamma(\alpha)}$$

$$= \alpha\beta$$

(c) Since $E(X^2) = \mu'_2$, we have

$$E(X^2) = \frac{\beta^2 \Gamma(\alpha + 2)}{\Gamma(\alpha)}$$

$$= \frac{\beta^2 (\alpha + 1)\Gamma(\alpha + 1)}{\Gamma(\alpha)}$$

$$= \alpha(\alpha + 1)\beta^2 = \alpha^2\beta^2 + \alpha\beta^2$$

Therefore

$$\text{Var}(X) = (\alpha^2\beta^2 + \alpha\beta^2) - (\alpha\beta)^2 = \alpha\beta^2$$

(d) Finally,

$$M_X(t) = \int_0^\infty e^{xt} \frac{1}{\beta^\alpha \Gamma(\alpha)} x^{\alpha-1} e^{-x/\beta} \, dx$$

$$= \frac{1}{\beta^\alpha \Gamma(\alpha)} \int_0^\infty x^{\alpha-1} \exp\left[-x\left(\frac{1}{\beta} - t\right)\right] dx$$

If we use the change of variables

$$y = x\left(\frac{1}{\beta} - t\right)$$

$$dy = \left(\frac{1}{\beta} - t\right) dx$$

we obtain

$$M_X(t) = \frac{1}{\beta^\alpha \Gamma(\alpha)[(1/\beta) - t]^\alpha} \int_0^\infty y^{\alpha-1} e^{-y} \, dy$$

and therefore,

$$M_X(t) = \frac{\Gamma(\alpha)}{\beta^\alpha \Gamma(\alpha)[(1/\beta) - t]^\alpha} = \frac{1}{(1 - \beta t)^\alpha}$$

Corollary 8.13

If X has exponential distribution with parameter β, then

(a) $E(X) = \beta$
(b) $\text{Var}(X) = \beta^2$
(c) $M_X(t) = 1/(1 - \beta t)$

Corollary 8.14

If X has χ^2 distribution with k degrees of freedom, then

(a) $E(X) = k$
(b) $\text{Var}(X) = 2k$
(c) $M_X(t) = (1 - 2t)^{-k/2}$

Example 1 Suppose that family income in New York City (in units of $10,000) has gamma distribution with parameters $\alpha = 2$ and $\beta = \frac{1}{2}$ (Figure 8.9). By Proposition 8.12, we see that the average family income in New York City is $\alpha\beta = 2 \cdot \frac{1}{2} = 1$ unit, or $10,000 dollars. We wish to calculate the probability that a family's income will exceed $20,000. We let X denote

Figure 8.9

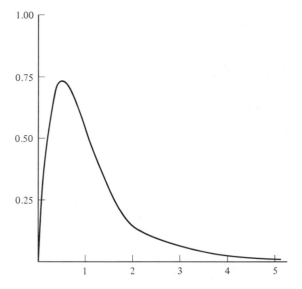

the income, in units of \$10,000, of a family chosen at random, and must calculate $P(X > 2)$. But this equals

$$P(X > 2) = \frac{1}{(\frac{1}{2})^2 \Gamma(2)} \int_2^\infty x^{2-1} e^{-2x} \, dx$$

$$= 4 \int_2^\infty x e^{-2x} \, dx$$

$$= 4 \left(\frac{-x}{2} e^{-2x} \right) \Big|_2^\infty + 2 \int_2^\infty e^{-2x} \, dx$$

$$= (-2x \cdot e^{-2x} - e^{-2x}) \Big|_2^\infty$$

$$= -(2x + 1)e^{-2x} \Big|_2^\infty$$

$$= 5e^{-4} = 0.0916$$

The continuous exponential distribution has an important connection with the discrete Poisson distribution. Under the assumptions stated at the end of Chapter 6, the probability $p(0; t)$ that a (rare) event will not occur over a time interval of length $t > 0$ equals $e^{-\alpha t}$, where α represents the expected number of occurrences of the event over a time interval of length 1. If we define a random variable Y to represent the waiting time until the event first (or next) occurs, then

$$F_Y(t) = P(Y \le t) = 1 - e^{-\alpha t}$$

Consequently,

$$f_Y(t) = \frac{d[F_Y(t)]}{dt} = \alpha e^{-\alpha t}$$

so Y has an exponential distribution with parameter $1/\alpha$.

Example 2   The time T (in hours) between automobile accidents at a particularly dangerous intersection may be assumed to have exponential distribution, according to the discussion above. Statistics reveal that $E(T) = 10$ hours, and so our exponential has parameter $\alpha = 10$. The probability that there will be no accidents at this intersection on a given day equals

$$P(T \ge 24) = 1 - P(0 \le T \le 24)$$

$$= 1 - \int_0^{24} \frac{1}{10} e^{-t/10} \, dt$$

$$= 1 + e^{-t/10} \Big|_0^{24}$$

$$= 1 + e^{-24/10} - e^{-0}$$

$$= e^{-24/10} = 0.091$$

Example 3 Let T represent the time (in minutes) between the arrival of cars at the tollgate on the Southern State Parkway during the rush hour. According to the discussion above, it is reasonable to assume that T has exponential distribution. Statistics indicate the $E(T) = 10$ seconds, so our exponential has parameter $\alpha = \frac{1}{6}$ minute ($= 10$ seconds). The probability that the time between two successive arrivals at the tollgate will be between 1 and 2 minutes equals

$$P(1 \le T \le 2) = \int_1^2 6e^{-6t}\, dt = -e^{-6t}\Big|_1^2$$

$$= e^{-6} - e^{-12}$$

$$= 0.0025$$

Example 4 The time T (in hours) between successive failures of an electronic computer part can also be regarded as an exponential distribution. If statistics reveal that $E(T) = 36$ hours, then the probability that the component will not fail during a two day period equals

$$P(T \ge 48) = 1 - P(0 \le T \le 48)$$

$$= 1 - \int_0^{48} \tfrac{1}{36} e^{-t/36}\, dt$$

$$= 1 + e^{-48/36} - e^{-0}$$

$$= e^{-4/3}$$

$$= 0.264$$

A machine, such as a computer or a rocket engine, that is dependent on electronic components which are subject to a reasonably high rate of failure (wearing out) are often designed to contain several spare parts to be used once the original component fails. We assume that a machine contains m identical electronic components and define

$T_i = $ "time until component i fails," $i = 1, 2, \ldots, 10$

Then we may consider the $\{T_i\}$ to be independent exponential random variables with equal parameter α. If we let

$$T = T_1 + T_2 + \cdots + T_m$$

then T is the time until the machine fails because all of the identical components have failed. According to Exercise 1, T has gamma distribution with parameters $\alpha = m$ and $\beta = \alpha$.

Example 5 Suppose that a rocket engine has been designed to contain a system of three identical electronic components, which are to be used one at a time. Suppose that the average lifespan of such components is one day (24 hours). What is the probability that these components will last through a proposed three-day space flight? If we let T represent the time (in days) until the system of three components fails, then T has gamma distribution with parameters $\alpha = 3$ and $\beta =$ one day. Proposition 8.12 indicates that $E(T) = 3 \cdot 1 = 3$ days. We wish to calculate

$$P(T \geq 3) = 1 - P(0 \leq T \leq 3)$$

$$= 1 - \frac{1}{\Gamma(3)} \int_0^3 x^2 e^{-x}\, dx$$

$$= 1 - \left(-e^{-x} \left(\frac{1 + x + x^2}{2} \right) \right) \Big|_0^3$$

$$= 1 - \left(\frac{-8.5}{e^3} \right) + 1$$

$$= 0.423$$

The range of values that may be assumed by a continuous random variable having gamma distribution is unbounded above. Such random variables may assume any value between zero and positive infinity, although it is true that the probability that such a random variable will assume a value in the interval $[a, \infty)$ diminishes very rapidly as the value of a increases. When we study the normal probability distribution in the next section, we will see another example of a continuous random variable whose range is unbounded (in this case, in both directions). This is the rule rather than the exception. However, there are many examples of random variables, or probability distributions, which assume values that are restricted to a certain bounded interval. Random variables which express proportions or percentages are one example. We now study the only family of continuous probability distributions whose values are restricted to a bounded interval.

DEFINITION

A function of the form

$$f(x) = \begin{cases} \dfrac{\Gamma(\alpha + \beta)}{\Gamma(\alpha)\Gamma(\beta)} x^{\alpha - 1}(1 - x)^{\beta - 1} & \text{if } 0 < x < 1 \\ 0 & \text{otherwise} \end{cases}$$

is known as a *beta-density function*. The corresponding probability distribution is called the *beta distribution*, with parameters α and β.

Proposition 8.15

The beta distribution with parameters α and β is, in fact, a probability distribution, and

(a) $E(X) = \alpha/(\alpha + \beta)$

(b) $\text{Var}(X) = \alpha\beta/(\alpha + \beta)^2(\alpha + \beta + 1)$

Proof: Verification is similar to that for the gamma distribution, and is left as an exercise.

The graphs of the beta densities when $\alpha = 2$, $\beta = 4$, $\alpha = 3$, $\beta = 3$, and $\alpha = 4$, $\beta = 2$ can be found in Figure 8.10.

Notice that the beta densities as defined above are all standardized to the unit interval $[0, 1]$. Should we wish a beta density defined on a more general interval $[a, b]$, we would transform the original beta random variable X into $(b - a)X + a$.

Example 6 The percentage of bettors who make money at the race track on a given day may be considered to have a beta distribution with parameters $\alpha = 1$ and $\beta = 5$. On the average, $1/(1 + 5) = \frac{1}{6}$ of the bettors come out winners on any given day. The probability that fewer than ten percent come out winners on a given day is

$$\frac{\Gamma(6)}{\Gamma(5)\Gamma(1)} \int_0^{0.1} x^0 (1 - x)^4 \, dx = 5 \int_0^{0.1} (1 - x)^4 \, dx$$

$$= -(1 - x)^5 \Big|_0^{0.1} = 0.6723$$

Figure 8.10

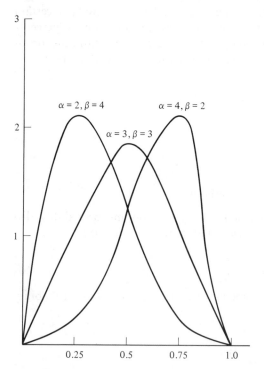

Example 7 The percentage of the working force that is unemployed on any given day may be assumed to have beta distribution with parameters $\alpha = 2$ and $\beta = 18$. On the average, $2/(2 + 18) = 0.10$, or 10 percent, of the working force is unemployed. The probability that more than 20 percent of the working force is unemployed on a given day is

$$\frac{\Gamma(20)}{\Gamma(2)\Gamma(18)} \int_{0.2}^{1} x(1 - x)^{17} \, dx$$

$$= (19 \cdot 18) \int_{0.2}^{1} x(1 - x)^{17} \, dx$$

$$= (19 \cdot 18)\left\{\frac{-x(1 - x)^{18}}{18}\right\}\Big|_{0.2}^{1} + \frac{1}{18} \int_{0.2}^{1} (1 - x)^{18} \, dx$$

$$= (19 \cdot 18)\left\{\frac{-x(1 - x)^{18}}{18} - \frac{(1 - x)^{19}}{(18 \cdot 19)}\right\}\Big|_{0.2}^{1}$$

$$= \{-19x(1 - x)^{18} - (1 - x)^{19}\}\Big|_{0.2}^{1}$$

$$= 0.0829$$

Notice that when $\alpha = \beta = 1$,

$$\frac{\Gamma(\alpha + \beta)}{\Gamma(\alpha)\Gamma(\beta)} \cdot x^{\alpha - 1} \cdot (1 - x)^{\beta - 1} = \frac{\Gamma(2)}{\Gamma(1)\Gamma(1)} x^{0} \cdot (1 - x)^{0} = 1$$

so that in this case, the beta density reduces to

$$f(x) = \begin{cases} 1 & \text{if } 0 < x < 1 \\ 0 & \text{otherwise} \end{cases}$$

We extend this to the following definition:

DEFINITION

A continuous random variable with density function

$$f(x) = \begin{cases} k & \text{if } a < x < b \\ 0 & \text{otherwise} \end{cases}$$

is said to be *uniformly distributed* on the interval $[a, b]$.

The density function of a uniform random variable therefore is constant (of value k) over its interval of definition, and zero elsewhere. Since

$$\int_{a}^{b} f(x) \, dx = \int_{a}^{b} k \, dx = 1$$

it is easy to see that the constant k must equal $1/(b - a)$, the reciprocal of the length of the interval.

As we previously noted, the uniform density over the unit interval is just one member of the family of beta densities defined on this interval. In general, the collection of uniform densities is a subset of the collection of beta densities.

What we have done here with uniform random variables is extend to continuous random variables a concept that has been with us throughout this book. The graph of a uniform density (Figure 8.11) suggests that such a random variable assumes each value in its interval of definition with equal likelihood. When we studied finite sample spaces, we found it to be a commonplace occurrence that each element of the sample space be equally likely. This concept of uniformity extended to random variables with finite range, where all possible values assumed by the random variable could, at times, be assumed to be equally likely. The idea of uniformity was also basic in our definition of probability in nondiscrete sample spaces. We have now extended this concept to continuous random variables.

The distribution function of the uniform random variable (Figure 8.12) is derived as follows:

$$F(x) = \int_{-\infty}^{x} f(t)\, dt = \int_{a}^{x} \frac{1}{b-a}\, dt$$

Figure 8.11

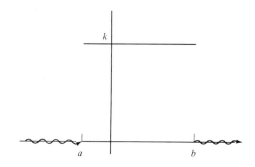

Figure 8.12

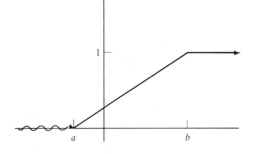

provided that $a \leq x \leq b$. Therefore,

$$F(x) = \begin{cases} 0 & \text{if } x \leq a \\ (x-a)/(b-a) & \text{if } a \leq x \leq b \\ 1 & \text{if } x \geq b \end{cases}$$

Proposition 8.16

If X is a uniformly distributed continuous random variable defined on the interval $[a, b]$, then

(a) $E(X) = (a + b)/2$
(b) $\text{Var}(X) = (a - b)^2/12$

Proof: We prove this result directly, although we could simply have invoked Proposition 8.15 with $\alpha = \beta = 1$.

(a) $E(X) = \int_{-\infty}^{\infty} x \cdot f(x) \, dx = \int_{a}^{b} \frac{x}{b-a} \, dx$

$$= \left(\frac{1}{b-a} \right) \left(\frac{x^2}{2} \right) \Big|_{a}^{b}$$

$$= \frac{b^2 - a^2}{2(b-a)}$$

$$= \frac{(b-a)(b+a)}{2(b-a)}$$

$$= \frac{a+b}{2}$$

(b) $E(X^2) = \int_{-\infty}^{\infty} x^2 \cdot f(x) \, dx = \int_{a}^{b} \frac{x^2}{b-a} \, dx$

$$= \left(\frac{1}{b-a} \right) \left(\frac{x^3}{3} \right) \Big|_{a}^{b}$$

$$= \frac{b^3 - a^3}{3(b-a)}$$

$$= \frac{b^2 + ab + a^2}{3}$$

Therefore,

$$\text{Var}(X) = \frac{b^2 + ab + a^2}{3} - \frac{(a+b)^2}{4} = \frac{(a-b)^2}{12}$$

Notice that the mean of the uniform distribution is the midpoint of the interval over which the random variable is defined. This is certainly logical, since all points in the interval are supposedly assumed with equal likelihood.

The two examples that follow could be solved by the elementary methods discussed in Section 2.5. In a sense, we actually were using a uniform distribution then, although we did not call it such.

Example 8 The time it takes a man to walk from his home to the railroad station is uniformly distributed between 15 and 20 minutes. The man leaves home at 7:30 to catch a train which departs at 7:48, and we wish to calculate the probability that he will catch the train. If we let T denote the time it will take for the man to arrive at the station, then T is a continuous random variable, uniform over the interval $[15, 20]$. We wish to calculate the probability $P(15 \leq T \leq 18)$, since the train leaves the station 18 minutes after the man leaves his home. Since $b - a = 20 - 15 = 5$, we have $k = \frac{1}{5}$, and so

$$P(15 \leq T \leq 18) = \int_{15}^{18} \frac{1}{5} \, dx = \frac{x}{5}\Big|_{15}^{18} = \frac{3}{5}$$

Essentially all we have done is divide the length of the interval $[15, 18]$—the event—by the length of the interval $[15, 20]$—the sample space. (See Example 2 in Section 2.5.)

Example 9 Suppose that the value g in the quadratic $x^2 + gx + g$ varies uniformly over the interval $[0, 6]$. We wish to calculate the probability that the quadratic has real roots. Now a quadratic $ax^2 + bx + c$ has real roots when the discriminant $b^2 - 4ac \geq 0$. In our case, then, we must have $g^2 - 4g \geq 0$. Since g is chosen from the interval $[0, 6]$, this is equivalent to $g - 4 \geq 0$, or $4 \leq g \leq 6$. Now g can be considered a uniform random variable over the interval $[0, 6]$, and since the interval has length 6, $k = \frac{1}{6}$. Therefore,

$$P(4 \leq g \leq 6) = \int_{4}^{6} \frac{1}{6} \, dx = \frac{x}{6}\Big|_{4}^{6} = \frac{1}{3}$$

Most computing centers have available what is called a *random number generator*. This is a program which generates decimal numbers in the interval $[0, 1]$ of the form $0.a_1 a_2 a_3 \ldots$, in such a way that each decimal place assumes each of the values $0, 1, 2, \ldots, 9$ with equal likelihood, and so that the values assumed by a decimal place a_i in no way effects the value assumed by another decimal place a_j, $i \neq j$. That is, the numbers generated are totally random, there is no pattern in the occurrence of the digits $0, 1, 2, \ldots, 9$ in any of the decimal places. As a result, a random number generator gives a uniform distribution over the interval $[0, 1]$. It would be more accurate, actually, to say that it almost gives a uniform continuous distribution. A computer with fixed length words can only give k-decimal place accuracy, where k depends on the word length, so we actually get a very large (10^k) finite distribution rather than a continuous distribution. But in any case, the distribution obtained is uniform.

EXERCISES

1. Suppose that $X_1, X_2, \ldots, X_n$ are independent exponential random variables with parameter α. Prove that $X_1 + X_2 + \cdots + X_n$ has gamma distribution with parameters n and α.
2. Prove Proposition 8.15.
3. Suppose that X and Y are independent uniform distributions defined on the unit interval $[0, 1]$. Find the density function of the random variable $X + Y$.
4. Prove that $\Gamma(\alpha) = (\alpha - 1)\Gamma(\alpha - 1)$.
5. Suppose that X has gamma distribution with parameters α_1 and β, and Y has gamma distribution with parameters α_2 and β. Prove that

   (a) $X + Y$ has gamma distribution with parameters $\alpha_1 + \alpha_2$ and β, and
   (b) $X/(X + Y)$ has beta distribution with parameters α_1 and α_2, provided that X and Y are independent.

6. Find the moment-generating function of the uniform continuous distribution.
7. Suppose that X has uniform density on the unit interval, and that $Y = X^3$. Prove that the density of Y is

$$g(x) = \begin{cases} (\tfrac{1}{3})x^{-2/3} & \text{if } 0 < x < 1 \\ 0 & \text{otherwise} \end{cases}$$

   Also, determine the density of Y if

   (a) $Y = cX$
   (b) $Y = 1/X$
   (c) $Y = \sqrt{X}$
   (d) $Y = X/(1 + X)$
   (e) $Y = -\log_e X$

8. Suppose that family income (in units of $\$1,000$) has gamma distribution with parameters $\alpha = 2$ and $\beta = 4$. Find the probability that a family randomly selected will have income less than $\$4,000$; greater than $\$20,000$; between $\$6,000$ and $\$12,000$.
9. Suppose that the annual proportion of erroneous income tax returns has beta distribution with $\alpha = 2$ and $\beta = 9$. Find the probability that in a given year the percentage of erroneous income tax returns will be less than ten percent.
10. The number of cars arriving at a traffic light is Poisson distributed with parameter $\alpha = 10$ per minute. A car has just passed the light. What is the probability that it will be at least 30 seconds before another car arrives at the light?
11. Suppose that X has uniform distribution on the unit interval $[0, 1]$. Prove that $-\lambda^{-1} \log(1 - X)$ has exponential distribution.
12. Suppose that buses run from New York to a certain town in New Jersey every half hour. What is the probability that a man arriving at the bus terminal will have to wait less than 20 minutes for a bus?
13. A traffic light is red for 30 seconds, green for 25 seconds, and then yellow for 5 seconds. What is the probability that a car arriving at the light will not have to wait? (Assume the car does not go through a yellow light.)
14. Suppose that the life of a television tube is exponentially distributed with mean 2000 hours. What is the probability the tube will survive 2500 hours?
15. Find the mode (maximum point) of the gamma and beta densities.

8.4 The Normal Distribution

Certainly the most important and useful probability distribution is the normal distribution. First of all, many experiments result in random variables which are normally distributed. Secondly, the normal distribution is very useful as an approximation to many other distributions. Most importantly, the normal distribution is the basis for many of the techniques in statistical inference. Without the normal distribution, many of the numerical calculations that arise in statistical inference problems would be virtually impossible.

DEFINITION

A function

$$n_{\mu, \sigma}(x) = \frac{1}{\sigma\sqrt{2\pi}} \exp\left[-\frac{1}{2}\left(\frac{x - \mu}{\sigma}\right)^2 \right]$$

is called a *normal density function* with parameters μ and σ; its (cumulative) distribution function

$$N_{\mu, \sigma}(x) = \int_{-\infty}^{x} \frac{1}{\sigma\sqrt{2\pi}} \exp\left[-\frac{1}{2}\left(\frac{t - \mu}{\sigma}\right)^2 \right] dt$$

is called a *normal distribution function* with parameters μ and σ.

Notice that the normal density function $n_{\mu, \sigma}(x)$ (Figure 8.13) is symmetric about the point $x = \mu$. The function is continuous and bell shaped, with its mode (peak, absolute maximum) at $x = \mu$. The value σ determines the height of the function at its peak: $n_{\mu, \sigma}(\mu) = 1/\sigma\sqrt{2\pi}$. The size of $n_{\mu, \sigma}(x)$ at $x = \mu$ is therefore inversely proportional to the value of σ.

Figure 8.13

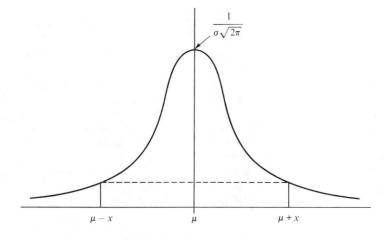

$$\frac{1}{\sigma\sqrt{2\pi}}$$

$\mu - x$ μ $\mu + x$

Notice that if we make the change of variables

$$z = \frac{t - \mu}{\sigma}$$

in the expression

$$N_{\mu, \sigma}(x) = \frac{1}{\sigma\sqrt{2\pi}} \int_{-\infty}^{x} \exp\left[-\frac{1}{2}\left(\frac{t - \mu}{\sigma}\right)^2 \right] dt$$

we obtain

$$N_{\mu, \sigma}(x) = \frac{1}{\sigma\sqrt{2\pi}} \int_{-\infty}^{(x - \mu)/\sigma} e^{-(1/2)z^2} \sigma \, dz$$

$$= \frac{1}{\sqrt{2\pi}} \int_{-\infty}^{(x - \mu)/\sigma} e^{-(1/2)z^2} \, dz$$

But this latter expression is simply $N_{0, 1}((x - \mu)/\sigma)$. So we have proved

Proposition 8.17

For any choice of parameters μ and σ, we have

$$N_{\mu, \sigma}(x) = N_{0, 1}\left(\frac{x - \mu}{\sigma}\right)$$

That is, any normal calculation $N_{\mu, \sigma}(x)$ can be done using one specific member of the normal family, $N_{0, 1}(x)$, after a suitable change of variables has been made.

Proposition 8.18

The normal density $n_{\mu, \alpha}$ is a continuous probability density. That is,

$$\frac{1}{\sigma\sqrt{2\pi}} \int_{-\infty}^{\infty} \exp\left[-\frac{1}{2}\left(\frac{t - \mu}{\sigma}\right)^2 \right] dt = 1$$

Proof: According to Proposition 8.17, it is sufficient to show that

$$I = \frac{1}{\sqrt{2\pi}} \int_{-\infty}^{\infty} e^{-(1/2)t^2} \, dt = 1$$

Since the integrand $e^{-(1/2)t^2}$ is always nonnegative, this is equivalent to showing that $I^2 = 1$, but

$$I^2 = \frac{1}{2\pi} \int_{-\infty}^{\infty} e^{-(1/2)t^2} \, dt \int_{-\infty}^{\infty} e^{-(1/2)s^2} \, ds$$

$$= \frac{1}{2\pi} \int_{-\infty}^{\infty} \int_{-\infty}^{\infty} \exp\left[-\tfrac{1}{2}(t^2 + s^2)\right] \, ds \, dt$$

Introducing polar coordinates

$$s = r \cos \theta$$
$$t = r \sin \theta$$
$$ds \, dt = r \, dr \, d\theta$$

we obtain

$$I^2 = \frac{1}{2\pi} \int_0^{2\pi} \left(\int_0^\infty r e^{-(1/2)r^2} \, dr \right) d\theta$$

$$= \frac{1}{2\pi} \int_0^{2\pi} \left. (-e^{-(1/2)r^2}) \right|_0^\infty d\theta$$

$$= \frac{1}{2\pi} \int_0^{2\pi} d\theta = 1$$

Proposition 8.19

If X is a continuous random variable having normal distribution with parameters μ and σ, then

(a) $E(X) = \mu$
(b) $\mathrm{Var}(X) = \sigma^2$

Proof:

(a) If X has normal density $n_{\mu, \sigma}(x)$, then

$$E(X) = \frac{1}{\sigma \sqrt{2\pi}} \int_{-\infty}^\infty t \exp\left[-\frac{1}{2} \left(\frac{t - \mu}{\sigma} \right)^2 \right] dt$$

Making the change of variables $z = (t - \mu)/\sigma$, we obtain

$$E(X) = \frac{1}{\sqrt{2\pi}} \int_{-\infty}^\infty (\sigma z + \mu) e^{-(1/2)z^2} \, dz$$

$$= \frac{\sigma}{\sqrt{2\pi}} \int_{-\infty}^\infty z e^{-(1/2)z^2} \, dz + \frac{\mu}{\sqrt{2\pi}} \int_{-\infty}^\infty e^{-(1/2)z^2} \, dz$$

Since $z e^{-(1/2)z^2}$ is an odd function, the first of these integrals is zero. Applying Proposition 8.18 to the second integral, we obtain

$$E(X) = \frac{\sigma}{\sqrt{2\pi}} \cdot 0 + \mu \cdot 1 = \mu$$

(b) Likewise,

$$E(X^2) = \frac{1}{\sigma \sqrt{2\pi}} \int_{-\infty}^\infty t^2 \exp\left[-\frac{1}{2} \left(\frac{t - \mu}{\sigma} \right)^2 \right] dt$$

and again setting $z = (t - \mu)/\sigma$, we obtain

$$E(X^2) = \frac{1}{\sqrt{2\pi}} \int_{-\infty}^{\infty} (\sigma z + \mu)^2 e^{-(1/2)z^2} \, dz$$

$$= \frac{\sigma^2}{\sqrt{2\pi}} \int_{-\infty}^{\infty} z^2 e^{-(1/2)z^2} \, dz + \frac{2\sigma\mu}{\sqrt{2\pi}} \int_{-\infty}^{\infty} z e^{-(1/2)z^2} \, dz$$

$$+ \frac{\mu^2}{\sqrt{2\pi}} \int_{-\infty}^{\infty} e^{-(1/2)z^2} \, dz$$

$$= \frac{\sigma^2}{\sqrt{2\pi}} \int_{-\infty}^{\infty} z^2 e^{-(1/2)z^2} \, dz + \mu^2$$

Integration by parts (let $u = z$ and $dv = z e^{-(1/2)z^2}$) gives

$$\frac{1}{\sqrt{2\pi}} \int_{-\infty}^{\infty} z^2 e^{-(1/2)z^2} \, dz = 1$$

and so $E(X^2) = \sigma^2 + \mu^2$. As a result,

$$\text{Var}(X) = (\sigma^2 + \mu^2) - \mu^2 = \sigma^2$$

Proposition 8.17 now takes on added significance. The normal density function $n_{0,1}(x)$ gives rise to a probability distribution which has mean 0 and standard deviation 1—in other words, it is a standardized distribution. And Proposition 8.17 indicates that the calculations (integrations) involving *any* normal distribution $N_{\mu,\sigma}(x)$ can be performed using the standard normal distribution $N_{0,1}(x)$, after a suitable change of variables has been made.

Unfortunately, the function $e^{-(1/2)x^2}$ cannot be integrated in closed form—there is no antiderivative, at least not one that can be expressed by some functional rule. Therefore, the values of $N_{0,1}(x)$ must be obtained by some method of integral approximation. The table of values in Appendix C, which are accurate to four decimal places, were obtained by means of Simpson's rule for integration.* The values of x are indicated in the left-hand column of the table, with increments of 0.01 indicated across the top of the table. Notice that only positive values of x are included in the table. The value in the table corresponding to x is the area under the standard normal density $n_{0,1}$ between 0 and x. According to Proposition 8.17, the area under this curve from $-\infty$ to $+\infty$ is 1, and since the curve is symmetric about $x = 0$, the area from $-\infty$ to 0 is $\frac{1}{2}$, as is the area from 0 to $+\infty$. Therefore, the value of $N_{0,1}(x)$, for $x \geq 0$, can be obtained by adding 0.5000 to the value found in the table. (Figure 8.14).

Example 1 $N_{0,1}(2.48) = 0.4934 + 0.5000 = 0.9934$ (Figure 8.15)

* G. B. Thomas, *Calculus and Analytic Geometry* (Reading, Massachussets: Addison-Wesley 1968).

Example 2 Suppose that X is a random variable with standard normal density and we wish to calculate $P(0.48 \leq X \leq 1.77)$. Using the normal table, we find that

$$P(X \leq 0.48) = 0.1844 + 0.5000 = 0.6844$$

and

$$P(X \leq 1.77) = 0.4616 + 0.5000 = 0.9616$$

From Figure 8.16, it is evident that

$$\{0.48 \leq X \leq 1.77\} = \{X \leq 1.77\} - \{X \leq 0.48\}$$

and so

$$P(0.48 \leq X \leq 1.77) = P(X \leq 1.77) - P(X \leq 0.48)$$
$$= 0.9616 - 0.6844 = 0.2772$$

Figure 8.14

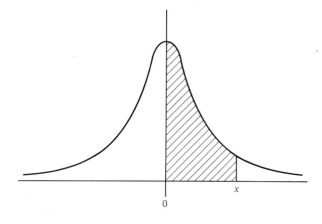

Figure 8.15

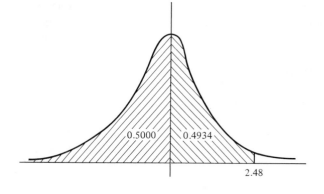

Example 3   If X has standard normal density and we wish to calculate $P(-1.33 \leq X \leq 0.76)$, then we proceed as follows (Figure 8.17):

$$P(X \leq 1.33) = 0.4082 + 0.5000 = 0.9082$$

so by symmetry,

$$P(X \geq -1.33) = 0.9082$$

and so

$$P(X \leq -1.33) = 0.0918$$

Since

$$P(X \leq 0.76) = 0.2764 + 0.5000 = 0.7764$$

we have

$$P(-1.33 \leq X \leq 0.76) = 0.7764 - 0.0918 = 0.6846$$

Example 4   Suppose that the temperature T in New York during May is normally distributed with mean $68°$ and standard deviation $6°$. We wish to calculate the probability that, at any given moment, the temperature is between $70°$ and $80°$. We therefore must calculate $P(70 \leq T \leq 80)$ but have no means to do so, since T is not standard normal, and therefore, the table of normal probabilities does not apply to T. What we must do is convert T

Figure 8.16

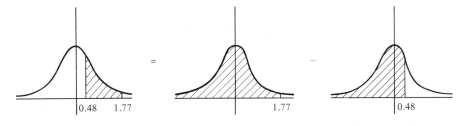

Figure 8.17

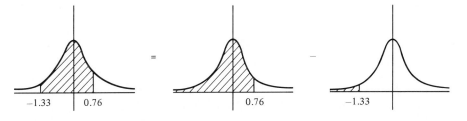

and the two values 70° and 80° to standard normal units. To do this, we make the change of variables

$$T' = \frac{T - 68}{6}$$

Then

$$70' = \frac{70 - 68}{6} = \frac{1}{3}$$

and

$$80' = \frac{80 - 68}{6} = 2$$

By Proposition 8.17,

$$P(70 \leq T \leq 80) = P(\tfrac{1}{3} \leq T' \leq 2)$$

Using the table of normal probabilities, we find

$$P(T' \leq 2) = 0.9772$$

and

$$P(T' \leq \tfrac{1}{3}) = 0.6293$$

and so

$$P(\tfrac{1}{3} \leq T' \leq 2) = P(T' \leq 2) - P(T' \leq \tfrac{1}{3})$$
$$= 0.9772 - 0.6293 = 0.3479.$$

Example 5 Suppose that heights H of high-school students are normally distributed with mean 66 in. and standard deviation 5 in. We wish to estimate the number of students in a group of 800 students who are likely to have a height of at least six feet. In standard units,

$$72 \text{ in.} \longleftrightarrow \frac{72 - 66}{5} = 1.2$$

and so we seek the probability (Figure 8.18)

$$P(H' \geq 1.2)$$

From the table of normal probabilities, we have

$$P(H' \leq 1.2) = 0.3849 + 0.5000 = 0.8849$$

and so

$$P(H' \geq 1.2) = 1 - P(H' \leq 1.2) = 0.1151$$

Therefore, we approximate that 11.51 percent of the 800 students, or approximately 92 students, will have a height of at least six feet.

Example 6 A machine produces items which are supposed to measure 2 in. in diameter, with a tolerance of 0.01 in. in either direction. If it is known that these measurements are normally distributed with mean 2 in. and standard deviation 0.005 in., what percentage of the items produced by the machine are regarded as being "defective." If we let X denote the diameter of an item chosen at random from the produce of the machine, then we wish to calculate $P(1.99 \text{ in.} \leq X \leq 2.01 \text{ in.})$. In standard normal units, we have

$$1.99 \longleftrightarrow \frac{1.99 - 2.00}{0.005} = -2$$

and

$$2.01 \longleftrightarrow \frac{2.01 - 2.00}{0.005} = +2$$

From our table of normal probabilities, we find that

$$P(0 \leq X' \leq 2) = 0.4773$$

and so

$$P(-2 \leq X' \leq +2) = 0.4773 + 0.4773 = 0.9546$$

Thus, more than 95 percent of the items produced by the machine are acceptable—they measure within 0.01 in. of 2 in.

We conclude this section by determining the moment-generating function of normal random variables.

Proposition 8.20

If X has normal distribution with parameters μ and σ, then

$$M_X(t) = \exp(\mu t + \tfrac{1}{2}t^2\sigma^2)$$

In particular, if X has standard normal distribution, then $M_X(t) = e^{(1/2)t^2}$.

Figure 8.18

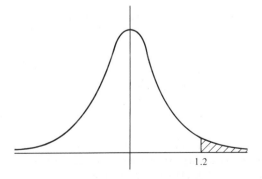

1.2

Proof: By definition,

$$M_X(t) = \int_{-\infty}^{\infty} e^{xt} \frac{1}{\sigma\sqrt{2\pi}} \exp\left[-\frac{1}{2}\left(\frac{x-\mu}{\sigma}\right)^2\right] dx$$

$$= \frac{1}{\sigma\sqrt{2\pi}} \int_{-\infty}^{\infty} \exp\{-(\tfrac{1}{2}\sigma^2)[-2xt\sigma^2 + (x-\mu)^2]\} \, dx$$

since

$$xt - \frac{1}{2}\left(\frac{x-\mu}{\sigma}\right)^2 = \frac{2xt\sigma^2 - (x-\mu)^2}{2\sigma^2}$$

but

$$(x-\mu)^2 - 2xt\sigma^2 = [x - (\mu + t\sigma^2)]^2 - 2\mu t\sigma^2 - t^2\sigma^4$$

and so

$$M_X(t) = \frac{1}{\sigma\sqrt{2\pi}} \int_{-\infty}^{\infty} \exp\{-(\tfrac{1}{2}\sigma^2)[x - (\mu + t\sigma^2)]^2 - 2\mu t\sigma^2 - t^2\sigma^4\} \, dx$$

$$= \exp(2\mu t\sigma^2 + t^2\sigma^4/2\sigma^2)/\sigma\sqrt{2\pi} \int_{-\infty}^{\infty} \exp\{-(\tfrac{1}{2}\sigma^2)[x - (\mu + t\sigma^2)]^2\} \, dx$$

$$= \exp(\mu t + \tfrac{1}{2}t^2\sigma^2)/\sigma\sqrt{2\pi} \int_{-\infty}^{\infty} \exp\left\{-\frac{1}{2}\left[\frac{x - (\mu + t\sigma^2)}{\sigma}\right]^2\right\} \, dx$$

$$= \exp(\mu t + \tfrac{1}{2}t^2\sigma^2)N_{\mu + t\sigma^2,\,\sigma}(\infty)$$

and since the latter term equals 1, we have

$$M_X(t) = \exp(\mu t + \tfrac{1}{2}t^2\sigma^2)$$

as required. Substituting $\mu = 0$ and $\sigma = 1$ into this equation, we find that the moment-generating function for a standard normal random variable is

$$M_X(t) = \exp(0t + \tfrac{1}{2}t^2 1^2) = \exp(\tfrac{1}{2}t^2)$$

EXERCISES

1. Suppose that X has standard normal distribution. Find

   $$P(X \le 1.32), \, P(0.33 \le X \le 1.35), \, P(X \le -1.56), \, P(-1.29 \le X \le 0.34)$$

   and $P(-1.80 \le X \le -0.45)$.
2. Suppose that X has normal distribution with parameters $\mu = 10$ and $\sigma = 2$. Find $P(X \le 12), P(X \ge 11.5), P(9.24 \le X \le 10.90), P(X \le 9.00)$, and $P(7.50 \le X \le 9.50)$.
3. Suppose that X_1 and X_2 are independent normal random variables with means μ_1 and μ_2, and standard deviations σ_1 and σ_2, respectively. Prove that $X_1 \pm X_2$ is normal with mean $\mu_1 \pm \mu_2$ and standard deviation $\sqrt{\sigma_1^2 + \sigma_2^2}$. (*Hint*: See Lemma 8.21.)
4. Suppose that grade-point averages are normally distributed with mean 2.6 and standard deviation 0.45. A student is placed on probation if his average is below 2.0. What is the probability that a student will be placed on probation?

5. Suppose that the annual snowfall in Albany is normally distributed with an average of 46 in. and a standard deviation of 4 in. Over a span of 50 years, how many years can be expected to have a total snowfall of between 40 and 50 in.?
6. The top 15 percent of the students in a class will get A's, while the bottom 10 percent will fail. If the grades are normally distributed with mean 75 and standard deviation 10, determine the lowest grade to receive an A and the lowest grade that passes.
7. Golf ball diameters are normally distributed with mean 1.96 in. and standard deviation 0.04 in. A ball fails to meet USGA specifications if its diameter is less than 1.90 in. or more than 2.02 in. What percentage of golf balls fail to meet specifications?
8. Suppose that X_1 and X_2 are independent normal random variables with means μ_1 and μ_2 both zero, and standard deviations σ_1 and σ_2, respectively. Find the density function of X_1/X_2.
9. Suppose that petal lengths for plants of a certain species are normally distributed and average 3.2 cm with a standard deviation of 1.2 cm. What is the probability that a plant of the species will have petal length exceeding 5 cm?
10. Suppose that the scores on an IQ exam are normally distributed with mean 100 and standard deviation 15. What percentage score higher than 120 on this exam?

8.5 The Central Limit Theorem

We come now to a remarkable and highly important result which is the cornerstone of statistical inference. It is called the central limit theorem, and without it many of the calculations required in this field would be virtually impossible.

The proof of the central limit theorem is rather involved. We will present here a proof which makes use of the moment-generating functions. However, we will have to assume, without proof, the following two results about moment-generating functions, both of which should appear quite plausible to the reader.

Lemma 8.21

When the moment-generating function of a random variable exists, the probability distribution of the random variable is completely determined. In other words, the existence of all the moments of a random variable completely determines the structure of the random variable.

Thus, if we are able to determine and recognize the moment-generating function of a random variable, we have also identified the random variable.

Lemma 8.22

If $X_1, X_2, \ldots, X_n, \ldots$ is a sequence of random variables such that, for any t,

$$\lim_{n \to \infty} M_{X_i}(t) = M_Y(t)$$

for some random variable Y, then the probability distribution of the X_i's approaches the probability distribution of Y.

Theorem 8.23 Central Limit Theorem

Suppose that $X_1, X_2, \ldots, X_n$ are independent, identically distributed random variables with common mean μ, common standard deviation σ, and common moment-generating function $M_X(t)$, and suppose that $\bar{X} = X_1 + X_2 + \cdots + X_n$. Then for sufficiently large n the probability distribution of $\bar{X}$ is approximately equal to the normal distribution with mean $n\mu$ and standard deviation $\sigma\sqrt{n}$, or equivalently, the standardized random variable

$$X' = \frac{\bar{X} - n\mu}{\sigma\sqrt{n}}$$

is approximately standard normal.

Proof: By Proposition 4.17(c), we have

$$M_{X'}(t) = M_{(\bar{X} - n\mu)/\sigma\sqrt{n}}(t) = \exp\left(\frac{-n\mu t}{\sigma\sqrt{n}}\right) M_X\left(\frac{t}{\sigma\sqrt{n}}\right)$$

which, in turn, by Corollary 5.6, gives

$$M_{X'}(t) = \exp\left(\frac{-n\mu t}{\sigma\sqrt{n}}\right)\left(M_X\left(\frac{t}{\sigma\sqrt{n}}\right)\right)^n$$

Applying the natural logarithm to both sides of this equation, we obtain

$$\log_e[M_{X'}(t)] = \frac{-n\mu}{\sigma\sqrt{n}}t + n\log_e\left(M_X\left(\frac{t}{\sigma\sqrt{n}}\right)\right)$$

By Proposition 4.15, we can write $M_X(t/\sigma\sqrt{n})$ as a power series in t, which gives us

$$\log_e[M_{X'}(t)] = \frac{-n\mu}{\sigma\sqrt{n}}t + n\log_e\left(\sum_{k=0}^{\infty} \mu_k' \frac{(t/\sigma\sqrt{n})^k}{k!}\right)$$

where μ_k' is the kth moment about the origin of the identically distributed random variables $X_1, X_2, \ldots, X_n$. We rewrite this in the form

$$\log_e[M_{X'}(t)] = \frac{-n\mu}{\sigma\sqrt{n}}t + n\log_e\left(1 + \sum_{k=1}^{\infty} \frac{\mu_k' t^k}{\sigma^k n^{k/2} k!}\right)$$

and then use the Maclaurin expansion

$$\log_e(1 + x) = \sum_{k=1}^{\infty} (-1)^{k+1} \frac{x^k}{k}$$

which converges when $|x| < 1$, to obtain

$$\log_e[M_{X'}(t)] = \frac{-n\mu}{\sigma\sqrt{n}}t$$

$$+ n\left[\sum_{k=1}^{\infty} \frac{\mu_k' t^k}{\sigma^k n^{k/2} k!} - \frac{1}{2}\left(\sum_{k=1}^{\infty} \frac{\mu_k' t^k}{\sigma^k n^{k/2} k!}\right)^2 + \frac{1}{3}\left(\sum_{k=1}^{\infty} \frac{\mu_k' t^k}{\sigma^k n^{k/2} k!}\right)^3 - + - + \cdots\right]$$

provided that n is chosen large enough so that the term

$$\left| \sum_{k=1}^{\infty} \frac{\mu_k' t^k}{\sigma^k n^{k/2} k!} \right| < 1$$

Rearranging these terms in increasing powers of t, we find that $\log_e[M_{X'}(t)]$ equals

$$\left(\frac{-\sqrt{n}\mu}{\sigma} + \frac{\sqrt{n}\mu_1'}{\sigma} \right) t + \left[\frac{\mu_2'}{2\sigma^2} - \frac{(\mu_1')^2}{2\sigma^2} \right] t^2$$

$$+ \left[\frac{\mu_3'}{6\sigma^3\sqrt{n}} - \frac{\mu_1'\mu_2'}{2\sigma^3\sqrt{n}} + \frac{(\mu_1')^3}{3\sigma^3\sqrt{n}} \right] t^3 + - \cdots$$

Since $\mu_1' = \mu$, the coefficient of t reduces to zero, and since $\mu_2' - (\mu_1')^2 = \sigma^2$, we find that $\log_e[M_{X'}(t)]$ equals

$$\frac{t^2}{2} + \left(\frac{\mu_3' - 3\mu_1'\mu_2' + 2(\mu_1')^3}{6\sigma^3\sqrt{n}} \right) t^3 + \cdots$$

Notice that the coefficient of t^3 contains the term $\sqrt{n}$ in its denominator. In general, the coefficient of t^k will contain the term $n^{(k-2)/2}$ in its denominator. Consequently,

$$\lim_{n \to \infty} \log_e[M_{X'}(t)] = \frac{t^2}{2}$$

with the higher power terms in t vanishing in the limit. By the continuity of the logarithm function, we then have

$$\lim_{n \to \infty} M_{X'}(t) = e^{(1/2)t^2}$$

which, by Proposition 8.20, is the moment-generating function of the standard normal distribution. Combining the results of Lemmas 8.21 and 8.22, we find that the probability distribution of the random variable X' approaches that of the standard normal distribution as n grows large.

The most remarkable thing about the central limit theorem is the fact that the probability distribution of the random variables $X_1, X_2, \ldots, X_n$ is not important. Whether this distribution be uniform, binomial, Poisson, normal, etc., the distribution of their sum, $\bar{X}$, begins to take on aspects of the bell-shaped normal curve after n is only moderately large. As a matter of fact, the resemblance is quite good, even when n is as low as 10.

Example 1 The heights of 100 people are recorded, each being rounded off to the nearest inch. We wish to calculate the probability that the total roundoff error will not exceed 5 in. We define 100 random variables.

X_k = roundoff error on kth person's height

$k = 1, 2, \ldots, 100$. We may assume that each of these random variables has uniform distribution on the interval $[-\frac{1}{2}, \frac{1}{2}]$, and so $E(X_k) = 0$ and

$\text{Var}(X_k) = \frac{1}{12}$, for $k = 1, 2, \ldots, 100$. The total roundoff error is then

$$\bar{X} = X_1 + X_2 + \cdots + X_{100}$$

which, according to the central limit theorem, is approximately normal with mean $100 \cdot 0 = 0$ and standard deviation $\sqrt{100} \cdot \sqrt{1/12} = 5/\sqrt{3} = 2.887$. The probability we must calculate is $P(-5 \leq \bar{X} \leq 5)$. In standard normal units,

$$5 \longleftrightarrow \frac{5 - 0}{2.887} = 1.732$$

and so we must calculate

$$P(-1.732 \leq X' \leq 1.732)$$

where $X' = (X - 0)/2.887$. From the table of normal probabilities, we find that $P(0 \leq X' \leq 1.732) = 0.4584$, and so $P(-1.732 \leq X' \leq 1.732) = 0.9168$. Therefore, there is a better than 91 percent chance that the total roundoff error will not exceed 5 in.

Table 8.1
Poisson Simulation

Value of X	Frequency	Value of X	Frequency
19	1	42	573
20	2	43	566
21	4	44	476
22	5	45	454
23	12	46	331
24	18	47	309
25	26	48	248
26	44	49	206
27	89	50	167
28	114	51	133
29	125	52	110
30	183	53	83
31	236	54	71
32	319	55	48
33	333	56	29
34	442	57	21
35	508	58	19
36	558	59	5
37	598	60	10
38	579	61	3
39	627	62	1
40	664	63	1
41	648	64	1
		65	1

Example 2 Suppose that $X = X_1 + X_2 + \cdots + X_{20}$, where, for each $k = 1, 2, \ldots, 20$, the random variable X_k has Poisson distribution with parameter $\lambda = 2$. According to the central limit theorem, X should have a probability distribution that is approximately normal with mean $20 \cdot 2 = 40$ (since the mean of each Poisson random variable is 2). A computer simulation of 10,000 different values of X gave the results listed in Table 8.1. A histogram, rounding these frequencies to the nearest multiple of 30, is shown in Figure 8.19. Notice that the histogram is shaped very much like a normal distribution. Also note that the peak of the histogram occurs at 40, as would be expected.

The central limit theorem can be applied to the sample mean (for sampling *with* replacement) because, in this case, the sampling random variables $X_1, X_2, \ldots, X_n$ are independent and identically distributed.

Proposition 8.24

If $\bar{X}$ is the sample mean for samples of size n taken with replacement from a population with mean μ and standard deviation σ, then $\bar{X}$ is closely approximated by the normal distribution with mean μ and standard deviation $\sigma/\sqrt{n}$. Equivalently, the standardized sample mean $\bar{X}' = (\bar{X} - \mu)/(\sigma/\sqrt{n})$ is approximately standard normal.

Proof: We have

$$\bar{X} = \frac{X_1 + X_2 + \cdots + X_n}{n}$$

where the $\{X_i\}$ are independent and identically distributed since the sampling

Figure 8.19

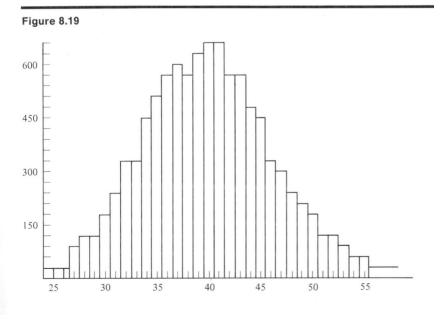

is done with replacement. By the central limit theorem,

$$\frac{(X_1 + X_2 + \cdots + X_n) - n\mu}{\sigma\sqrt{n}}$$

is approximately standard normal. However,

$$\frac{\bar{X} - \mu}{\sigma/\sqrt{n}} = \frac{(X_1 + X_2 + \cdots + X_n)/n - \mu}{\sigma/\sqrt{n}}$$

$$= \frac{(X_1 + X_2 + \cdots + X_n) - n\mu}{\sigma\sqrt{n}}$$

and therefore, the standardized sample mean is approximately standard normal.

It is this fact, that the sample mean is approximately normal, that we shall exploit in our study of statistical inference.

Example 3 Referring back to Example 8 in Section 8.3, suppose that we observe this man for a period of 100 days, and record his arrival time at the railroad station on each of these days. We wish to determine the probability that his average arrival time for the 100 days will lie between 7:47 and 7:48. This average arrival time is the sample mean $\bar{X}$ for 100 trials of the uniform distribution over the interval $[15, 20]$ (or equivalently, $[7:45, 7:50]$). By Proposition 8.16, $E(X) = 17.5$ (or 7:45.5) and $\text{Var}(X) = \frac{25}{12}$ minutes. The standardized sample mean

$$\bar{X}' = \frac{\bar{X} - 17.5}{\dfrac{\sqrt{25/12}}{\sqrt{100}}}$$

$$= \frac{\bar{X} - 17.5}{0.144}$$

is approximately standard normal. In standard units,

$$17 \longleftrightarrow \frac{17 - 17.5}{0.144} = -3.464$$

and

$$18 \longleftrightarrow \frac{18 - 17.5}{0.144} = 3.464$$

and so

$$P(17 \le \bar{X} \le 18) = P(-3.464 \le \bar{X}' \le 3.464)$$

$$= 2 \cdot P(0 \le \bar{X}' \le 3.464)$$

$$= 2 \cdot 0.4997$$

$$= 0.9994$$

Example 4 Packages that are delivered to a certain factory have mean weight 300 pounds and standard deviation 50 pounds. Suppose that 25 packages are loaded on a truck for delivery to this factory. We wish to calculate the probability that the total weight of the 25 packages exceeds the 8200 pound capacity of the truck. This problem can be restated in terms of the average of the 25 packages as follows: What is the probability that the average weight of the sample exceeds $8200/25 = 328$ pounds per package? Our problem therefore concerns the sample mean for samples of size 25 taken from a population with mean $\mu = 300$ pounds and standard deviation $\sigma = 50$ pounds. Hence, the sample mean $\bar{X}$ has mean $\mu = 300$ pounds and standard deviation $\sigma/\sqrt{n} = 10$ pounds, and is approximately normal. We wish to determine the probability $P(\bar{X} > 328)$. In standard units,

$$328 \longleftrightarrow \frac{328 - 300}{10} = 2.8$$

and so

$$P(\bar{X} > 328) = P(X' > 2.8) = 0.0026$$

If a computer programmer wishes to generate a population that is uniformly distributed, he simply uses a random number generator. If he should wish to generate a population that is normally distributed, he might, using a random number generator, generate samples of size n, for some specific value of n. According to the central limit theorem, the averages of these samples are normally distributed, and so they (the averages) can be used as the desired normal population. The population discussed in Section 8.1, the heights of 10,000 people, was generated in exactly this way, using $n = 8$.

We have already seen that the Poisson distribution can be used to approximate the binomial distribution when the binomial parameter p is close to either 0 or 1. In any other case, we have the following:

Proposition 8.25

If n is large and neither p nor q are close to 0, then the binomial distribution $b(k; n, p)$ is closely approximated by the normal distribution with mean np and standard deviation $\sqrt{npq}$.

Proof: We define n indicator random variables X_k as follows:

$$X_k = \begin{cases} 1 & \text{if a success occurs on the } k\text{th trial} \\ 0 & \text{otherwise} \end{cases}$$

If X is binomial with parameters n and p, then

$$X = X_1 + X_2 + \cdots + X_n$$

and the $\{X_i\}$ are independent and identically distributed. By the central limit theorem, then X is approximately normal with mean $nE(X_k) = np$ and variance $n \operatorname{Var}(X_k) = npq$, and, therefore, standard deviation $\sqrt{npq}$.

It was in this form that the central limit theorem was first "discovered" (not proved) by de Moivre in 1793.

According to Proposition 8.25, a binomial probability of the form

$$\sum_{k=a}^{b} b(k; n, p)$$

can be approximated by the integral

$$\frac{1}{\sqrt{2\pi}} \int_{(a-np)/\sqrt{npq}}^{(b-np)/\sqrt{npq}} e^{-(1/2)x^2} \, dx$$

Notice that if either p or q were close to zero, the denominators $\sqrt{npq}$ in the two limits of integration would be very close to zero. Both terms then become exceedingly large, and the normal approximation to the binomial, in this case, becomes very poor.

Example 5 A fair coin is tossed 12 times, and we wish to determine the probability that the number of heads that occur is between 4 and 7. Using the binomial distribution with parameters $n = 12$ and $p = \frac{1}{2}$, we can calculate the exact probability:

$$P(X = 4) = b(4; 12, \tfrac{1}{2}) = \frac{495}{4096}$$

$$P(X = 5) = b(5; 12, \tfrac{1}{2}) = \frac{792}{4096}$$

$$P(X = 6) = b(6; 12, \tfrac{1}{2}) = \frac{924}{4096}$$

$$P(X = 7) = b(7; 12, \tfrac{1}{2}) = \frac{792}{4096}$$

Adding, we find

$$P(4 \leq X \leq 7) = \sum_{k=4}^{7} b(k; 12, \tfrac{1}{2}) = 0.7332$$

If we wish to use the normal probability table to approximate this probability, we must first decide which normal curve to use. According to Proposition 8.25, we should use the normal distribution which has the same mean and standard deviation as the binomial. Hence, we will use the normal curve with parameters $\mu = 12 \cdot \frac{1}{2} = 6$ and $\sigma = \sqrt{12 \cdot \frac{1}{2} \cdot \frac{1}{2}} = 1.732$. This normal curve is going to be used to approximate the histogram in Figure 8.20, where the cross-hatched area represents the probability that $4 \leq X \leq 7$. This area is approximated by the area under $n_{6, 1.732}$ between $x = 3.5$ and $x = 7.5$ (rather than between $x = 4$ and $x = 7$) because the four bars of the histogram representing $P(4 \leq X \leq 7)$ extend from $x = 3.5$ to $x = 7.5$. In standard units

$$3.5 \longleftrightarrow \frac{3.5 - 6}{1.732} = -1.45$$

and

$$7.5 \longleftrightarrow \frac{7.5 - 6}{1.732} = 0.87$$

and so

$$
\begin{aligned}
P(4 \leq X \leq 7) &= P(-1.45 \leq X' \leq 0.87) \\
&= 0.8078 - (1.0 - 0.9265) \\
&= 0.7345
\end{aligned}
$$

Thus, our approximation contains an error of 0.0011, which is really not very large, especially considering that the sample size $n = 12$ was quite small.

It should be emphasized that a normal approximation to a binomial distribution entails the use of a normal curve as an approximation to a histogram. Care should be taken that the limits of integration correspond to the extremities of the bars whose area is to be approximated.

Example 6 We wish to estimate the probability that digit 7 occurs at most 950 times among 10,000 random digits. Here we have $n = 10,000$ Bernoulli trials, each with probability of success (occurrence of digit 7) being $p = \frac{1}{10}$. If X denotes the number of 7's occurring in 10,000 trials, then we must calculate

$$P(X \leq 950) = \sum_{k=0}^{950} b(k; 10,000, 0.10)$$

Figure 8.20

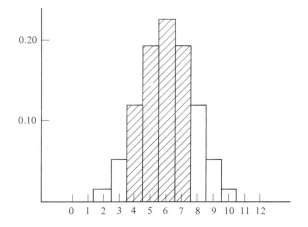

It would indeed be a time consuming task to calculate the exact value of this probability using the binomial distribution. With the number of trials being so large, we would expect the normal approximation to be very accurate. We shall approximate the area in the bars of the histogram from $k = 0$ to $k = 950$ by the area from 0 to 950.5 under the normal curve with mean $10{,}000 \cdot \frac{1}{10} = 1{,}000$ and standard deviation $\sqrt{10{,}000 \cdot \frac{1}{10} \cdot \frac{9}{10}} = 30$. In standard units,

$$0 \longleftrightarrow \frac{0 - 1000}{30} = \frac{-100}{3}$$

and

$$950.5 \longleftrightarrow \frac{950.5 - 1000}{30} = -1.65$$

and so we approximate

$$P(X \leq 950) = P(0 \leq X \leq 950.5)$$

$$= P\left(\frac{-100}{3} \leq X' \leq -1.65\right)$$

$$= P(X' \leq -1.65) = 0.0495$$

Thus, the probability that digit 7 occurs at most 950 times among 10,000 random digits is approximately 0.0495.

Example 7 An airline has statistics which indicate that five percent of the people who book a flight will fail to show up for the flight. For a flight booked full for 300 people, the airline sells 20 standby tickets. We wish to calculate the probability that all 20 standbys get seated on the flight. The 20 standbys will be seated if 20 or more of the original 300 ticket holders fail to make the flight. The probability that this will happen is

$$\sum_{k=20}^{300} b(k; 300, 0.05)$$

which we will approximate as the area under the normal curve with mean $300 \cdot 0.05 = 15$ and standard deviation $\sqrt{300 \cdot 0.05 \cdot 0.95} = 3.775$ between 19.5 and 300.5 (or, essentially, between 19.5 and positive infinity). In standard units

$$19.5 \longleftrightarrow \frac{19.5 - 15}{3.775} = 1.19$$

and so the probability we seek, $P(X \geq 20)$, where X denotes the number of ticket holders who fail to make the flight, equals

$$P(X \geq 20) = P(X' \geq 1.19) = 0.1170$$

Another consequence of the two lemmas that were crucial in the proof of the central limit theorem is the following:

Proposition 8.26

If X denotes a random variable having gamma distribution with parameters α and β, then, as $\alpha \to \infty$ (while β remains constant), the distribution of X becomes approximately normal.

Proof: We will prove that the moment-generating function of the standardized gamma distribution approaches the moment-generating function of the standard normal distribution as $\alpha \to \infty$. The desired result will then follow from Lemmas 8.21 and 8.22. If X has gamma distribution with parameters α and β, then $E(X) = \alpha\beta$, $\text{Var}(X) = \alpha\beta^2$, and $M_X(t) = (1 - \beta t)^{-\alpha}$. Consequently, the standardized form of X is

$$X' = \frac{X - \alpha\beta}{\sqrt{\alpha\beta^2}}$$

By Proposition 4.17(c),

$$M_{X'}(t) = \exp\left[-\left(\frac{\alpha\beta}{\sqrt{\alpha}\,\beta}\right)t\right] M_X\left(\frac{t}{\sqrt{\alpha}\,\beta}\right)$$

$$= e^{-\sqrt{\alpha}\,t} M_X\left(\frac{t}{\beta\sqrt{\alpha}}\right)$$

and so

$$\log_e M_{X'}(t) = -\sqrt{\alpha}\,t + \log_e M_X\left(\frac{t}{\beta\sqrt{\alpha}}\right)$$

$$= -\sqrt{\alpha}\,t - \alpha \log_e\left(1 - \frac{t}{\sqrt{\alpha}}\right)$$

Using the Maclaurin expansion

$$\log_e(1 + x) = \sum_{k=1}^{\infty} (-1)^{k+1} \frac{x^k}{k}$$

which converges when $|x| < 1$, we obtain

$$\log_e M_{X'}(t) = -\sqrt{\alpha}\,t - \alpha\left[\frac{-t}{\sqrt{\alpha}} - \frac{1}{2}\left(\frac{t}{\sqrt{\alpha}}\right)^2 - \left(\frac{1}{3}\right)\left(\frac{t}{\sqrt{\alpha}}\right)^3 - \cdots\right]$$

$$= -\sqrt{\alpha}\,t + \sqrt{\alpha}\,t + \frac{t^2}{2} + \frac{t^3}{3\sqrt{\alpha}} + \frac{t^4}{4\alpha} + \cdots$$

where α is chosen large enough so that $|t/\sqrt{\alpha}| < 1$. Hence,

$$\log_e M_{X'}(t) = \frac{t^2}{2} + R$$

where the term R contains α in its denominator. Consequently,

$$\lim_{\alpha \to \infty} \log_e M_{X'}(t) = \frac{t^2}{2} + \lim_{\alpha \to \infty} R = \frac{t^2}{2}$$

and therefore,

$$\lim_{\alpha \to \infty} M_{X'}(t) = e^{(1/2)t^2}$$

the moment-generating function of the standard normal distribution.

Example 8   Suppose that family income (in units of $1000) in a certain community has gamma distribution with parameters $\alpha = 100$ and $\beta = \frac{1}{10}$. We wish to calculate the probability that a family living in this community will have a total income in excess of $15,000. The average family in this community has an income of $\alpha\beta = 100 \cdot \frac{1}{10} = 10$ units, or $10,000. If we let X represent the income of a family chosen at random, then the probability we seek is $P(X > 15)$. Using the gamma distribution, this would involve evaluating the integral

$$\int_{15}^{\infty} x^{99} e^{-10x} \, dx$$

which would be quite tedious. However, the value of α is fairly large, so we can use the normal curve with parameters $\mu = \alpha\beta = 100 \cdot \frac{1}{10} = 10$ and $\sigma^2 = \alpha\beta^2 = 100 \cdot (\frac{1}{10})^2 = 1$ to approximate the gamma distribution above. In standard units,

$$15 \longleftrightarrow \frac{15 - 10}{1} = 5$$

and according to the normal probability tables, the probability that $X' > 5$ is negligible. Therefore, it is almost impossible for a family in this community to make in excess of $15,000 per year.

We conclude this chapter by considering a special instance of the sample mean. Suppose that we have a population, a certain proportion p of which are of a certain type (possess a certain characteristic, or set of characteristics). Such a population is called a Bernoulli population with parameter p. Now suppose that we take a sample of n items with replacement from this population. We define n identically distributed random variables

$$X_k = \begin{cases} 1 & \text{if the } k\text{th item is of the given type} \\ 0 & \text{otherwise} \end{cases}$$

each of which has mean p and variance pq (where $q = 1 - p$). The sample mean

$$\bar{X} = \frac{X_1 + X_2 + \cdots + X_n}{n}$$

is actually just the proportion of elements of the given type in the sample, and for this reason, the sample mean, in this case, is called the *sample proportion* and is denoted $\bar{p}$.

Proposition 8.27

If $\bar{p}$ is the sample proportion for samples of size n taken with replacement from a Bernoulli population with parameter p, then

(a) $\bar{p}$ is approximately normal, for sufficiently large n;
(b) $E(\bar{p}) = p$
(c) $\text{Var}(\bar{p}) = pq/n$

Proof:

(a) By Proposition 8.24, the sample mean is approximately normal, and by definition, the sample proportion is, in fact, a sample mean.
(b) By Proposition 5.20(a), $E(\bar{p}) = E(\bar{X}) = E(X_k) = p$
(c) By Proposition 5.20(b),

$$\text{Var}(\bar{p}) = \text{Var}(\bar{X}) = \frac{\text{Var}(X_k)}{n} = \frac{pq}{n}$$

Example 9 Election returns showed that a particular candidate received 46 percent of the votes. We would like to estimate, in a preelection sample of 200 voters, the probability that this candidate actually received a majority. Our population is Bernoulli—each voter either voted for or against this candidate—and the parameter p is known to be 0.46. We are inquiring about the chances of obtaining a sample of size $n = 200$ with sample proportion $\bar{p} \geq 100.5/200 = 0.5025$. Note that we use 100.5 rather than 101 because, once again, we are using a normal curve to approximate a histogram. Since the sample size $n = 200$ is fairly large, the sample proportion is approximately normal, and it has mean $E(\bar{p}) = 0.46$ and standard deviation

$$\sigma_{\bar{p}} = \sqrt{\frac{(0.46)(0.54)}{200}} = 0.0352$$

In standard units

$$0.5025 \longleftarrow \frac{0.5025 - 0.4600}{0.0352} = 1.21$$

and so

$$P(\bar{p} \geq 0.5025) = P(\bar{p}' \geq 1.21) = 0.1131$$

Thus, there is only a small probability (11 percent) that a preelection sample of 200 voters would be so misleading as to suggest the candidate has a majority of the voters in his corner. If, instead, we had chosen our sample size to be $n = 1000$ voters, then the standard deviation of the sample proportion would be

$$\sigma_{\bar{p}} = \sqrt{\frac{(0.46)(0.54)}{1000}} = 0.0158$$

and in standard units,

$$0.5005 \longleftrightarrow \frac{0.5005 - 0.4600}{0.0158} = 2.56$$

where $0.5005 = 500.5/1000$, rather than $501/1000$. Hence, in this case,

$$P(\bar{p} \geq 0.5005) = P(\bar{p}' \geq 2.56) = 0.0052$$

In other words, a sample of size 1000 makes the difference between 46 percent and a majority appear much larger than would a sample of size 200.

Example 10 It is known that two percent of the items produced by a machine are defective. We wish to find the probability that in a shipment of 400 items more than three percent are defective. Our population is Bernoulli— the items produced by the machine are either defective or not defective— and the population parameter $p = 0.02$. Consequently,

$$\sigma_{\bar{p}} = \sqrt{\frac{(0.02)(0.98)}{400}} = 0.007$$

Since three percent of 400 items is 12 items, we seek the probability that $\bar{p} \geq 12.5/400 = 0.0313$, rather than $12/400 = 0.03$. In standard units,

$$0.0313 \longleftrightarrow \frac{0.0313 - 0.0200}{0.007} = 1.607$$

and so

$$P(\bar{p} \geq 0.0313) = P(\bar{p}' \geq 1.607) = 0.0541$$

EXERCISES

1. A lot of 1000 fried chicken dinners have mean weight of 12 ounces with a standard deviation of 0.6 ounces. What is the probability that the total weight of a sample of 100 of these dinners will exceed 1195 ounces?

2. A sample of 100 is taken from a large group of items of which five percent are defective. What is the probability that fewer than 5 in the sample are defective?

3. A sample of size 10 is taken from a population which is normally distributed with mean 12 and variance 16. What is the probability that the mean of the sample is within 1 of 12?

4. Suppose that the weight of airline passengers with their luggage averages 205 pounds with standard deviation of 20 pounds. What is the probability that the weight of 144 passengers and their luggage will exceed 30,000 pounds?

5. A runner paces off 100 meters for a race. If his paces are independent and average 0.97 meters with a standard deviation of 0.1 meters, what is the probability that his 100 paces differ from 100 meters by no more than 5 meters?

6. Suppose that it is a known fact that 0.1 percent of all babies are born with birth defects. Suppose that 20,000 babies were born in a given city during a 12-month period. What is the probability that 10 or fewer were born with birth defects?

7. The mortality rate for 41-year-olds is known to be one percent. A large insurance company has issued 50,000 term (one-year) policies on people aged 41 years. What is the probability that the insurance company will have to pay off on more than 550 of these policies?

8. Suppose that the lifetime of a particular brand of flashlight battery has gamma distribution with parameters $\alpha = 25$ and $\beta = 2$ hours. Estimate the probability that one of these flashlights will last longer than 60 hours.

9. The results of an IQ test are known to be normally distributed with mean 100 and standard deviation 16. What is the probability a group of ten people will average 110 or higher?

10. Suppose that the incomes of people living in a certain community are normally distributed with mean $15,000 and standard deviation $2,000. What is the probability that the 20 people living in a certain apartment building average more than $20,000 per year?

ADDITIONAL PROBLEMS

1. A point is chosen at random in the rectangle $0 \le x \le 2, 0 \le y \le 3$. If X equals the product of the coordinates of the point, then find $P(X < 1)$.

2. A circular dart board is divided into three circular regions of radii 1, 2, and 3. The inner circle is worth five points, the middle annulus is worth three points, and the outer annulus is worth one point. Suppose that the place the dart hits the board is a continuous random variable with density function

$$f(x) = \begin{cases} xe^{-x} & \text{if } x > 0 \\ 0 & \text{otherwise} \end{cases}$$

Let X denote the value of a hit. Find $E(X)$.

3. A point is selected from the interval $[0, 2]$, and X denotes the distance from the point to the origin. If the point is twice as likely to be chosen from the interval $[0, 1]$ than from the interval $[1, 2]$, find the probability $P(X \le a)$.

4. The life span in hours of a light bulb is a random variable X with density function

$$f(x) = \begin{cases} A/x^3 & \text{if } 1500 \le x \le 2500 \\ 0 & \text{otherwise} \end{cases}$$

Determine A. If five of these bulbs are installed, what is the probability that at most one needs replacing after 2000 hours.

5. The life span of light bulbs is a continuous random variable with density function

$$f(x) = \begin{cases} e^{-x} & \text{if } x \ge 0 \\ 0 & \text{otherwise} \end{cases}$$

where x is in units of 100 hours. If it costs $3.00 to manufacture one bulb, which will be sold for $7.00 with a money back guarantee if $x < 0.5$, find the expected profit per bulb.

6. The number of inoperative milk trucks each day at a dairy has Poisson distribution with an average of 2 per day. What is the probability that in 30 days, chosen at random, the total number of inoperative milk trucks is between 50 and 60?

7. A polling organization samples 900 voters to estimate the proportion of voters planning to vote for candidate A. How large would the true proportion p have to be for candidate A to be 95 percent sure that the majority of those sampled will vote for him?

8. An urn contains 99 red balls and one black ball. A large number n of balls are drawn with replacement. How large must n be so that we can be 90 percent confident that the proportion of blacks in the sample differs from 0.01 by less than 0.05?

9

Statistical Inference

9.1 Introduction

We have already mentioned that the procedures of mathematical statistics are, in a sense, inverse to the procedures of probability theory which we have been studying. In both cases, we work with a population whose distribution is determined by a certain parameter or set of parameters. For example, the population might be Bernoulli with parameter p, or the population might be normally distributed with parameters μ and σ.

Problems in probability theory work from knowledge of the parameter(s), and therefore, complete knowledge of the distribution of the population, to statements about the relative frequencies of samples of various types chosen from the population. Problems in statistical theory, on the other hand, work from one particular sample chosen at random from the population to an estimation of the values of the parameters specifying the population. The parameters in this case are assumed to be unknown.

The following examples are typical of the type of problems we will be discussing in our study of statistical inference.

Example 1   A teacher has just experimented with a different method of teaching a particular subject. In order to determine if the new method is more effective than his previous method, the teacher compares the class's average on a certain advanced placement examination with the averages of previous classes on similar examinations. Should the present class show a higher average on the exam than did previous classes, the teacher must be able to determine if their average is *significantly* higher than previous classes', and therefore, provides solid evidence that the new teaching method is really an improvement over the older method.

Example 2   A political pollster wishes to estimate the percentage p of voters favoring a certain candidate. By interviewing a sample of N people, the pollster finds that k of them favor the candidate. With the formula k/N an estimate for p is obtained.

Example 3   A quality control inspector must estimate the true percentage p of defectives in a lot based upon the number k of defectives that are found in a sample of N items that are inspected. In this case, too, the estimate of the true value of p will be based on the quotient k/N.

The values that are obtained from samples, such as the means and proportions discussed above, are called *statistics*, or *sample statistics*. These are to be contrasted with the population parameters. The purpose of the sample statistics, of course, is to allow us to make reasonable deductions concerning the population parameters.

We have not yet mentioned how we will use the sample statistics to form an estimate for the population parameter(s). At times, we may wish to determine one value as our estimate of a population parameter, to be used in place of the parameter in all necessary calculations. An estimate of this type is called a *point estimate*; we will discuss point estimates in Section 9.2. On other occasions, we may prefer a range of values in which we may be fairly certain that the population parameter falls. An estimate of this type is called an *interval estimate*; we will discuss interval estimates in Section 9.3.

An alternate approach that we will discuss in Section 9.4 is known as *hypothesis testing*. Before the sample is taken, an educated guess (hypothesis) is made concerning the true value of the population parameter. Information from the sample is then used as evidence either supporting or refuting the hypothesis.

9.2 Point Estimators

Suppose that we have a population that is specified by a random variable X whose probability distribution (or density function) depends on an unknown parameter π. In order to estimate the value of π, we take a sample of size n, with replacement, from the population. That is, we sample n values of the random variable X, thereby defining n independent random variables $X_1, X_2, \ldots, X_n$ distributed identically to X. In this context, we make the following definition:

DEFINITION

A *point estimator* of the unknown population parameter π is any function $e(X_1, X_2, \ldots, X_n)$.

That is, given sample values $a_1, a_2, \ldots, a_n$ for the random variables $X_1, X_2, \ldots, X_n$, we simply calculate the functional value $e(a_1, a_2, \ldots, a_n)$ and use this value as our estimate for the unknown population parameter π.

Example 1 We wish to estimate the mean μ of a population. For a point estimator of μ, we define

$$e(X_1, X_2, \ldots, X_n) = (X_1 + X_2 + \cdots + X_n)/n$$

the average of the n sample values. Alternatively, we could define $e(X_1, X_2, \ldots, X_n)$ to be the median of the n sample values $X_1, X_2, \ldots, X_n$.

Thus, our point estimator function e is defined on the sample space S^n, where S represents the population from which we are sampling.

Although it is unreasonable to expect that a point estimator function will be perfect, it is not unreasonable to demand that a point estimator function do its job well on the average. So we make the following definition:

DEFINITION

A point estimator function e is called *unbiased* if

$$E\big[\{e(a_1, a_2, \ldots, a_n) : (a_1, a_2, \ldots, a_n) \in S^n\}\big] = \pi$$

An unbiased estimator, then, is one for which the average of all possible point estimates, corresponding to all possible samples of size n, is equal to the unknown population parameter.

Proposition 5.20(a) indicates that the sample mean $\bar{x}$ is an unbiased estimator for the population mean μ, and Proposition 8.27(b) indicates that the sample proportion $\bar{p}$ is an unbiased estimator for the population proportion p. On the other hand, using the definition of variance as stated in Chapter 4, we shall find that the variance of a sample will not be an unbiased estimator for the variance of the population. However, if we make the following slight modification in the definition, we will obtain an unbiased estimator.

DEFINITION

If $\{x_1, x_2, \ldots, x_n\}$ is a sample of size n taken from a population S, then we define the *sample variance* $\bar{s}^2$ to equal

$$\bar{s}^2 = \frac{\sum\limits_{i=1}^{n} (x_i - \bar{x})^2}{n - 1}$$

where $\bar{x}$ is the sample mean $(\sum_{i=1}^{n} x_i)/n$.

The sample variance $\bar{s}^2$ is related to the variance of the sample s^2 by

$$\bar{s}^2 = \left(\frac{n}{n - 1}\right)s^2$$

Proposition 9.1

The sample variance $\bar{s}^2$ is an unbiased estimator of the population variance σ^2.

Proof: By definition,

$$E(\bar{s}^2) = E\left(\frac{\sum\limits_{i=1}^{n} (x_i - \bar{x})^2}{n - 1}\right)$$

$$= \left(\frac{1}{n - 1}\right) E\left[\sum_{i=1}^{n} (x_i - \mu + \mu - \bar{x})^2\right]$$

$$= \left(\frac{1}{n - 1}\right) E\left[\sum_{i=1}^{n} (x_i - \mu)^2 - 2(x_i - \mu)(\bar{x} - \mu) + (\bar{x} - \mu)^2\right]$$

We consider these three summands one at a time. First, we find that

$$E\left[\sum_{i=1}^{n} (x_i - \mu)^2\right] = \sum_{i=1}^{n} E[(x_i - \mu)^2]$$

$$= \sum_{i=1}^{n} \sigma^2 = n\sigma^2$$

Next,

$$E\left\{\sum_{i=1}^{n} \left[-2(x_i - \mu)(\bar{x} - \mu)\right]\right\} = -2(\bar{x} - \mu) E\left[\sum_{i=1}^{n} (x_i - \mu)\right]$$

$$= -2(\bar{x} - \mu) \sum_{i=1}^{n} E(x_i - \mu) = 0$$

since $E(X - \mu_X) = 0$. Finally,

$$E\left[\sum_{i=1}^{n} (\bar{x} - \mu)^2\right] = \sum_{i=1}^{n} E[(\bar{x} - \mu)^2]$$

$$= \sum_{i=1}^{n} \left(\frac{\sigma^2}{n}\right) \qquad [\text{Proposition 5.20(b)}]$$

$$= \sigma^2$$

Therefore,

$$E(\bar{s}^2) = \frac{n\sigma^2 - \sigma^2}{n - 1} = \sigma^2$$

As the method of the proof indicates, the variance s^2 of the sample would not be unbiased because

$$E(s^2) = \left(\frac{n - 1}{n}\right)\sigma^2$$

Another highly desirable quality of a point estimator is that it have as small a variance as possible. Among several unbiased estimators of a population parameter available, one would naturally choose the estimator which has the smallest variance, reasoning that this estimator would be the least likely to give an estimate far from the true value of the population parameter.

DEFINITION

An unbiased estimator $\bar{\pi}$ is called a *best estimator* (or a *minimum variance estimator*) for a population parameter π_0 if $\text{Var}(\bar{\pi}) \leq \text{Var}(\pi)$ for any other unbiased estimator π of π_0.

Two best estimators for the same parameter π are essentially the same since they have identical means and variances (see Exercise 1 in Section 4.4).

The following result is a step in the direction of determining when an unbiased estimator is, in fact, a best estimator. The proof is quite complicated and nonintuitive, and so is omitted. For more details, see an advanced text.*

Proposition 9.2 The Cramer-Rao Inequality

Suppose that S is a population specified by a random variable X whose distribution (or density function) is determined, at least partially, by a

---

* C. R. Rao, *Advanced Statistical Methods in Biometric Research* (New York, New York: John Wiley and Sons 1952).

parameter π. If $\bar{\pi}$ is an unbiased estimator of π, then

$$\text{Var}(\bar{\pi}) \geq nE\left[\left(\frac{\delta \log_e f(x)}{\delta \pi}\right)^2\right]^{-1}$$

In the discrete case,

$$E\left(\frac{\delta \log_e f(x)}{\delta \pi}\right)^2 = \sum_{i=1}^{n} \left(\frac{\delta \log_e f(x_i)}{\delta \pi}\right)^2 f(x_i)$$

while, in the continuous case, this term equals

$$\int_{-\infty}^{\infty} \left(\frac{\delta \log_e f(x)}{\delta \pi}\right)^2 f(x) \, dx$$

The following result is an immediate consequence of the Cramer-Rao inequality:

Corollary 9.3

Under the assumptions of Proposition 9.2, if $\bar{\pi}$ is an unbiased estimator of π and if

$$\text{Var}(\bar{\pi}) = nE\left[\left(\frac{\delta \log_e f(x)}{\delta \pi}\right)^2\right]^{-1}$$

then $\bar{\pi}$ is a best estimator for π.

Example 2 Suppose that S is a Bernoulli population with parameter p. Therefore, $S = \{0, 1\}$, with $f(0) = 1 - p$ and $f(1) = p$. Consequently $\log_e f(0) = \log_e(1 - p)$ and $\log_e f(1) = \log_e p$, and so

$$\frac{\delta \log_e f(0)}{\delta p} = \frac{-1}{1 - p}$$

and

$$\frac{\delta \log_e f(1)}{\delta p} = \frac{1}{p}$$

As a result,

$$E\left[\left(\frac{\delta \log_e f(x)}{\delta p}\right)^2\right] = \left(\frac{-1}{1 - p}\right)^2 (1 - p) + \left(\frac{1}{p}\right)^2 p$$

$$= \frac{1}{1 - p} + \frac{1}{p}$$

$$= \frac{1}{p(1 - p)}$$

and so

$$\frac{1}{nE[(\delta \log_e f(x)/\delta p)^2]} = \frac{p(1-p)}{n}$$

This is just the variance of the sample proportion $\bar{p}$, which we already know to be an unbiased estimator of p. Therefore, $\bar{p}$ is a best estimator for p.

Example 3 Suppose that a population S is normally distributed with parameters μ and σ. Working with the density function

$$f(x) = \left(\frac{1}{\sigma\sqrt{2\pi}}\right) \exp\left[-\frac{1}{2}\left(\frac{x-\mu}{\sigma}\right)^2\right]$$

we have

$$\log_e f(x) = \log_e\left(\frac{1}{\sigma\sqrt{2\pi}}\right) + \log_e\left[\exp\left(-\frac{1}{2}\left(\frac{x-\mu}{\sigma}\right)^2\right)\right]$$

$$= \log_e\left(\frac{1}{\sigma\sqrt{2\pi}}\right) - \frac{1}{2}\left(\frac{x-\mu}{\sigma}\right)^2$$

Therefore,

$$\frac{\delta(\log_e f(x))}{\delta\mu} = \frac{x-\mu}{\sigma^2}$$

$$= \left(\frac{1}{\sigma}\right)\left(\frac{x-\mu}{\sigma}\right)$$

and so

$$E\left[\left(\frac{\delta \log_e f(x)}{\delta\mu}\right)^2\right] = E\left[\left(\frac{1}{\sigma^2}\right)\left(\frac{x-\mu}{\sigma}\right)^2\right]$$

$$= \left(\frac{1}{\sigma^2}\right) E\left[\left(\frac{x-\mu}{\sigma}\right)^2\right]$$

$$= \frac{1}{\sigma^2}$$

with the last equality following because the random variable $(X-\mu)/\sigma$ is in standardized form. As a result,

$$\frac{1}{nE[(\delta \log_e f(x)/\delta\mu)^2]} = \frac{1}{(n/\sigma^2)} = \frac{\sigma^2}{n}$$

which is the variance of $\bar{x}$, the sample mean. Since $\bar{x}$ is an unbiased estimator of μ, $\bar{x}$ is a best estimator of μ.

A third desirable quality that a good point estimator should have is that it do a better and better job as the sample size gets larger and larger. We therefore make the following definition, which should be highly reminiscent of Chebyshev's theorem:

DEFINITION

An estimator $\bar{\pi}$ of a population parameter π is called *consistent* if

$$\lim_{n \to \infty} P(|\bar{\pi} - \pi| > c) = 0$$

where n is the sample size.

We have the following sufficient conditions that an unbiased estimator be a consistent estimator of π:

Proposition 9.4

If $\bar{\pi}$ is an unbiased estimator of π and

$$\lim_{n \to \infty} \text{Var}(\bar{\pi}) = 0$$

then $\bar{\pi}$ is a consistent estimator of π.

Proof: It is an immediate consequence of Chebyshev's theorem that

$$P\left[|\pi - E(\bar{\pi})| > c\right] < \frac{\text{Var}(\bar{\pi})}{c^2}$$

But $E(\bar{\pi}) = \pi$ since $\bar{\pi}$ is unbiased, so that

$$0 \le P(|\bar{\pi} - \pi| > c) < \frac{\text{Var}(\bar{\pi})}{c^2}$$

and since $\text{Var}(\bar{\pi}) \to 0$ as $n \to \infty$, the result follows.

Since $\text{Var}(\bar{x}) = \sigma^2/n \to 0$ as $n \to \infty$, the sample mean $\bar{x}$ is a consistent estimator of μ. Also, since $\text{Var}(\bar{p}) = pq/n \to 0$ as $n \to \infty$, $\bar{p}$ is a consistent estimator of p. We shall see in Chapter 10 that $\bar{s}^2$ is a consistent estimator of the variance σ^2 of a normal population.

One final quality that we would like a point estimator to have is that it make what has been observed in the sample as highly probable as possible. This can be done by choosing that value of the parameter π which maximizes the joint probability or joint density function arising from the sample.

Since we are sampling with replacement, the random variables $\{X_k\}$ that result from our sampling are independent and identically distributed. Therefore, we have

$$P\left(\bigcap_{k=1}^{n} \{X_k = a_k\}\right) = f_1(a_1) f_2(a_2) \cdots f_n(a_n)$$

where $f_k(a_k) = P(X_k = a_k)$ is the probability function of X_k. Note that since each $f_k(a_k)$ is a function of the unknown parameter π, the probability

$\prod_{k=1}^{n} f_k(a_k)$ is also a function of π. We shall denote this probability by
$f(a_1, a_2, \ldots, a_n | \pi)$
and call it the *likelihood function* of π. In this context, we define:

DEFINITION

The *maximum likelihood estimator* of π is that value $e(a_1, a_2, \ldots, a_n)$ that maximizes the probability $f(a_1, a_2, \ldots, a_n | \pi)$.

Note that the values $a_1, a_2, \ldots, a_n$ are known. A sample of size n has been taken, and these are the values that have occurred. We now wish to determine which value of π best explains the occurrence of the event

$$\bigcap_{k=1}^{n} \{X_k = a_k\}$$

Example 4 Suppose that we have a Bernoulli population with parameter p, which we wish to estimate. To do so, we take n samples of size m, with replacement, from this population, and denote by $a_1, a_2, \ldots, a_n$ the number of "successes" (items with the prescribed set of characteristics) occurring in the n samples. Since each trial of the experiment is binomial with parameters m and p, we have

$$f_k(a_k) = C_{a_k}^{m} \cdot p^{a_k} \cdot (1 - p)^{m - a_k}$$

and hence,

$$f(a_1, a_2, \ldots, a_n | p) = \prod_{k=1}^{n} C_{a_k}^{m} \cdot p^{a_k} \cdot (1 - p)^{m - a_k}$$

$$= \prod_{k=1}^{n} C_{a_k}^{m} \cdot p^{\sum_{k=1}^{n} a_k} \cdot (1 - p)^{mn - \sum_{k=1}^{n} a_k}$$

In order to maximize $f(a_1, a_2, \ldots, a_n | p)$, it is sufficient to maximize $\log_e f(a_1, a_2, \ldots, a_n | p)$, since the logarithm function is strictly increasing. But

$$\log_e f(a_1, a_2, \ldots, a_n | p) = \log_e \prod_{k=1}^{n} C_{a_k}^{m}$$

$$+ \log_e p^{\sum_{k=1}^{m} a_k}$$

$$+ \log_e (1 - p)^{mn - \sum_{k=1}^{m} a_k}$$

which equals

$$\log_e \prod_{k=1}^{n} C_{a_k}^{m} + \left(\sum_{k=1}^{n} a_k \right) \log_e p + \left(mn - \sum_{k=1}^{n} a_k \right) \log_e (1 - p)$$

Therefore,

$$\frac{\delta}{\delta p} \log_e f(a_1, a_2, \ldots, a_n | p) = \frac{\sum_{k=1}^{n} a_k}{p} - \frac{mn - \sum_{k=1}^{n} a_k}{1 - p}$$

Therefore, the probability $f(a_1, a_2, \ldots, a_n | p)$ will assume its maximum value (with respect to p) when

$$\frac{\sum_{k=1}^{n} a_k}{p} = \frac{mn - \sum_{k=1}^{n} a_k}{1 - p}$$

Solving this equation for p, we find that

$$p = \frac{\sum_{k=1}^{n} a_k}{mn} = \left(\frac{1}{n}\right) \frac{\sum_{k=1}^{n} a_k}{m}$$

or, putting it in another form,

$$p = \left(\frac{1}{n}\right)\left(\frac{a_1}{m} + \frac{a_2}{m} + \cdots + \frac{a_n}{m}\right)$$

The maximum likelihood estimator for p is therefore the average of the proportions obtained from the n samples taken. Notice that if we had taken just one sample ($n = 1$) of m elements, the maximum likelihood estimator for p would be a_1/m, where a_1 is the number of "successes" occurring in the sample. This is simply saying that the sample proportion $\bar{p} = a_1/m$ is, in some sense, the most likely estimate for the population proportion. It is the estimate that makes what happened in the sample—a_1 successes—appear most likely. That is, it is this value of $p(a_1/m)$ that gives a_1 successes in m trials its highest probability.

Example 5   Suppose that we have a population that is normally distributed with parameters μ and σ which we wish to estimate. We take a sample of n elements and obtain the n values $a_1, a_2, \ldots, a_n$. Since we are dealing with the continuous normal distribution, the probability $f(a_1, a_2, \ldots, a_n | \pi)$ is defined in terms of the normal density function:

$$f(a_1, a_2, \ldots, a_n | \pi) = \prod_{k=1}^{n} \left(\frac{1}{\sigma\sqrt{2\pi}}\right) \exp\left[-\frac{1}{2}\left(\frac{a_k - \mu}{\sigma}\right)^2\right]$$

$$= \left(\frac{1}{2\pi}\right)^{n/2} \left(\frac{1}{\sigma}\right)^n \exp\left[-\frac{1}{2}\left(\frac{1}{\sigma^2}\right) \sum_{k=1}^{n} (a_k - \mu)^2\right]$$

and therefore,

$$\log_e f(a_1, a_2, \ldots, a_n | \pi) = \log_e\left(\frac{1}{2\pi}\right)^{n/2} + \log_e\left(\frac{1}{\sigma^n}\right)$$

$$+ \log_e \exp\left[-\frac{1}{2\sigma^2} \sum_{k=1}^{n} (a_k - \mu)^2\right]$$

$$= -\frac{n}{2} \log_e 2\pi - n \log_e \sigma - \frac{\sum_{k=1}^{n} (a_k - \mu)^2}{2\sigma^2}$$

Now

$$\frac{\delta}{\delta\mu} \log_e f(a_1, a_2, \ldots, a_n | \pi) = \frac{\sum\limits_{k=1}^{n} (a_k - \mu)}{\sigma^2}$$

and

$$\frac{\delta}{\delta\sigma} \log_e f(a_1, a_2, \ldots, a_n | \pi) = \frac{-n}{\sigma} + \frac{\sum\limits_{k=1}^{n} (a_k - \mu)^2}{\sigma^3}$$

Setting the first of these equations equal to zero yields

$$\frac{\left(\sum\limits_{k=1}^{n} a_k\right)}{n}$$

the sample mean. So the sample mean is the maximum likelihood estimator for the parameter μ in a normal population. Setting the second equation equal to zero yields

$$\sqrt{\frac{\sum\limits_{k=1}^{n} (a_k - \mu)^2}{n}}$$

the standard deviation of the sample, which is therefore the maximum likelihood estimator for the parameter σ in a normal population. Notice that the maximum likelihood estimator for the mean μ is unbiased, while the maximum likelihood estimator for the deviation σ is biased.

We close this section by pointing out a major weakness inherent in point estimators. The one value chosen as the point estimate for the population parameter π is very likely to be incorrect, and we have no way (yet) to determine *how* incorrect it may be. The value of the point estimate is highly dependent on the one sample chosen, and there is no guarantee that *that* sample will be representative of the population from which it was chosen.

EXERCISES

1. Prove Corollary 9.3.
2. Show that the sample mean is a maximum likelihood estimator for the defining parameter λ in a Poisson distribution. Is the sample mean an unbiased estimator? Is it a consistent estimator? Is it a best estimator?
3. Show that the sample mean is a maximum likelihood estimator for the defining parameter in an exponential distribution. Is the sample mean an unbiased estimator? Is it a consistent estimator? Is it a best estimator?
4. According to the *method of moments*, estimations can be obtained for unknown population parameters by equating the first few moments of the population with the corresponding moments of the sample. If there are k population parameters to be

determined, then the first k moments (about the origin) must be used, giving k equations in the k unknowns (the parameters). Suppose that $x_1, x_2, \ldots, x_n$ is a sample of n items taken from a Poisson population with parameter λ. Use the method of moments to estimate λ. (This involves μ_1', the first moment about the origin, and one equation in the one unknown, λ).

5. Use the method of moments to estimate the parameters α and β in a population which has gamma distribution.

6. Use the method of moments to estimate the parameter α in a population having beta distribution with $\beta = 1$.

9.3 Interval Estimates

Instead of approximating a population parameter with a single value, with no method of determining how accurate this estimate might be, it would be much more informative if we could determine a range of values in which we could say, with a certain prescribed amount of confidence, that the parameter will lie. In this section, we discuss a method for determining such a range for both population means and population proportions.

Before proceeding any farther, we introduce the following notation. Given any fraction α between 0 and 1 [that is, $\alpha \in (0, 1)$], we denote by $z_{\alpha/2}$ that point on the x axis such that the area under the standard normal curve between $-z_{\alpha/2}$ and $+z_{\alpha/2}$ equals $1 - \alpha$ (Figure 9.1). That is,

$$\frac{1}{\sqrt{2\pi}} \int_{-z_{\alpha/2}}^{z_{\alpha/2}} e^{-(1/2)x^2} \, dx = 1 - \alpha$$

Since the standard normal curve is symmetric about the line $x = 0$, the remaining area, which equals α, is divided equally between the two tails of

Figure 9.1

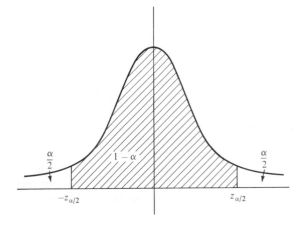

the curve. That is, the area under the curve to the right of $z_{\alpha/2}$ equals $\alpha/2$, and the area under the curve to the left of $-z_{\alpha/2}$ also equals $\alpha/2$.

The following table gives the values of $z_{\alpha/2}$ corresponding to frequently used values of α:

α	$1 - \alpha$	$z_{\alpha/2}$
0.10	0.90	1.64
0.05	0.95	1.96
0.02	0.98	2.33
0.01	0.99	2.58

Therefore, the probability that a random variable with standard normal distribution assumes a value between -1.96 and $+1.96$ equals 0.95. Recalling that the mean and standard deviation of the standard normal distribution are 0 and 1, respectively, we see that the probability that such a random variable assumes a value within two deviations of its mean is slightly higher than 0.95. This result can be extended to normally distributed random variables in general. Suppose that X is such a random variable, with parameters μ and σ. Under the change of variables $z = (x - \mu)/\sigma$, we find that

$$\frac{1}{\sigma\sqrt{2\pi}} \int_{\mu - z_{\alpha/2}\sigma}^{\mu + z_{\alpha/2}\sigma} \exp\left[-\frac{1}{2}\left(\frac{x - \mu}{\sigma}\right)^2\right] dx = \frac{1}{\sqrt{2\pi}} \int_{-z_{\alpha/2}}^{z_{\alpha/2}} e^{-(1/2)z^2} dz$$

Therefore, the probability that X assumes a value within 1.96 deviations σ of its mean μ equals 0.95. Compare this with the information obtained from Chebyshev's inequality: The probability that *any* random variable assumes a value within two deviations of its mean is at least $1 - (\frac{1}{2})^2 = \frac{3}{4}$.

Now suppose that we wish to estimate the mean μ of a population. We will assume for the moment that the standard deviation σ of the population is known. (We will discuss what happens when σ is unknown in Chapter 10.) We take a sample of size n, with replacement, and find that our sample has mean $\bar{x}$. Our sample's average is but one value of the sample mean, a random variable which itself has mean μ and, for samples of size n, deviation $\sigma/\sqrt{n}$. According to the central limit theorem, the distribution of the sample mean can be closely approximated by a normal curve (with parameters μ and $\sigma/\sqrt{n}$), and therefore, we can say that

$$P\left[|\bar{x} - \mu| \leq z_{\alpha/2}\left(\frac{\sigma}{\sqrt{n}}\right)\right] = 1 - \alpha$$

That is, we can assert with probability $1 - \alpha$ that the value of our sample's average (the value assumed by the sample mean, in this case) will differ from the population's average (which is also the average of the sample mean) by no more than $z_{\alpha/2}$ (sample mean) deviations. Since

$$|\bar{x} - \mu| \leq z_{\alpha/2}\left(\frac{\sigma}{\sqrt{n}}\right)$$

is equivalent to

$$-z_{\alpha/2}\left(\frac{\sigma}{\sqrt{n}}\right) \leq \bar{x} - \mu \leq z_{\alpha/2}\left(\frac{\sigma}{\sqrt{n}}\right)$$

or

$$\mu - z_{\alpha/2}\frac{\sigma}{\sqrt{n}} \leq \bar{x} \leq \mu + z_{\alpha/2}\frac{\sigma}{\sqrt{n}}$$

we can assert with probability $1 - \alpha$ that our sample average falls in the interval

$$\left(\mu - z_{\alpha/2}\frac{\sigma}{\sqrt{n}}, \mu + z_{\alpha/2}\frac{\sigma}{\sqrt{n}}\right)$$

Unfortunately, μ is unknown, and so it is impossible to determine the end points of this interval. But this interval is important in that it does contain a certain percentage of the sample averages. That is what we mean when we say "we can assert with probability $1 - \alpha$." We are saying that a certain proportion $1 - \alpha$ of all possible sample averages will fall in the above interval. A small percentage α of the samples will be so unrepresentative of the population that they will produce sample averages that will fall outside this interval.

However, we are not interested in what happens to the majority of the sample averages. We are interested only in the one sample average $\bar{x}$ that we have observed, and what this sample average allows us to conclude about the population average μ. Now the inequality

$$|\bar{x} - \mu| \leq z_{\alpha/2}\frac{\sigma}{\sqrt{n}}$$

can be restated as

$$|\mu - \bar{x}| \leq z_{\alpha/2}\frac{\sigma}{\sqrt{n}}$$

which is equivalent to

$$-z_{\alpha/2}\frac{\sigma}{\sqrt{n}} \leq \mu - \bar{x} \leq z_{\alpha/2}\frac{\sigma}{\sqrt{n}}$$

or

$$\bar{x} - z_{\alpha/2}\frac{\sigma}{\sqrt{n}} \leq \mu \leq \bar{x} + z_{\alpha/2}\frac{\sigma}{\sqrt{n}}$$

Therefore, we can assert with probability $1 - \alpha$ that the population average μ falls in the interval

$$\left(\bar{x} - z_{\alpha/2}\frac{\sigma}{\sqrt{n}}, \bar{x} + z_{\alpha/2}\frac{\sigma}{\sqrt{n}}\right)$$

This interval is called a *confidence interval* for μ, while the endpoints of the interval, $\bar{x} \pm z_{\alpha/2}(\sigma/\sqrt{n})$, are called the *confidence limits* for μ. The value (probability) $1 - \alpha$ is called the *degree of confidence*.

Notice that the confidence interval is very much dependent on the mean $\bar{x}$ of the sample that we have taken. The value $\bar{x}$ is the midpoint of the confidence interval. Different samples, which probably will have different means, will lead to different confidence intervals for μ. However, we can assert with degree of confidence $1 - \alpha$ that the true value of μ will fall somewhere in the confidence interval. What this means is that most $[(1 - \alpha) \text{ percent}]$ of the possible samples of size n will have sample average close enough to the true value of μ that the confidence interval determined by that sample average will contain μ. Only α percent of such samples have a mean $\bar{x}$ so different from μ that the confidence interval about $\bar{x}$ fails to include μ.

The length of the confidence interval is controlled by three different factors, as is suggested by the formula for the length, $2z_{\alpha/2}(\sigma/\sqrt{n})$:

1. One factor is the degree of confidence $1 - \alpha$, which determines the size of $z_{\alpha/2}$. A higher degree of confidence corresponds to a larger value of $z_{\alpha/2}$, and therefore, to a wider confidence interval. Hence, the more confidence we desire, the less we will have to be confident about. A larger confidence interval gives less information about the true value of μ. A smaller confidence interval pinpoints the value of μ much better than does a larger confidence interval.

2. A second factor is the standard deviation of the population. A population with large deviation will give rise to samples with large deviations and widely varying means. On the other hand, a population with small deviation will give rise to samples with small deviations and quite similar means. In other words, the larger the value of σ, the more we would expect the sample averages to differ from the true population average which they are supposed to approximate. For small values of σ, we would expect the sample averages to concentrate around μ. Thus, we would expect the length of the confidence interval to be directly proportional to the size of σ.

3. The final factor that is of importance is the size of the sample. We would naturally expect that the average of a large sample would give us a better approximation to the population average than would the average of a smaller sample. Consequently, we would expect that the length of the confidence interval would be inversely proportional to the size of the sample.

It is customary to require that the sample size n be at least 30 in order that the normal approximation to the sample mean be sufficiently accurate. In the case $n \geq 30$, the value of the sample deviation s can be substituted in the formula for the confidence interval for the population deviation σ, should the latter be unknown. We will discuss this point in more detail in Chapter 10.

Example 1   A random sample of 100 brand A cigarettes yields an average of $\bar{x} = 26$ milligrams of nicotine. It is known that $\sigma = 8$ milligrams of nicotine. With degree of confidence $1 - \alpha = 0.99$, we find that the average nicotine content μ for this brand of cigarettes satisfies

$$26 - 2.58 \cdot \frac{8}{\sqrt{100}} \leq \mu \leq 26 + 2.58 \cdot \frac{8}{\sqrt{100}}$$

which is equivalent to $23.94 \leq \mu \leq 28.06$

Example 2 A sample of 40 exam scores taken from a large population of exam scores is found to have an average of $\bar{x} = 77.125$. The standard deviation for this type of exam score is known to be $\sigma = 13.36$. With degree of confidence 0.95, we estimate that the average score μ for this exam lies in the interval

$$77.125 - 1.96 \cdot \frac{13.36}{\sqrt{40}} \leq \mu \leq 77.125 + 1.96 \cdot \frac{13.36}{\sqrt{40}}$$

which is equivalent to $72.985 \leq \mu \leq 81.265$. Of course, there is a five percent chance that our sample will mislead us, and the true population mean μ will not lie in this interval.

If we were to use our sample mean $\bar{x}$ as a point estimate for the population mean μ, then we could say with degree of confidence $1 - \alpha$ that the error we make will not exceed

$$E = z_{\alpha/2} \frac{\sigma}{\sqrt{n}}$$

This is because the true population mean will fall, with probability $1 - \alpha$, in an interval of length $2E$ centered about $\bar{x}$. Should we wish to decrease the size of this error term, while at the same time holding the degree of confidence constant (and working with a known σ), our only choice would be to increase the sample size n. If, for example, we wish to have $E \leq E_0$, for some specific value of E_0, then solving the equation

$$z_{\alpha/2} \frac{\sigma}{\sqrt{n}} \leq E_0$$

for n, we find that we must choose n so that

$$n \geq \left(\frac{z_{\alpha/2} \cdot \sigma}{E_0} \right)^2$$

Example 1 (continued) Suppose that we wish to reduce the length of the confidence interval to 2. That is, we want $E = 1$. We must therefore choose a sample of size

$$n \geq \left(\frac{2.58 \cdot 8}{1} \right)^2 = 424.8$$

Hence, a sample of size 425 will suffice. This means that, regardless of which sample of size 425 we take, the confidence interval will extend one unit above and one unit below the sample's average, and 99 percent of the time, this interval will contain the true population mean.

Example 2 (continued) If we wish to reduce our error to at most 2, with 95 percent degree of confidence, then we must choose our sample size

$$n \geq \left(\frac{1.96 \cdot 13.36}{2}\right)^2 = 171.42$$

That is, a sample of size 172 will suffice.

Example 3 A computer simulation of 100 samples of size 20 from a population uniformly distributed on the interval $[50, 100]$ yielded the following results. Table 9.1 gives the distribution of the 100 sample averages. Recall that the uniform distribution has mean $\mu = 75$ and standard deviation

$$\sigma = \sqrt{\frac{(b-a)^2}{12}} = \sqrt{\frac{50^2}{12}} = 14.434$$

Table 9.1
Distribution of 100 Sample Averages

Sample average	Frequency
65	1
66	0
67	0
68	1
69	1
70	4
71	6
72	11
73	13
74	17
75	16
76	9
77	5
78	10
79	5
80	0
81	0
82	0
83	1

The 95 percent "confidence interval" about $\mu = 75$ is

$$75 - 1.96 \cdot \frac{14.434}{\sqrt{20}} \leq \bar{x} \leq 75 + 1.96 \cdot \frac{14.434}{\sqrt{20}}$$

which is equivalent to $68.482 \leq \bar{x} \leq 81.518$. Table 9.1 indicates that exactly 97 of our 100 samples had averages which fell in this interval. On the other hand, exactly 98 of the 100 samples were such that the population mean 75 actually fell in the 95 percent confidence interval centered about the sample average. That is,

$$75 \in \left(\bar{x} - 1.96 \cdot \frac{14.434}{\sqrt{20}}, \bar{x} + 1.96 \cdot \frac{14.434}{\sqrt{20}} \right)$$

or

$$75 \in (\bar{x} - 6.518, \bar{x} + 6.518)$$

for 98 of the 100 sample averages $\bar{x}$. We point out that the average of the 100 sample averages was 74.816. Recall that the average of the averages of all possible samples of size 20 taken from this population would be 75.

Next, suppose that we wish to estimate the proportion p of the elements in a population that have a certain characteristic. We take a sample of size n, with replacement, and find that the proportion of elements with the given characteristic in the sample is $\bar{p}$. Now our particular sample proportion is just one particular value assumed by the random variable known as the sample proportion. We have seen that this random variable has mean p and standard deviation $\sqrt{pq/n}$, and that its distribution is approximately that of the normal distribution with the same mean and standard deviation. As a result, we can say that

$$P\left(|\bar{p} - p| \leq z_{\alpha/2} \sqrt{\frac{pq}{n}} \right) = 1 - \alpha$$

and consequently, we can assert with probability $1 - \alpha$ that

$$\bar{p} - z_{\alpha/2} \sqrt{\frac{pq}{n}} \leq p \leq \bar{p} + z_{\alpha/2} \sqrt{\frac{pq}{n}}$$

We have therefore determined the form of the confidence interval for the unknown parameter p.

In practice, the unknowns p and q in the expression for the confidence limits can be replaced by $\bar{p}$ and $\bar{q} = 1 - \bar{p}$. Thus, in practice, the $1 - \alpha$ percent degree of confidence interval for p is

$$\bar{p} - z_{\alpha/2} \sqrt{\frac{\bar{p}\bar{q}}{n}} \leq p \leq \bar{p} + z_{\alpha/2} \sqrt{\frac{\bar{p}\bar{q}}{n}}$$

Example 4 A political pollster wishes to estimate the percentage of the voting public which favors a particular candidate. He takes a sample of

100 voters and finds that 55 of them favor the candidate. What can he conclude about the candidate's chances of winning the election? Since $n = 100$, $\bar{p} = 0.55$, and $\bar{q} = 0.45$, the 95 percent confidence interval is

$$0.55 - 1.96 \sqrt{\frac{0.55 \cdot 0.45}{100}} \leq p \leq 0.55 + 1.96 \sqrt{\frac{0.55 \cdot 0.45}{100}}$$

which is equivalent to $0.45 \leq p \leq 0.65$. Hence, although 55 percent of the pollster's sample favored the candidate, the pollster cannot conclude, at a 95 percent level of confidence, any more than that the true percentage of voters favoring the candidate lies in the interval between 45 and 65 percent. Most importantly, the pollster cannot conclude that a majority of the voters favor the candidate.

Example 5 A large urn contains white and red balls. In order to estimate the proportion p of white balls in the urn, a sample of 49 balls is drawn with replacement and found to contain 35 white balls. As a result, with 95 percent confidence, we can say that p lies in the following confidence interval about $\bar{p} = \frac{35}{49} = 0.714$:

$$0.714 - 1.96 \sqrt{\frac{0.714 \cdot 0.286}{49}} \leq p \leq 0.714 + 1.96 \sqrt{\frac{0.714 \cdot 0.286}{49}}$$

which is equivalent to $0.588 \leq p \leq 0.841$. Once again, we have a rather large confidence interval, in this case, one that covers all of 25 percent. Obviously, we must take larger samples if we wish to obtain more precise confidence intervals. If, for example, we had taken a sample of $n = 400$ balls, and found that 289 were white, we would have obtained the following confidence interval about $\bar{p} = \frac{289}{400} = 0.7225$:

$$0.7225 - 1.96 \sqrt{\frac{0.7225 \cdot 0.2775}{400}} \leq p \leq 0.7225 + 1.96 \sqrt{\frac{0.7225 \cdot 0.2775}{400}}$$

which is equivalent to $0.6786 \leq p \leq 0.7664$.

If we were to use the sample proportion as a point estimate for the population proportion, we would make an error that would not exceed

$$E = z_{\alpha/2} \sqrt{\frac{pq}{n}}$$

Should we wish to control this error term, we would be faced with a problem that did not arise in connection with the error term for sample means. In the case of sample proportions, the error term contains the unknown population proportion p. To alleviate this problem, we note that the function $x(1 - x)$ assumes its maximum over the interval $[0, 1]$ at the point $x = \frac{1}{2}$, and so $x(1 - x) \leq \frac{1}{4}$ over the interval $[0, 1]$. Consequently, $0 \leq pq \leq \frac{1}{4}$, and so

$$E = z_{\alpha/2} \sqrt{\frac{pq}{n}} \leq \frac{z_{\alpha/2}}{\sqrt{4n}}$$

Should we wish to make $E \le E_0$, for some specific value of E_0, it would suffice to make

$$\frac{z_{\alpha/2}}{\sqrt{4n}} \le E_0$$

or

$$n \ge \left(\frac{z_{\alpha/2}}{2E_0}\right)^2$$

Example 4 (*continued*)   Should the pollster wish to obtain a sample proportion that he could be 95 percent certain was within 2 percent of the true population proportion—that is, should he desire an error term of not more than 2 percent or a confidence interval of length 4 percent—then he would have to take a sample of size

$$n \ge \left(\frac{1.96}{2 \cdot 0.02}\right)^2 = 49^2 = 2401$$

Roughly speaking, a sample of size 2400 would be necessary.

Example 6   A computer simulation of 100 samples of size 25 taken from a Bernoulli population with parameter $p = 0.40$ yielded the following results. Table 9.2 gives the distribution of the 100 sample proportions. The average proportion over the 100 samples was 0.4028, contrasted to $p = 0.40$, the average of the proportions of all possible samples of size 25 taken from this

---

Table 9.2
Distribution of 100 Sample Proportions

Sample proportion	Frequency
0.12	1
0.16	0
0.20	1
0.24	3
0.28	11
0.32	13
0.36	13
0.40	18
0.44	13
0.48	8
0.52	7
0.56	8
0.60	4

population. The 95 percent "confidence interval" about $p = 0.40$ is

$$0.40 - 1.96 \sqrt{\frac{0.4 \cdot 0.6}{25}} \leq p \leq 0.40 + 1.96 \sqrt{\frac{0.4 \cdot 0.6}{25}}$$

which is equivalent to $0.208 \leq p \leq 0.592$. Of the 100 samples generated, exactly 94 had sample proportions which fell within this interval. On the other hand, each of the 100 sample proportions $\bar{p}$ determines a confidence interval

$$\bar{p} - 1.96 \sqrt{\frac{\bar{p}\bar{q}}{n}} \leq p \leq \bar{p} + 1.96 \sqrt{\frac{\bar{p}\bar{q}}{n}}$$

Of the 100 samples generated, once again exactly 94 had sample proportions which determined confidence intervals which actually did contain $p = 0.40$.

EXERCISES

1. A sample of 100 light bulbs had an average life span of 950 hours with a standard deviation of 100 hours. Determine a 95 percent confidence interval for the true average life span of these light bulbs. How large a sample would be required if we desired a confidence interval of length 40 hours?

2. A sample of 50 twelve-ounce cans of peanuts had an average weight of 11.9 ounces with a standard deviation of 0.1 ounce. Determine a 99 percent confidence interval for the true weight of the contents of these cans.

3. A sample of 64 items produced by a certain machine revealed an average diameter of 8 in. with a standard deviation of 0.3 in. Determine a 95 percent confidence interval for the true average diameter of the items produced by this machine.

4. A machine produces 42 defective items in a sample of 600 items. Determine a 95 percent confidence interval for the true percentage of defectives produced by this machine.

5. In a random sample of 1000 viewers, 184 were found to be watching a particular television show. Determine a 98 percent confidence interval for the true percentage of viewers who were watching that particular show. If you wanted to estimate this percentage to within 1 percent, how large a sample would you have to take?

6. A sample of 250 households revealed 175 which had pets. Determine a 95 percent confidence interval for the true proportion of households that have pets.

7. A survey of 200 women aged 20 years and up revealed 85 who used the birth control pill. Find a 99 percent confidence interval for the true proportion of women (20 and up) who use the pill.

8. How large a sample is necessary to estimate a Bernoulli parameter p to within 1 percent? If it is known for a fact that p does not exceed 25 percent? If it does not exceed 10 percent?

9.4 Hypothesis Testing

Suppose again that we have a population whose distribution is determined by an unknown parameter π, and suppose that we have good reason to believe that $\pi = \pi_0$. Rather than using the information obtained from a sample to specify either a point estimate or a confidence interval estimate for π, we shall instead use the information contained in the sample to either support or deny the hypothesis that $\pi = \pi_0$. In other words, we will use the sample information as a basis for deciding between two possible courses of action. Either we will accept the hypothesis that $\pi = \pi_0$, or at least subject it to further testing, or reject it completely and look for a more reasonable estimate for the parameter π.

We begin our discussion of this procedure, called *hypothesis testing*, by considering the following example:

Example 1   A manufacturer wishes to improve his productivity. He has recently read an advertisement about a new machine which claims that 90 percent of the items produced by the machine are nondefective. The advertisement also states that prospective buyers may observe the machine in operation and inspect its produce.

Our friend, realizing that such a machine would be of great value to his business, decides to observe the production of 25 items and, on the basis of this sample, decide whether or not to purchase the machine. He must therefore decide whether or not to accept the hypothesis that $p = 0.90$, where p is the probability that an item produced by the machine is nondefective. All he will have to base his decision on is the number of nondefectives in the sample of size 25 that he observes—that, and a little probabilistic reasoning.

If the hypothesis that $p = 0.90$ is true, and if k denotes the number of non-defectives in the observed sample, then

$$P(k \le 19) = \sum_{k=0}^{19} b(k; 25, 0.90) = 0.0334$$

Thus, it is not very likely that a machine producing 90 percent nondefectives will produce only 19 or fewer nondefectives in a sample of size 25, although it nevertheless is possible. (We point out that the value 19 was determined by trial and error. The probabilities $P(k \le c)$ were calculated, for $c = 25$, 24, 23, ... , until the value of $P(k \le c)$ became exceptionally small.)

Our manufacturer decides upon a rule for making his decision. He will accept (believe) the advertiser's claim that $p = 0.90$, and purchase the machine, provided that the observed value of k is at least 20. We emphasize that this rule depends only on the value of k, and not at all (except as it effects the value of k) on the actual value of p. Regardless of what the actual value of p is, the manufacturer will follow the above rule because he does not know what the value of p really is. Remember, all that the manufacturer knows is the observed value of k.

Now there is a chance that the manufacturer will make an error. It is possible that the hypothesis $p = 0.90$ is true, and still the machine produces $k \le 19$ nondefectives in a sample of 25. But such an observed value of k will lead the manufacturer to reject the true hypothesis that $p = 0.90$, and not buy a machine he probably should buy. Such an error—rejecting a true hypothesis—is called a *type 1 error*. The probability of making such an error, which we shall denote by α, is called the *level of significance* of the test (rule). In our example, $\alpha = 0.0334$. The set of outcomes that lead to the rejection of the hypothesis is called the *critical region* of the test. The critical region for the test above is $k \le 19$.

If the advertiser's claim is, in fact, a lie, the manufacturer would hope that his rule would lead him to reject the hypothesis that $p = 0.90$, and not buy the machine. Suppose that, in fact, the true value of p is 0.80. Then the probability that the hypothesis $p = 0.90$ will be rejected is

$$P(k \le 19) = \sum_{k=0}^{19} b(k; 25, 0.80) = 0.3833$$

(Notice that, in calculating $P(k \le 19)$ here, the value $p = 0.80$ is used, rather than the value $p = 0.90$.) Therefore, the probability that the false hypothesis $p = 0.90$ will be accepted is 0.6167. An error of this type—accepting a false hypothesis—is called a *type 2 error*. The probability of making such an error depends on the actual true value of the parameter p. We define the *power* of the test to be the function which gives the probability of rejecting the hypothesis $p = 0.90$ for all possible values $0 \le p \le 1$. If, for example, $p = 0.90$, the power is 0.0334, the level of significance of the test. For any other value of p, the power $P(k \le 19)$ can be determined from a table of binomial probabilities. Several values of the power function for the test above are given in the table below:

Value of p	Power
0.80	0.3833
0.75	0.6217
0.70	0.8065
0.60	0.9264
0.50	0.9979

The graph of this power function is shown in Figure 9.2.

The example above is typical of many problems in statistics, although the procedure followed above is not in the usual order. The general problem involved is whether or not to reject a hypothesis concerning some population parameter. Since most of our parameters are closely related to the mean, we now discuss a general procedure to be followed in deciding whether or not to reject the hypothesis that $\mu = \mu_0$. The hypothesis $\mu = \mu_0$ is commonly known as the *null hypothesis*, and is denoted H_0. One reason for the negative aspect of this name is that it is common practice to set up such a hypothesis in the hope of rejecting it.

If the null hypothesis were false, we would want to accept one of the alternatives $\mu \neq \mu_0$, $\mu < \mu_0$, or $\mu > \mu_0$. Whichever alternative hypothesis fits the particular problem shall be denoted H_1. If H_1 is $\mu = \mu_0$, it is called a *two-sided alternative*; if H_1 is either $\mu < \mu_0$ or $\mu > \mu_0$, it is called a *one-sided alternative*.

After having decided upon the null and alternative hypotheses, we must then decide upon the level of significance of the test. The deciding factor here should be the seriousness of making a type 1 error. We might also wish to control the power function of the test, that is, control the chances of making a type 2 error. Thus, we might simultaneously wish to keep α near zero and the value of the power function near one. The normal way to do this is by means of a wise choice of sample size, although this at times might require such a large sample as to become economically not feasible. At times a compromise between the level of significance, the value of the power function, and the size of the sample becomes necessary.

We then take a sample of size n, with replacement, calculate the sample average $\bar{x}$, and consider the statistic

$$z = \frac{\bar{x} - \mu_0}{\sigma/\sqrt{n}}$$

(If the standard deviation σ of the population is unknown, it may be replaced by the standard deviation s of the sample, provided that the sample size $n \geq 30$. We will discuss this in more detail in Chapter 10.) Notice that z is the standardized form of the sample mean $\bar{x}$, under the assumption that $\mu = \mu_0$; that is, under the assumption that our null hypothesis is true. Consequently, should the null hypothesis be true, the statistic z would have standard normal distribution.

The values of $\bar{x}$ will, naturally, vary from being fairly close to μ_0 to being either well above or well below μ_0, provided, of course, that μ_0 is the true population mean. Should the logical alternative hypothesis be one-sided, these extreme values of $\bar{x}$ would tend to be on just the one side of μ_0. The problem facing us is how to interpret the values of $\bar{x}$ (or z), particularly the extreme values. This is done by means of the critical region of the test, which is determined by the level of significance α and the nature of the alternative

Figure 9.2

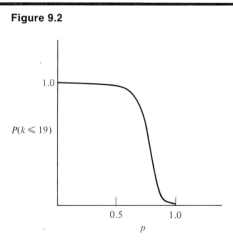

hypothesis H_1. The following table will prove helpful in determining critical regions in the examples that follow:

α	One-sided z_α	Two-sided $z_{\alpha/2}$
0.05	1.64	1.96
0.01	2.33	2.58

To explain what these numbers mean, suppose that we consider the case $\alpha = 0.05$. The values ± 1.96, recall, are such that 95 percent of the area under the standard normal curve lies in the interval between them, with the remaining 5 percent of the area equally distributed in the two tails of the curve, as indicated in Figure 9.3. Thus, these two tails contain those values of z that correspond to the most unlikely 5 percent of all possible sample averages.

On the other hand, the value 1.64 is such that 95 percent of the area under the standard normal curve lies to its left, with the remaining 5 percent to its right, as indicated in Figure 9.4. In this case, the most unlikely 5 percent of

Figure 9.3

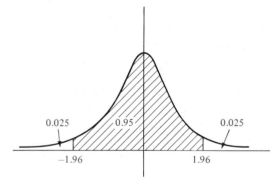

Figure 9.4

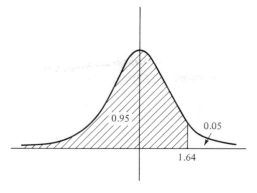

all possible sample averages correspond to the tail to the right. Reversing the roles of right and left, similar remarks could be made concerning the value -1.64.

In general then, the critical region for our test is defined to be

$|z| > z_{\alpha/2}$ if H_1 is two-sided

$z > z_\alpha$ if H_1 is $\mu > \mu_0$

$z < -z_\alpha$ if H_1 is $\mu < \mu_0$

The critical region is simply the tail (or tails) corresponding to the extreme values of $\bar{x}$, the most unlikely α percent of all possible sample averages.

We then set up the following criterion for rejecting or not rejecting the null hypothesis H_0:

"If the statistic z, defined above, falls in the critical region, reject the null hypothesis; otherwise, either accept H_0 or reserve judgement on H_0."

The logic behind our criterion is as follows: If we are working at level of significance $\alpha = 0.05$, for example, then the critical region is defined in such a manner that five percent of our possible samples will have means which fall in the critical region, provided that H_0 is true, that is, that $\mu = \mu_0$. If the sample that we take has a mean which falls in the critical region, then we are faced with the dilemma of either accepting the fact that *our* sample is that one bad sample in 20 whose mean lies in the critical region (how could *we* have such bad luck?), or trying to explain why this unusual sample has occurred. The obvious explanation, of course, is that the null hypothesis is not true after all, that the true value of the population mean is some value other than μ_0, and that in light of the true value of the population mean, the occurrence of the "unusual" sample would no longer seem so unlikely. Indeed, we shall adopt this latter approach.

Recall that in Section 9.3, we found that we could assert with probability $1 - \alpha$ (in our case, $1 - \alpha = 0.95$) that our sample average lies in the interval

$$\mu - z_{\alpha/2} \frac{\sigma}{\sqrt{n}} \leq \bar{x} \leq \mu + z_{\alpha/2} \frac{\sigma}{\sqrt{n}}$$

centered around the true population mean μ. If our sample value $\bar{x}$ does happen to lie in the two-sided critical region at level of significance $\alpha = 0.05$, then $\bar{x}$ does not lie in the interval

$$\mu_0 - z_{\alpha/2} \frac{\sigma}{\sqrt{n}} \leq \bar{x} \leq \mu_0 + z_{\alpha/2} \frac{\sigma}{\sqrt{n}}$$

centered around the suspected mean μ_0. Our only possible reaction is that since 95 percent of all possible sample averages should lie in the former interval, and since our sample average did *not* fall in the latter interval, the "only" possible conclusion is that these two intervals are not the same. That is, we must conclude that $\mu \neq \mu_0$.

Example 2   When operating properly, a machine will produce items that measure 2 inches in diameter, with a deviation of 0.05 inch. To insure that the machine remains in proper operating condition, at regular intervals samples of size 49 are checked. Suppose that such a sample produced an average diameter of 1.98 inches. What can be concluded about the condition of the machine? The statistical hypothesis that we wish to test here is

$H_0: \mu = 2$ inches

Hopefully, we will not be able to reject this hypothesis, because to do so would mean incurring the expense of having the machine repaired. If the machine were operating improperly, the items it produces may be either too small in diameter, or too large in diameter. (It could also happen that the machine is producing some items too large and others too small. This would produce an unusually large sample deviation, and we will discuss methods for testing for such an occurrence in Chapter 10.) Consequently, we must use the two-sided alternative

$H_1: \mu \neq 2$ inches

A type 1 error here would mean having a properly functioning machine sent in for repairs, while a type 2 error would mean not having an improperly functioning machine sent in for repairs. Since a type 1 error would mean nothing more than an unnecessary expense, we will set a rather high level of significance $\alpha = 0.05$. This gives the critical region $|z| \geq 1.96$. Since our statistic

$$z = \frac{1.98 - 2.00}{0.05/\sqrt{49}} = -2.8$$

does fall in this critical region, we must reject our null hypothesis that μ equals 2 inches, and have the machine repaired.

Example 3   Suppose that it is a known fact that if a brand of cigarettes averages 30 milligrams (or more) of nicotine per cigarette, then lung cancer is almost certain to develop in the user. On the other hand, if the average nicotine content is less than 30 mg per cigarette, then it is relatively safe to continue smoking this particular brand of cigarettes. A random sample of 100 brand X cigarettes yields an average of 27 mg of nicotine per cigarette. Can a person safely continue to smoke this brand of cigarettes? What we wish to do here is test the hypothesis

$H_0: \mu = 30$ mg

against the one-sided alternative

$H_1: \mu < 30$ mg

Actually, our null hypothesis is that the cigarettes are dangerous ($\mu \geq 30$), while the alternative (hypothesis) is that they are not dangerous. We wish to test our hypothesis at a very low level of significance, since making a type 1 error means continuing to smoke a potentially dangerous brand of cigarettes, one which does average 30 mg (or more) of nicotine per cigarette. Thus, making a type 1 error would almost certainly lead to the development of lung cancer. We choose $\alpha = 0.01$ to simplify our calculations, although it may be desirable to choose α much smaller. Our critical region is therefore $z < -2.33$. Our statistic

$$z = \frac{27 - 30}{8/\sqrt{100}} = -3.75$$

(where we assume, as we did in Example 1 of the preceding section, that $\sigma = 8$ mg), which clearly falls in our critical region. Therefore, we must reject the null hypothesis (that the cigarettes are potentially dangerous), and may safely continue to smoke this brand of cigarettes.

Example 4   The contents of a medium-sized box of Kellogg's corn flakes are supposed to weigh 15 ounces, according to the statement on the package. In order to avoid problems with the government, the people at Kelloggs must be careful that the contents of these boxes do weigh 15 ounces. To keep check that this is, in fact, the case, they might proceed as follows. At regular intervals, the contents of 50 boxes are weighed, and the sample average is used to test the null hypothesis

$H_0: \mu = 15$ ounces

against the one-sided alternative

$H_1: \mu < 15$ ounces

In this case, making a type 1 error means concluding that the contents of the boxes are deficient when, in fact, they are not. On the other hand, making a type 2 error means incorrectly concluding that the boxes weigh the required 15 ounces when, in fact, they weigh less. Obviously, a type 2 error is more critical, since it is this type of error that will get Kelloggs in trouble with the government. Therefore, we use the moderately high level of significance $\alpha = 0.05$, which gives the critical region $z < -1.64$. Suppose a sample of 50 boxes revealed a sample average of 14.9 ounces per box. If we accept the fact that such measurements produce a standard deviation of $\sigma = 0.5$ ounces per box, then our statistic

$$z = \frac{14.9 - 15.0}{0.5/\sqrt{50}} = -1.43$$

does not fall in the critical region. Therefore, unless other tests prove differently, Kelloggs has no cause for concern, at least not on the basis of this particular sample. Since making a type 2 error is very important in this example, let's evaluate the power of this test at $\mu = 14.9$ ounces. That is, let's calculate the probability that the hypothesis $\mu = 15$ ounces will be rejected if, in fact, $\mu = 14.9$ ounces. By definition, the power is the probability

$$P\left(\frac{\bar{x} - 15}{0.5/\sqrt{50}} < -1.64\right)$$

However, the term $(\bar{x} - 15)/(0.5/\sqrt{50})$ is no longer standard normal, since 14.9, rather than 15, is the true population mean. Therefore, the term $(\bar{x} - 14.9)/(0.5/\sqrt{50})$ is standard normal, and our calculation of the power of the test at $\mu = 14.9$ proceeds as follows:

$$P\left(\frac{\bar{x} - 15}{0.5/\sqrt{50}} < -1.64\right)$$

$$= P\left(\frac{\bar{x} - 14.9}{0.5/\sqrt{50}} + \frac{14.9 - 15}{0.5/\sqrt{50}} < -1.64\right)$$

$$= P\left(\frac{\bar{x} - 14.9}{0.5/\sqrt{50}} - 1.43 < -1.64\right)$$

$$= P\left(\frac{\bar{x} - 14.9}{0.5/\sqrt{50}} < -0.21\right) = 0.4168$$

So the probability of rejecting $\mu = 15$, if, in fact, $\mu = 14.9$, equals 0.4168. In similar fashion, we find the power at $\mu = 14.8$ to be 0.8888 and the power at $\mu = 14.7$ to be 0.9960. Now suppose that we wanted both $\alpha = 0.05$ and the power of the test at $\mu = 14.9$ to be 0.90. This could be achieved by taking the sample size n large enough so that

$$P\left(\frac{\bar{x} - 15}{0.5/\sqrt{n}} < -1.64\right) = 0.90$$

when, in fact, $\mu = 14.9$. Notice that the level of significance $\alpha = 0.05$ appears in this expression in the value -1.64 on the right-hand side of the inequality. In the case that $\mu = 14.9$, the term $(\bar{x} - 14.9)/(0.5/\sqrt{n})$ is standard normal, and so

$$0.90 = P\left(\frac{\bar{x} - 15}{0.5/\sqrt{n}} < -1.64\right)$$

$$= P\left(\frac{\bar{x} - 14.9}{0.5/\sqrt{n}} + \frac{14.9 - 15}{0.5/\sqrt{n}} < -1.64\right)$$

$$= P\left(\frac{\bar{x} - 14.9}{0.5/\sqrt{n}} - \frac{\sqrt{n}}{5} < -1.64\right)$$

$$= P\left(\frac{\bar{x} - 14.9}{0.5/\sqrt{n}} < \frac{\sqrt{n}}{5} - 1.64\right)$$

Since $(\bar{x} - 14.9)/(0.5/\sqrt{n})$ is standard normal, we have

$$\frac{\sqrt{n}}{5} - 1.64 = 1.281$$

(since, if y is standard normal, $P(y < c) = 0.90$ implies that $c = 1.281$) which is equivalent to $n = 213.306$. Therefore, a sample of size 214 would suffice at level of significance 0.05 to guarantee that the power of the test at $\mu = 14.9$ is 0.90.

Should we wish to test a hypothesis concerning a population proportion, we would use almost the identical procedure. The null hypothesis would be of the form

$$H_0 : p = p_0$$

with either a one-sided or a two-sided alternative. The statistic z tested would be the standardized form of $\bar{p}$, the sample proportion, standardized under the assumption that the null hypothesis $p = p_0$ is true; that is,

$$z = \frac{\bar{p} - p_0}{\sqrt{p_0 q_0 / n}}$$

Example 1 (continued)   Suppose that we wish to test the hypothesis

$$H_0 : p = 0.90$$

that the advertiser's claim is legitimate, at the five percent level of significance. The alternative hypothesis here would be

$$H_1 : p < 0.90$$

since values of $p > 0.90$ would be just as acceptable as the value $p = 0.90$. The one-sided critical region, together with the five percent level of significance, define the critical region $z < -1.64$. If, in the observed sample of 25 items, we found that 19 were nondefective, then $\bar{p} = 0.72$, and

$$z = \frac{0.72 - 0.90}{\sqrt{0.90 \cdot 0.10/25}} = -3$$

which is certainly critical. If, however, we had found that 20 of the 25 items observed were nondefective, we would have $\bar{p} = 0.80$, and

$$z = \frac{0.80 - 0.90}{\sqrt{0.90 \cdot 0.10/25}} = -1.67$$

This value of z lies just slightly over the borderline in the critical region. Thus, at the five percent level of significance, at least 21 of the 25 items observed must be nondefective to support the advertiser's claim. Recall that our rule previously had been to accept his claim unless the observed number of nondefectives was 19 or fewer, but this was at level of significance $\alpha = 0.0334$. The one-sided critical region corresponding to this value of would be $z < -1.83$, which would not contain the statistic $z = -1.67$ corresponding to $k = 20$ nondefectives in the sample observed.

Example 5   A production process is regarded as being under control if it produces five percent or fewer defectives. Suppose that a sample of 100 such items contained eight defectives. Is this cause for concern that the process is no longer under control? To answer this question, we test the hypothesis

$H_0: p = 0.05$

against the one-sided alternative

$H_1: p > 0.05$

at the five percent level of significance. (Do you understand why we choose a one-sided critical region?) The critical region is therefore, equal to $z > 1.64$. Since the sample proportion $\bar{p} = 0.08$, the statistic

$$z = \frac{0.08 - 0.05}{\sqrt{0.05 \cdot 0.95/100}} = 1.38$$

does not fall in the critical region, and so we have no reason, based on this test performed on this sample, to believe that the process is no longer under control.

Example 6 It is known that approximately 1 out of every 10 people use a certain product. After a concentrated advertising campaign in a certain area, a sample of 400 people contained 60 who were using the product. We wish to determine if the advertising campaign has had a significant effect on consumer preference, so we test the null hypothesis

$H_0: p = 0.10$

that the advertising campaign has had *no* effect (this is one reason for the term *null* hypothesis), hoping that the evidence available in the sample will allow us to reject H_0. The alternative hypothesis would be

$H_1: p > 0.10$

and at level of significance $\alpha = 0.01$, we have the critical region $z > 2.33$. Our sample proportion was found to be $\bar{p} = \frac{60}{400} = 0.15$, and so our sample statistic

$$z = \frac{0.15 - 0.10}{\sqrt{0.10 \cdot 0.90/400}} = \frac{10}{3}$$

which is resoundingly a critical value. Hence, we may reject the null hypothesis, and feel confident that the advertising campaign has significantly raised the percentage of people using this product.

Example 7 Suppose that we wish to estimate the number of fish in a certain lake. Rather than catching all the fish and counting them (and then eating them), we adopt the following plan. On one day, we catch 1000 fish, mark each with a red dot, and then release them. A few days later, we again catch 1000 fish (both days catching them from many different parts of the lake), and record the number of fish we have just caught that bear our red mark. This number, which we shall denote k, will be critical in our estimation of the size of the fish population of the lake. For suppose that there are N fish in the lake. Then, on the second day we go out to catch fish, the probability that a fish caught bears our red mark is $p = 1000/N$. But the proportion of fish bearing a red mark in our second sample is $\bar{p} = k/1000$. Our problem is to determine which values of N make this sample proportion seem not unreasonable. Therefore, at level of significance $\alpha = 0.05$, we test the hypothesis

$$H_0 : p = \frac{1000}{N}$$

against a two-sided alternative. We wish to determine for which values of N the sample proportion $\bar{p}$ is *not* critical, at a five percent level of significance. The sample statistic we must test is

$$z = \frac{k/1000 - 1000/N}{\sqrt{(1000/N)[(N - 1000)/N]/1000}}$$

$$= \frac{kN - 100{,}000}{1000\sqrt{N - 1000}}$$

Table 9.3
Estimation of the Size of a Fish Population

Value of k	Smallest possible N	Largest possible N
10	54,616	183,801
20	32,584	77,021
30	23,528	47,409
40	18,516	33,886
50	15,311	26,227
60	13,077	21,324
70	11,427	17,929
80	10,156	15,445
90	9,147	13,552
100	8,323	12,062
110	7,639	10,860
120	7,063	9,873
130	6,568	9,045
140	6,139	8,343
150	5,765	7,740
160	5,435	7,216
170	5,141	6,757
180	4,878	6,353
190	4,642	5,993
200	4,427	5,670

A computer program was written to determine the smallest and largest values of N, corresponding to any given value of k, for which the sample statistic z is not critical. In other words, we have calculated the values of N that correspond to the left- and right-hand boundaries of the critical region. The results can be found in Table 9.3 below. For example, if our second sample of 1,000 fish contained $k = 100$ with red marks, we could say, allowing ourselves a five percent chance of error, that the population of the lake was somewhere between 8,323 and 12,062 fish.

EXERCISES

1. A person claims to have ESP. In order to prove his contention, he is asked to identify the colors of 50 cards drawn with replacement from an ordinary deck. Suppose that he identifies 32 colors correctly. What can be concluded about his powers of ESP?

2. A manufacturer claims that his brand of light bulbs have an average life span of 1000 hours. A sample of 100 of his light bulbs show an average life span of 950 hours. Should we accept his claim? Use a five percent level of significance and a standard deviation of $\sigma = 100$ hours.

3. Scores on a college entrance examination in mathematics over the years have averaged 67 with a standard deviation of 7.5. A class of 35 high school students recently averaged 70 on this exam. Is this sufficient reason to believe that the true average is now higher than 67?

4. A coin is tossed 50 times, and comes up heads 28 times. Is this sufficient evidence to believe that the coin is biased in favor of heads?

5. A medicine is claimed to be 90 percent effective. In a sample of 200 patients, 170 were cured. At level of significance 0.01, can this claim be believed?

6. An advertising agency claims that 20 percent of all television viewers watch a particular TV program. In a random sample of 1000 viewers, only 184 were found to be watching the program. Is this sufficient evidence to dismiss the advertiser's claim?

7. If minority groups constitute 15 percent of the population in a city, but only 9 of the 85 workers in a factory were from minority groups, does this suggest discrimination in hiring?

8. Experience shows that five percent of a certain article are defectively produced. A new worker produces eight defective articles in a group of 100. At a five percent level of significance, can it be said that he is incompetent?

9. A politician aspiring to the presidency decides to initiate a campaign if samples indicate that 50 percent of the voters recognize his name. Suppose that a sample of 100 voters contained only 42 who recognized the politician's name. Is the result of this sample strong enough evidence to persuade the politician not to start a campaign, or should he seek more evidence in further samples?

9.5 Theory of Hypothesis Testing

In this section, we will discuss some of the theory of hypothesis testing, with our goal being to establish criteria that will tell us which is the "best" critical region for testing a particular null hypothesis H_0 against a given alternative

H_1 at some level of significance α. This section is, in essence, a theoretical justification of the techniques discussed in the preceding section.

In the preceding section, we discussed only one particular type of hypothesis test, wherein we test a *simple* null hypothesis against a *composite* alternative. A statistical hypothesis is called *simple* if it specifies all the undefined parameters which determine the distribution of a population; a hypothesis which leaves a parameter (or parameters) undefined is called *composite*. For example, a hypothesis of the type

$$H_0: p = 0.50$$

that specifies the parameter p of a Bernoulli population is simple, whereas the hypothesis

$$H_1: p > 0.50$$

is composite, since it does not specify an exact value for p. Likewise, the hypothesis

$$H_0: \mu = 2$$

specifying the parameter μ of a population which is normally distributed is composite because it fails to specify the parameter σ.

Although the test of a simple null hypothesis against a composite alternative is by far the most common, it is helpful, by way of preparation, to discuss first the test of a simple null hypothesis against a simple alternative hypothesis. Let us suppose that we have a population whose distribution is determined by a single parameter π, and that we wish to test the null hypothesis

$$H_0: \pi = \pi_0$$

against the alternative hypothesis

$$H_1: \pi = \pi_1$$

at level of significance α. Recall that α is the probability of making a type 1 error—accepting H_1 when, in fact, H_0 is true. In this situation, a type 2 error occurs if we accept H_0 when, in fact, it is H_1 that is true. We shall denote the probability of making a type 2 error by β.

We would, of course, like our test to be so efficient that both α and β are very small. But, unfortunately, it is a statistical fact of life that, for a fixed sample size n, there is an inverse type of relationship between α and β. That is to say, if we were in some way able to modify our test so as to decrease the value of the probability α, we would find that we have caused an increase in the value of the probability β, and vice-versa.

Any test of the hypothesis H_0 against the alternative H_1 specifies a critical region, a subset of the set of all sample statistics which leads to the rejection of the null hypothesis. In order to determine which test (critical region) is "best," with respect to the size of α and β, we will first restrict the size of α, thereby placing a restriction on how large the critical region may be, in relation to the set of all sample statistics. We will then determine which of the potential critical regions meeting our size requirements on α yields the smallest values of β. In other words, if we wish to test H_0 against H_1 at level of significance α, we will first restrict ourselves to possible critical regions which constitute at most α percent of the set of all sample statistics, and then determine which of these critical regions gives us the smallest chance of making a type 2 error. Such a critical region is said to be *most powerful* for testing H_0 against H_1. We shall also say, if our test is done at level of significance α, that our critical region is of *size α*.

We shall now discuss the Neyman-Pearson lemma, which will give us sufficient conditions for the existence of a most powerful critical region for testing a simple null hypothesis H_0 against a simple alternative hypothesis H_1. The Neyman-Pearson criterion is stated in terms of the likelihood function discussed in Section 9.2. Recall that if $\{x_1, x_2, \ldots, x_n\}$ is a random sample of size n taken with replacement from a population whose distribution is determined by the parameter π, then the likelihood function is defined by

$$L = f(x_1, x_2, \ldots, x_n \mid \pi) = \prod_{i=1}^{n} f(x_i \mid \pi)$$

where f is the probability function or density function of the population. This function calculates the likelihood (probability) that a particular sample occurs, given a particular value for the parameter π. In the case of testing a simple null hypothesis

$$H_0 \colon \pi = \pi_0$$

against a simple alternative hypothesis

$$H_1 \colon \pi = \pi_1$$

two of the likelihood functions are of special importance:

$$L_0 = f(x_1, x_2, \ldots, x_n \mid \pi_0)$$

which calculates the likelihood of the given sample occurring under the assumption that the parameter $\pi = \pi_0$, and

$$L_1 = f(x_1, x_2, \ldots, x_n \mid \pi_1)$$

which calculates the likelihood of the given sample occurring under the assumption that the parameter $\pi = \pi_1$.

If the null hypothesis were false, we would expect that L_1 would be considerably larger than L_0, and consequently, the quotient L_0/L_1, which is known as the *likelihood ratio*, would be rather small. This leads us to the Neyman-Pearson lemma:

Proposition 9.5 Neyman-Pearson Lemma

If R is a critical region of size α such that, for some constant λ,

$$\frac{L_0}{L_1} \leq \lambda$$

for all sample statistics which fall in R, and such that

$$\frac{L_0}{L_1} \geq \lambda$$

for all sample statistics which do not fall in R, then R is a most powerful critical region of size α for testing H_0 against H_1.

Proof: Suppose that R' is another critical region of size α. By the definition of "size α" (level of significance α), we have

$$\alpha = \int \cdots \int_R f(x_1, x_2, \ldots, x_n | \pi_0)\, dx_1\, dx_2 \cdots dx_n$$

$$= \int \cdots \int_{R'} f(x_1, x_2, \ldots, x_n | \pi_0)\, dx_1\, dx_2 \cdots dx_n$$

Using the notation L_0 and L_1, this becomes

$$\alpha = \int \cdots \int_R L_0\, d\bar{x} = \int \cdots \int_{R'} L_0\, d\bar{x}$$

where $d\bar{x} = dx_1\, dx_2 \cdots dx_n$. Since

$$R = (R \cap R') \cup [R \cap C(R')]$$

and

$$R' = (R \cap R') \cup [C(R) \cap R']$$

we have

$$\int \cdots \int_{R \cap R'} L_0\, d\bar{x} + \int \cdots \int_{R \cap C(R')} L_0\, d\bar{x} = \int \cdots \int_{R \cap R'} L_0\, d\bar{x} + \int \cdots \int_{C(R) \cap R'} L_0\, d\bar{x}$$

so that

$$\int \cdots \int_{R \cap C(R')} L_0\, d\bar{x} = \int \cdots \int_{C(R) \cap R'} L_0\, d\bar{x}$$

Since $R \cap C(R') \subseteq R$ and $C(R) \cap R' \subseteq R'$, we have

$$\int \cdots \int_{R \cap C(R')} L_1\, d\bar{x} \geq \int \cdots \int_{R \cap C(R')} \left(\frac{L_0}{\lambda}\right) d\bar{x}$$

$$= \int \cdots \int_{C(R) \cap R'} \left(\frac{L_0}{\lambda}\right) d\bar{x}$$

$$\geq \int \cdots \int_{C(R) \cap R'} L_1\, d\bar{x}$$

so that

$$\int \cdots \int_{R \cap C(R')} L_1\, d\bar{x} \geq \int \cdots \int_{C(R) \cap R'} L_1\, d\bar{x}$$

Therefore,

$$\int_R \cdots \int L_1 \, d\bar{x} = \int_{R \cap R'} \cdots \int L_1 \, d\bar{x} + \int_{R \cap C(R')} \cdots \int L_1 \, d\bar{x}$$

$$\geq \int_{R \cap R'} \cdots \int L_1 \, d\bar{x} + \int_{C(R) \cap R'} \cdots \int L_1 \, d\bar{x}$$

$$= \int_{R'} \cdots \int L_1 \, d\bar{x}$$

That is,

$$\int_R \cdots \int L_1 \, d\bar{x} \geq \int_{R'} \cdots \int L_1 \, d\bar{x}$$

but the multiple integral

$$\int_R \cdots \int L_1 \, d\bar{x}$$

is the probability that a given sample $\{x_1, x_2, \ldots, x_n\}$ will fall in the critical region (which leads to the rejection of $\pi = \pi_0$), given that $\pi = \pi_1$. In other words, it is the probability of *not* making a type 2 error, and the inequality

$$\int_R \cdots \int L_1 \, d\bar{x} \geq \int_{R'} \cdots \int L_1 \, d\bar{x}$$

is simply expressing the fact that the probability of not making a type 2 error is at its highest for the critical region R. Consequently, the probability of making a type 2 error is at its smallest for the critical region R, making R a most powerful region for testing $\pi = \pi_0$ against $\pi = \pi_1$.

Example 1   Suppose that we have a normal population with parameters μ and $\sigma = 1$, and wish to test the simple null hypothesis $\mu = \mu_0$ against the simple alternative hypothesis $\mu = \mu_1$, where we assume, for the sake of example, that $\mu_1 > \mu_0$. If $\{x_1, x_2, \ldots, x_n\}$ represents a random sample of size n taken from this population, then

$$L_0 = \prod_{i=1}^{n} \left(\frac{1}{\sqrt{2\pi}} \right) \exp\left[-\tfrac{1}{2}(x_i - \mu_0)^2 \right]$$

$$= \left(\frac{1}{\sqrt{2\pi}} \right)^n \exp\left[-\frac{1}{2} \sum_{i=1}^{n} (x_i - \mu_0)^2 \right]$$

and

$$L_1 = \prod_{i=1}^{n} \left(\frac{1}{\sqrt{2\pi}} \right) \exp\left[-\tfrac{1}{2}(x_i - \mu_1)^2 \right]$$

$$= \left(\frac{1}{\sqrt{2\pi}} \right)^n \exp\left[-\frac{1}{2} \sum_{i=1}^{n} (x_i - \mu_1)^2 \right]$$

Consequently,

$$\frac{L_0}{L_1} = \exp\left[+\frac{1}{2}\left(\sum_{i=1}^{n} (x_i - \mu_1)^2 - \sum_{i=1}^{n} (x_i - \mu_0)^2 \right) \right]$$

but

$$\sum_{i=1}^{n} (x_i - \mu_1)^2 - \sum_{i=1}^{n} (x_i - \mu_0)^2$$

$$= \sum_{i=1}^{n} (x_i^2 - 2x_i\mu_1 + \mu_1^2) - \sum_{i=1}^{n} (x_i^2 - 2x_i\mu_0 + \mu_0^2)$$

$$= n\mu_1^2 - n\mu_0^2 + 2(\mu_0 - \mu_1) \sum_{i=1}^{n} x_i$$

so that

$$\frac{L_0}{L_1} = \exp\left[\left(\frac{n}{2}\right)(\mu_1^2 - \mu_0^2) + (\mu_0 - \mu_1) \sum_{i=1}^{n} x_i \right]$$

To find a most powerful critical region R for testing $\mu = \mu_0$ against $\mu = \mu_1$, we must show that there exists a constant λ such that

$$\frac{L_0}{L_1} \le \lambda \quad \text{in } R \quad \text{and} \quad \frac{L_0}{L_1} \ge \lambda \quad \text{outside } R.$$

But

$$\exp\left[\left(\frac{n}{2}\right)(\mu_1^2 - \mu_0^2) + (\mu_0 - \mu_1) \sum_{i=1}^{n} x_i \right] \le \lambda$$

is equivalent to

$$\left(\frac{n}{2}\right)(\mu_1^2 - \mu_0^2) + (\mu_0 - \mu_1) \sum_{i=1}^{n} x_i \le \log \lambda$$

which, in turn, is equivalent to

$$(\mu_0 - \mu_1) \sum_{i=1}^{n} x_i \le \log \lambda - \left(\frac{n}{2}\right)(\mu_1^2 - \mu_0^2)$$

Dividing through by the negative value $n(\mu_0 - \mu_1)$, we obtain the equivalent form

$$\bar{x} = \frac{\displaystyle\sum_{i=1}^{n} x_i}{n} \ge \frac{\log \lambda - (n/2)(\mu_1^2 - \mu_0^2)}{n(\mu_0 - \mu_1)}$$

Hence, we can find a constant λ such that $L_0/L_1 \le \lambda$ inside R and $L_0/L_1 \ge \lambda$ outside R if and only if we can find a constant L, which is a function of λ, n, μ_0, and μ_1, such that

$$\bar{x} \ge L \quad \text{inside } R \quad \text{and} \quad \bar{x} \le L \quad \text{outside } R$$

Our work in Section 9.3 indicates that $L = \mu_0 + z_\alpha/\sqrt{n}$ defines a critical region of size α with the desired property, so

$$\bar{x} \geq \mu_0 + \frac{z_\alpha}{\sqrt{n}}$$

is a most powerful critical region for testing $\mu = \mu_0$ against $\mu = \mu_1$.

Example 2 Suppose that we wish to test the simple null hypothesis that the parameter p of a Bernoulli population equals p_0 against the simple alternative that $p = p_1 < p_0$. Suppose that x "successes" were found in a sample of size n taken from this population. The likelihood ratio then equals

$$\frac{L_0}{L_1} = \frac{b(x; n, p_0)}{b(x; n, p_1)} = \frac{C_x^{\,n} \cdot p_0^{\,x} \cdot (1 - p_0)^{n-x}}{C_x^{\,n} \cdot p_1^{\,x} \cdot (1 - p_1)^{n-x}}$$

$$= \left(\frac{p_0}{p_1}\right)^x \left(\frac{1 - p_0}{1 - p_1}\right)^{n-x}$$

Now $L_0/L_1 \leq \lambda$ is equivalent to

$$\left(\frac{p_0}{p_1}\right)^x \left(\frac{1 - p_0}{1 - p_1}\right)^{n-x} \leq \lambda$$

Since the logarithm is a strictly increasing function, this is equivalent to

$$\log\left[\left(\frac{p_0}{p_1}\right)^x \left(\frac{1 - p_0}{1 - p_1}\right)^{n-x}\right] \leq \log \lambda$$

which implies that

$$x \log\left(\frac{p_0}{p_1}\right) + (n - x) \log\left(\frac{1 - p_0}{1 - p_1}\right) \leq \log \lambda$$

But this implies that

$$x \log\left(\frac{p_0}{p_1}\right) - x \log\left(\frac{1 - p_0}{1 - p_1}\right) \leq \log \lambda - n \log\left(\frac{1 - p_0}{1 - p_1}\right)$$

or

$$x \log\left[\frac{p_0(1 - p_1)}{p_1(1 - p_0)}\right] \leq \log \lambda + \log\left(\frac{1 - p_1}{1 - p_0}\right)^n$$

Rearranging terms, we find this to be equivalent to

$$x \leq \frac{\log\left[\lambda\left(\dfrac{1 - p_1}{1 - p_0}\right)^n\right]}{\log\left[\dfrac{p_0(1 - p_1)}{p_1(1 - p_0)}\right]}$$

or

$$\bar{p} = \frac{x}{n} \leq \frac{\log\left[\lambda\left(\frac{1-p_1}{1-p_0}\right)^n\right]}{n\log\left[\frac{p_0(1-p_1)}{p_1(1-p_0)}\right]} = L$$

Therefore, we can find a constant λ such that

$$L_0/L_1 \leq \lambda \quad \text{inside } R \quad \text{and} \quad L_0/L_1 \geq \lambda$$

if and only if we can find a constant L, which is a function of λ, n, p_0, and p_1, such that

$$\bar{p} \leq L \quad \text{inside } R \quad \text{and} \quad \bar{p} \geq L \quad \text{outside } R$$

Again invoking our results from Section 9.3, we find that

$$L = p_0 + z_\alpha \sqrt{\frac{p_0(1-p_0)}{n}}$$

defines a critical region

$$\bar{p} \leq p_0 + z_\alpha \sqrt{\frac{p_0(1-p_0)}{n}}$$

of size α with the desired property.

We shall now attempt to extend the procedures we have just discussed to tests of a simple null hypothesis against a composite alternative hypothesis. The techniques we shall employ can also be used for tests of a composite null hypothesis against a composite alternative hypothesis, although we shall not discuss that topic in this text.

In the case of a composite alternative hypothesis, the power function of the test will play a key role in determining whether one critical region of size α is better than another. The power function for one test (critical region) is said to be *uniformly more powerful* than the power function for another test (critical region) if the values of the first power function are greater than or equal to the values of the second power function for all values of the parameter(s) specified by the alternative hypothesis. A test (critical region) of size α that is uniformly more powerful than any other test (critical region) of size α is said to be *uniformly most powerful* (UMP).

Since our alternative hypothesis is composite, we cannot use the Neyman-Pearson lemma to find a uniformly most powerful critical region. As a matter of fact, uniformly most powerful critical regions seldom exist. However, we can use an approach similar to the one used above to construct critical regions which have some nice properties.

Let us suppose that we have a population whose distribution (or density) function is determined by some unknown parameter π, and we wish to test the simple null hypothesis

$$H_0: \pi = \pi_0$$

against the composite alternative

$H_1: \pi \in P$

where P is some set of parameters not including π_0. For example, P might be $\{\pi : \pi \neq \pi_0\}$ or $\{\pi : \pi > \pi_0\}$. If $\{x_1, x_2, \ldots, x_n\}$ is a random sample taken from this population, then we can define likelihood functions

$L_0 = f(x_1, x_2, \ldots, x_n \mid \pi_0)$

and

$L_\pi = f(x_1, x_2, \ldots, x_n \mid \pi)$

for any $\pi \in P$, and we can define

$L_1 = \max_{\pi \in P} L_\pi$

Once again, we consider the likelihood ratio L_0/L_1, and reason as above that, should the null hypothesis be false, the value of L_0 will be small in comparison to the value of L_1. Consequently, it is the small values of the likelihood ratio that would tend to cast doubts on the validity of the null hypothesis, and so the critical region of our test could be defined as

$\dfrac{L_0}{L_1} \leq \lambda$

for some small λ to be determined. The problem then becomes one of attempting to equate this critical region to one stated more simply in terms of the parameter π and the information contained in the sample.

Example 3   Suppose, as in Example 1, that we have a normal population with parameters μ and $\sigma = 1$, and we wish to test the simple null hypothesis $\mu = \mu_0$ against the composite alternative hypothesis $\mu \neq \mu_0$. We have

$$L_0 = \left(\frac{1}{\sqrt{2\pi}}\right)^n \exp\left[-\frac{1}{2}\sum_{i=1}^{n}(x_i - \mu_0)^2\right]$$

as in Example 1. Using the results from Section 9.2 concerning maximum likelihood estimators (see Example 5 in Section 9.2), we have

$$L_1 = \left(\frac{1}{\sqrt{2\pi}}\right)^n \exp\left[-\frac{1}{2}\sum_{i=1}^{n}(x_i - \bar{x})^2\right]$$

Consequently,

$$\frac{L_0}{L_1} = \exp\left\{-\frac{1}{2}\left[\sum_{i=1}^{n}(x_i - \mu_0)^2 - \sum_{i=1}^{n}(x_i - \bar{x})^2\right]\right\}$$

but

$$(x_i - \mu_0)^2 - (x_i - \bar{x})^2 = x_i^2 - 2\mu_0 x_i + \mu_0^2 - (x_i^2 - 2\bar{x}x_i + \bar{x}^2)$$
$$= (2\bar{x} - 2\mu_0)x_i + \mu_0^2 - \bar{x}^2$$

Consequently,

$$\sum_{i=1}^{n} (x_i - \mu_0)^2 - \sum_{i=1}^{n} (x_i - \bar{x})^2 = \sum_{i=1}^{n} [(2\bar{x} - 2\mu_0)x_i + \mu_0^2 - \bar{x}^2]$$

$$= (2\bar{x} - 2\mu_0) \sum_{i=1}^{n} x_i + n\mu_0^2 - n\bar{x}^2$$

$$= (2\bar{x} - 2\mu_0)(n\bar{x}) + n\mu_0^2 - n\bar{x}^2$$

$$= 2n\bar{x}^2 - 2n\bar{x}\mu_0 + n\mu_0^2 - n\bar{x}^2$$

$$= n(\bar{x}^2 - 2\bar{x}\mu_0 + \mu_0^2)$$

$$= n(\bar{x} - \mu_0)^2$$

Therefore,

$$\frac{L_0}{L_1} = \exp\left[-\left(\frac{n}{2}\right)(\bar{x} - \mu_0)^2\right]$$

The critical region for the likelihood ratio is thus

$$\exp\left[-\left(\frac{n}{2}\right)(\bar{x} - \mu_0)^2\right] \leq \lambda$$

for some small λ, which we may assume is less than 1. Our formula defining the critical region is equivalent to

$$-\left(\frac{n}{2}\right)(\bar{x} - \mu_0)^2 \leq \log \lambda$$

or

$$(\bar{x} - \mu_0)^2 \geq \frac{-2 \log \lambda}{n}$$

Since $\lambda < 1$, we find that this is equivalent to

$$|\bar{x} - \mu_0| \geq \sqrt{\frac{-2 \log \lambda}{n}} = L$$

Using the results of Section 9.4, we see that the critical region

$$|\bar{x} - \mu_0| \geq \frac{z_{\alpha/2}}{\sqrt{n}} \qquad \text{(recall that } \sigma = 1)$$

is equivalent to the critical region determined by the likelihood ratio. That is, given level of significance α, we choose λ so that $-2 \log \lambda = (z_{\alpha/2})^2$.

Example 4 Suppose that we wish to test the simple null hypothesis that the parameter p of a Bernoulli population equals $p_0 = \frac{1}{2}$, against the composite alternative that $p \neq \frac{1}{2}$. If a sample of size n shows x "successes," then

$$L_0 = b(x; n, \tfrac{1}{2})$$

while

$$L_1 = b(x; n, x/n)$$

where the parameter x/n was determined using the maximum likelihood method discussed in Section 9.2 (see Example 4). Consequently,

$$\frac{L_0}{L_1} = \frac{b(x; n, \frac{1}{2})}{b(x; n, x/n)} = \frac{C_x{}^n(\frac{1}{2})^n}{C_x{}^n(x/n)^x[(n-x)/n]^{n-x}}$$

so that

$$\frac{L_0}{L_1} = \frac{(\frac{1}{2})^n}{(x/n)^x[(n-x)/n]^{n-x}}$$

The condition that $L_0/L_1 \le \lambda$ is equivalent to

$$\left(\frac{1}{2}\right)^n \le \lambda \left(\frac{x}{n}\right)^x \left(\frac{n-x}{n}\right)^{n-x}$$

which is equivalent to

$$-n \log 2 \le \log \lambda + x \log\left(\frac{x}{n}\right) + (n-x) \log\left(\frac{n-x}{n}\right)$$

or

$$-n \log 2 \le \log \lambda + x \log x - x \log n + (n-x)\log(n-x) - (n-x)\log n$$

which, in turn, is equivalent to

$$x \log x + (n-x)\log(n-x) \ge L$$

where

$$L = n \log\left(\frac{n}{2}\right) - \log \lambda$$

Notice that, since $0 < \lambda < 1$, $\log \lambda$ is negative, and so $L \ge n \log(n/2)$. Since the function $x \log x + (n-x)\log(n-x)$ is symmetric about $x = n/2$ and assumes its minimum value, $n \log(n/2)$, at $x = n/2$, the critical region determined by the likelihood ratio is equivalent to a critical region of the form

$$\left| x - \frac{n}{2} \right| > M$$

or to a critical region of the form

$$\left| \frac{x}{n} - \frac{1}{2} \right| > \frac{M}{n} = N$$

Since $\bar{p} = x/n$, this leaves us with the same critical region that we determined in Section 9.4:

$$\left| \bar{p} - \tfrac{1}{2} \right| > z_{\alpha/2} \sqrt{\frac{1}{4n}}$$

EXERCISES

1. Use the Neyman-Pearson lemma to construct a most powerful critical region in each of the following cases:
   (a) to test the hypothesis that $\sigma = \sigma_0$ against the hypothesis that $\sigma = \sigma_1 > \sigma_0$ for a normal population with parameter $\mu = 0$.
   (b) to test the hypothesis that $\beta = \beta_0$ against the hypothesis that $\beta = \beta_1$ for an exponential population with parameter β. (*Hint*: See Exercise 1, Section 8.3.)
2. An urn contains ten balls, a certain percentage p of which are red. In order to test the hypothesis $p = 0.40$ against the alternative that $p = 0.60$, three balls are drawn without replacement from the urn, and the null hypothesis is rejected unless the number of reds in the sample is 0 or 1. Find the probability of making a type 1 error, and the probability of making a type 2 error.
3. If a random sample of size n is used to test the null hypothesis that the parameter β of an exponential distribution equals β_0 against the alternative that it does not equal β_0, use the result of Exercise 3 in Section 9.2 to show that critical region determined by the likelihood ratio is of the form $xe^{-x/\beta} < L$. Analyze the graph of the function $xe^{-x/c}$ to determine an alternate form for this critical region.

ADDITIONAL PROBLEMS

1. The standard medication for a particular disease is effective 80 percent of the time. A new, supposedly superior medication is being tested. From a sample of 100 patients, how many must recover after using the new medication to prove its superiority, at a 1 percent level of significance?
2. A manufacturer claims that 20 percent of the public prefers his product. What percentage of a sample of 100 customers is necessary in order to refute this claim at a 5 percent level of significance?
3. In measuring reaction time, a psychiatrist estimates a standard deviation of 0.05 seconds. How large a sample must he take so that he can be 95 percent confident that his error in approximating reaction times will not exceed 0.01 seconds?
4. Suppose that we want to estimate the proportion of television owners with color sets. How large a sample must be taken so that the probability is 0.90 that the relative frequency obtained from the sample is within 0.05 of the true proportion?
5. If 2 percent of the items produced by a factory are defective, determine the largest shipment size that will have, with probability 0.95, less than five defectives.
6. Eight hundred guests are invited to a picnic under the assumption that each has probability $\frac{2}{3}$ of accepting the invitation. How many picnic baskets should be ordered if we wish to be 99 percent certain that there will be a picnic basket for each guest who comes to the picnic?
7. Refer to Exercise 5 in Section 9.4, where the hypothesis $p = 0.90$ was tested against the alternative $p < 0.90$. Calculate the power of the test if $p = 0.80$ and if $p = 0.70$. At level of significance 0.01, how large a sample must be taken so that the power at $p = 0.80$ will exceed 0.80?
8. Refer to Exercise 2 in Section 9.4, where the hypothesis $\mu = 1000$ was tested against the alternative $\mu < 1000$. Calculate the power of this test if, in fact, $\mu = 950$ and if $\mu = 990$. At level of significance 0.05, how large a sample must be taken so that the power at 975 will exceed 0.95?

10

Statistics
Revisited

10.1 Differences Between Two Population Means or Proportions

One of the most important and frequently occurring problems in statistics is to determine whether or not two populations share a common parameter. That parameter could be a mean, and we might be asking, for example, if two similar products have the same life span, or that parameter could be a proportion, and our problem might be to determine if the citizens of two cities feel the same way about a certain political issue. Perhaps the most popular variation of this type of problem is the determination of the impact of change. We might, for example, wish to determine if a new drug is more effective than a drug presently in use, or if a new product is, in some way, better than an older product.

Our approach to such problems will be to take a sample from each population, and then determine the sample parameter (perhaps a mean or a proportion) for each sample. Needless to say, our sample parameters will probably differ. We will then have to determine if this difference is statistically significant and reflects a true difference in the corresponding population parameters, or if the difference is small and can reasonably be attributed to chance.

Since we have already seen that sample proportions can be looked upon as a particular case of sample means, we shall set up the machinery for studying the differences between sample means, and then obtain results for studying the differences between sample proportions as corollaries.

Let us suppose that we have two populations, the first with mean μ_1 and standard deviation σ_1, and the second with mean μ_2 and standard deviation σ_2. (We will study the case σ_1 and σ_2 unknown in section 10.7.) Suppose that we take a sample of size n_1 from the first population and a sample of size n_2 from the second population, and obtain sample means of $\bar{x}_1$ and $\bar{x}_2$, respectively. This is equivalent to having set up n_1 independent, identically distributed random variables $X_1, X_2, \ldots, X_{n_1}$ for sampling from the first population, and n_2 independent, identically distributed random variables $Y_1, Y_2, \ldots, Y_{n_2}$ for sampling from the second population. Now

$$E(X_i) = \mu_1 \quad \text{and} \quad \text{Var}(X_i) = \sigma_1{}^2$$

for $i = 1, 2, \ldots, n_1$, and

$$E(Y_i) = \mu_2 \quad \text{and} \quad \text{Var}(Y_i) = \sigma_2{}^2$$

for $i = 1, 2, \ldots, n_2$. By definition, the two sample means equal

$$\bar{X} = \frac{\left(\sum_{i=1}^{n_1} X_i \right)}{n_1} \quad \text{and} \quad \bar{Y} = \frac{\left(\sum_{i=1}^{n_2} Y_i \right)}{n_2}$$

Under these conditions, we can prove:

Proposition 10.1

(a) $\bar{X} - \bar{Y}$ is approximately normal.
(b) $E(\bar{X} - \bar{Y}) = \mu_1 - \mu_2$
(c) $\text{Var}(\bar{X} - \bar{Y}) = \sigma_1^2/n_1 + \sigma_2^2/n_2$

Proof:

(a) Follows from the central limit theorem, which implies that both $\bar{X}$ and $\bar{Y}$ are normal, and consequently (see Exercise 3 in Section 8.4) their difference $\bar{X} - \bar{Y}$ is normal.

(b)
$$E(\bar{X} - \bar{Y}) = E\left(\frac{\sum_{i=1}^{n_1} X_i}{n_1} - \frac{\sum_{i=1}^{n_2} Y_i}{n_2}\right)$$

$$= \frac{1}{n_1}\sum_{i=1}^{n_1} E(X_i) - \frac{1}{n_2}\sum_{i=1}^{n_2} E(Y_i)$$

$$= \frac{n_1\mu_1}{n_1} - \frac{n_2\mu_2}{n_2}$$

$$= \mu_1 - \mu_2$$

(c) Using the fact that $X_1, X_2, \ldots, X_{n_1}$ are independent, that $Y_1, Y_2, \ldots, Y_{n_2}$ are independent, and that the samples taken from the two populations are independent, we have

$$\text{Var}(\bar{X} - \bar{Y}) = \text{Var}\left(\frac{\sum_{i=1}^{n_1} X_i}{n_1} - \frac{\sum_{i=1}^{n_2} Y_i}{n_2}\right)$$

$$= \left(\frac{1}{n_1}\right)^2 \sum_{i=1}^{n_1} \text{Var}(X_i) + \left(\frac{1}{n_2}\right)^2 \sum_{i=1}^{n_2} \text{Var}(Y_i)$$

$$= \frac{n_1\sigma_1^2}{n_1^2} + \frac{n_2\sigma_2^2}{n_2^2}$$

$$= \frac{\sigma_1^2}{n_1} + \frac{\sigma_2^2}{n_2}$$

Let us now apply these results to the case where the parameters being studied are proportions, rather than means. If $\bar{p}_1$ represents the proportion of items with a certain characteristic in a sample taken from the first population, and $\bar{p}_2$ the proportion of items with this characteristic in a sample taken from the second population, then we have seen that both $\bar{p}_1$ and $\bar{p}_2$ can be looked upon as sample means. Since, in this case, we have $E(X_i) = p_1$ (the true proportion of items with the given characteristic in the first population) and $\text{Var}(X_i) = p_1(1 - p_1)$, for $i = 1, 2, \ldots, n_1$, and $E(Y_i) = p_2$ (the true proportion for the second population) and $\text{Var}(Y_i) = p_2(1 - p_2)$, for $i = 1, 2, \ldots, n_2$, we have

Corollary 10.2

(a) $\bar{p}_1 - \bar{p}_2$ is approximately normal.

(b) $E(\bar{p}_1 - \bar{p}_2) = p_1 - p_2$

(c) $\text{Var}(\bar{p}_1 - \bar{p}_2) = p_1(1 - p_1)/n_1 + p_2(1 - p_2)/n_2$

The results we have just derived are usually put to use testing the hypothesis that the two populations do share a common parameter. For example, if we are attempting to determine if the two population means μ_1 and μ_2 are, in fact, the same, we would test the null hypothesis

$$H_0: \mu_1 = \mu_2$$

against one of the alternative hypotheses $\mu_1 \neq \mu_2$, $\mu_1 < \mu_2$, or $\mu_1 > \mu_2$ at a given level of significance. We would take samples of sizes n_1 and n_2 from the two populations, calculate sample means $\bar{x}_1$ and $\bar{x}_2$, and then consider the standard normal statistic

$$z = \frac{(\bar{x}_1 - \bar{x}_2) - E(\bar{x}_1 - \bar{x}_2)}{\sigma_{\bar{x}_1 - \bar{x}_2}}$$

$$= \frac{(\bar{x}_1 - \bar{x}_2) - (\mu_1 - \mu_2)}{\sqrt{\sigma_1^2/n_1 + \sigma_2^2/n_2}}$$

Under the assumption that the null hypotheses is, in fact, true, this statistic becomes

$$\frac{\bar{x}_1 - \bar{x}_2}{\sqrt{\sigma_1^2/n_1 + \sigma_2^2/n_2}}$$

We then either reject, or do not reject, the null hypothesis depending on whether the statistic z does or does not fall in the critical region.

Example 1 A company must decide whether to use brand A or brand B light bulbs, so a test is made of 100 light bulbs of each brand. Suppose that the 100 brand A light bulbs had an average life span of 985 hours, while the 100 brand B light bulbs had an average life span of 1004 hours. We assume it to be a known fact that the standard deviation for the life span of a light bulb is 80 hours. We wish to determine if the difference in life spans for the two samples is significant at the five percent level of significance, so we test the null hypothesis

$$H_0: \mu_A = \mu_B$$

against the alternative

$$H_1: \mu_A \neq \mu_B$$

where μ_A and μ_B are the actual life spans for brand A and brand B light bulbs. At the five percent level of significance, we have the two-sided critical region $|z| > 1.96$, and if we evaluate

$$z = \frac{985 - 1004}{\sqrt{(80)^2/100 + (80)^2/100}} = \frac{-19}{\sqrt{128}} = -1.68$$

we find that our sample statistic does not fall in the critical region. Therefore, we cannot conclude that the difference between the two sample means reflects a true difference in the two population means.

Example 1 (continued) If, in Example 1 above, we knew that brand A light bulbs were slightly less expensive than brand B light bulbs, then we would want to use brand A light bulbs unless the results of the sample showed that brand B was significantly better than brand A. Consequently, we would choose

$$H_1: \mu_A < \mu_B$$

as our alternative hypothesis, giving the critical region $z < -1.64$. We calculate the sample statistic z exactly as before, and find that z is now critical.

Example 2 Suppose that we wish to test whether the (proposed) new rotary engines for automobiles are cleaner, with respect to sulphur dioxide emissions, than the piston engines currently in use. Suppose that a sample of 50 rotary engines showed an emission level of 26 parts per million, while a sample of 50 piston engines showed an emission level of 29 parts per million. Suppose also that we can assume a standard deviation of 10 parts per million. We test the hypothesis

$$H_0: \mu_1 = \mu_2$$

that the average emission levels for the two types of engines are equal, against the alternative

$$H_1: \mu_1 < \mu_2$$

that the rotary engines are cleaner, at a one percent level of significance. We have the 1-sided critical region $z < -1.64$, and the sample statistic

$$z = \frac{26 - 29}{\sqrt{(10)^2/50 + (10)^2/50}} = -1.50$$

does not fall in this critical region. Therefore, on the basis of the evidence contained in these two small samples, we cannot conclude that the new rotary engine is cleaner than the old piston engine.

If the population parameter we are studying happens to be a proportion, we test the null hypothesis

$$H_0: p_1 = p_2$$

against one of the alternatives $p_1 \neq p_2$, $p_1 < p_2$, or $p_1 > p_2$ at the given level of significance, by determining whether or not the statistic

$$z = \frac{(\bar{p}_1 - \bar{p}_2) - (p_1 - p_2)}{\sqrt{p_1 q_1/n_1 + p_2 q_2/n_2}}$$

lies in the appropriate critical region. Under the assumption that the null hypothesis is true, that is, that $p_1 = p_2$, this statistic reduces to

$$z = \frac{\bar{p}_1 - \bar{p}_2}{\sqrt{p_1 q_1/n_1 + p_2 q_2/n_2}}$$

One problem emerges here, and that is that the values of the unknown, although perhaps equal, parameters p_1 and p_2 appear in the formula for z. To overcome this problem, we first merge the two samples into one sample, then calculate the pooled proportion

$$\bar{p} = \frac{n_1 p_1 + n_2 p_2}{n_1 + n_2}$$

the proportion, in the merged sample, of items with the given characteristic. Under the assumption that the null hypothesis is true, that the two populations have the same percentage of items with the given characteristic, then $\bar{p}$ is the best guess we can make concerning the true value of $p_1 = p_2$, based on the information contained in the two samples. Using $\bar{p}$ as a substitute for both p_1 and p_2, our statistic z becomes

$$z = \frac{\bar{p}_1 - \bar{p}_2}{\sqrt{\bar{p}\bar{q}/n_1 + \bar{p}\bar{q}/n_2}}$$

Example 3 We wish to determine whether the male and female voters in a certain city are of the same mind with respect to a certain issue. A sample of 144 men contained 70 (or 48.6 percent) in favor of the issue, while a sample of 100 women contained 55 (or 55 percent) in favor of the issue. We wish to determine if the sample proportions 0.486 and 0.550 indicate a substantial difference of opinion, or if the difference between these two samples can reasonably be attributed to chance. For this purpose, we evaluate first the pooled proportion

$$\bar{p} = \frac{144(0.486) + 100(0.550)}{144 + 100}$$

$$= \frac{125}{244} = 0.512$$

and then the statistic

$$z = \frac{0.486 - 0.550}{\sqrt{0.512 \cdot 0.488 \cdot [(1/144) + (1/100)]}}$$

$$= \frac{-0.064}{0.065}$$

$$= -0.985$$

which is clearly not large enough to fall in any one percent or five percent critical region. Hence, on the basis of this sample alone, we cannot say, for example, that the women feel more strongly about this issue than do the men.

Example 4 Suppose that we wish to test two different types of headache pills for quickness in providing relief. Suppose that, in a group of 100 people using the first type, 70 found that their headache was relieved within 15 minutes, while, in a group of 100 people using the second type, only 60 found relief within 15 minutes. Is this sufficient proof that the first type of pill provides relief faster than the second type? To answer this question, we test the null hypothesis

$$H_0: p_1 = p_2$$

that the two types of pills are equally effective (that is, the same percentage will find relief within 15 minutes using either brand) against the alternative

$$H_1: p_1 > p_2$$

that the first type will relieve a higher percentage within 15 minutes. At a one percent level of significance, we have the one-sided critical region $z > 2.33$, and since our sample statistic

$$z = \frac{0.70 - 0.60}{\sqrt{0.65 \cdot 0.35 \cdot (2/100)}} = 1.48$$

does not fall in this critical region, we cannot conclude, on the basis of these samples, that one pill is more effective than the other.

EXERCISES

1. To test the effectiveness of a vaccine, 150 animals were given the vaccine and 150 were not. All were then infected with the disease. Of those vaccinated, 10 died; among the others, 30 died. Was the vaccine effective?

2. A group of 900 people consisted of 433 women and 467 men. Eight of the men were color blind, while only two of the women were color blind. Does this suggest that men are more susceptible to color blindedness?

3. Fifty students from each of two colleges took an examination in mathematics, with the 50 from college A averaging 63, while the 50 from college B averaged 72. If the standard deviation for such exam scores is known to be 10, can we conclude that the mathematics department at college B is doing a better job than their counterparts at college A?

4. Suppose that 32 brand A flashlight batteries had an average life span of 32.4 hours, while 36 brand B flashlight batteries averaged 30.8 hours. Suppose also that the standard deviation in flashlight battery life spans is 2.2 hours. Can we conclude that brand A batteries last longer than brand B batteries?

5. A think tank asked 1000 men and 1000 women if they favored the president's policies against inflation. Sixty percent of the men and 52 percent of the women said they did not. Does this reflect a true difference of opinion between the men and women of the country?

6. Suppose that 70 percent of the 400 men surveyed and 75 percent of the 600 women surveyed favored stronger pollution control laws. Does this indicate that women feel more concerned about pollution than men?

7. A teacher wishes to test the effectiveness of two different textbooks, so he teaches two sections of 40 students each of the same subject, one using each textbook. Suppose that 85 percent of the students in the first section pass the course, but only 75 percent in the second section pass. Does this indicate that the first textbook is significantly more effective?

8. Two different manufacturing processes produce two percent and four percent defectives in samples of 250. Does this indicate that the first process is better than the second process?

9. Suppose that a group of 50 cars averaged 15 miles per gallon using one brand of gasoline, and the same group of cars then averaged 13 miles per gallon using a second brand of gasoline. Does this indicate that the first brand gives better mileage? Use a standard deviation of 2 miles per gallon.

10. Suppose that 1000 families surveyed in New York had an average income of $12,500, while 1000 families surveyed in Chicago had an average income of $11,900. Use a standard deviation of $2,500, and determine if average income per family is higher in New York than it is in Chicago.

10.2 The Chi-Square Distribution

The chi-square (χ^2) distribution has already been introduced in Section 8.3 as a special case of the gamma distribution ($\alpha = k/2$ and $\beta = 2$). The χ^2 distribution therefore depends only on the parameter k, called the *number of degrees of freedom*, has density function

$$f(x) = \begin{cases} [1/2^{k/2}\Gamma(k/2)]x^{(k-2)/2}e^{-x/2} & \text{for } x > 0 \\ 0 & \text{for } x \le 0 \end{cases}$$

and, as we have indicated in Corollary 8.14, has mean k, variance $2k$, and moment-generating function

$$M_X(t) = (1 - 2t)^{-k/2}$$

With the exception of the case $k = 1$, these χ^2 densities take somewhat the appearance of a normal curve that is skewed to the left, as indicated in Figure 10.1. The χ^2 density with one degree of freedom looks nothing like the other χ^2 curves.

For our work with the χ^2 distribution, we will adopt the following notation. We shall denote by $\chi^2_{\alpha, k}$ that point on the x-axis to the right of which lies α percent of the area under the χ^2 density with k degrees of freedom, as indicated in Figure 10.2. A table of values of the form $\chi^2_{\alpha, k}$, for $k = 1, 2, 3, \ldots, 30$ is included as Appendix D. Of particular interest to us will be the values $\chi^2_{0.05, k}$, those points on the x-axis to the right of which lie five percent of the area under the χ^2 density with k degrees of freedom, and the values $\chi^2_{0.95, k}$, those points on the x axis to the left of which lie five percent of the area under the χ^2 density with k degrees of freedom. We shall use these values in later sections of this chapter in the construction of 95 percent confidence intervals, and in testing hypotheses at the five percent level of significance.

The remainder of this section will be devoted to three important results concerning the χ^2 distribution. Then, in the following three sections, we will investigate some of the applications of the χ^2 distribution.

Proposition 10.3

If X has standard normal distribution, then X^2 has χ^2 distribution with one degree of freedom.

Proof: We consider the random variable X^2 in its equivalent form $|X|^2$, so that we will be able to use Proposition 8.3. Since X is standard normal, it

---

Figure 10.1

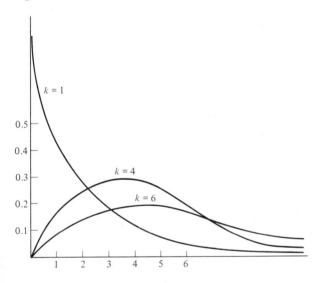

Figure 10.2

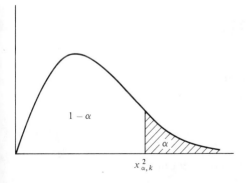

has density function

$$f(x) = \left(\frac{1}{\sqrt{2\pi}}\right)e^{-(1/2)x^2}$$

Consequently, $|X|$ has density function

$$g(x) = \begin{cases} (2/\sqrt{2\pi})e^{-(1/2)x^2} & \text{for } x > 0 \\ 0 & \text{otherwise} \end{cases}$$

We are now in a position to apply Proposition 8.3, because now the function $y = x^2$ is an increasing function over the range of the random variable $|X|$ on which it is defined. (This was not the case for the random variable X, which was also defined for negative values of x.) Since $dy = 2x\,dx$, we have $dx/dy = 1/2x = 1/2\sqrt{y}$, we obtain from Proposition 8.3 that the density function $h(y)$ of $|Y|^2$ (or Y^2) is

$$h(y) = \begin{cases} (2/\sqrt{2\pi})e^{-(1/2)y}(1/2\sqrt{y}) & \text{if } y > 0 \\ 0 & \text{otherwise} \end{cases}$$

That is,

$$h(y) = \begin{cases} (1/\sqrt{2\pi})e^{-(1/2)y}y^{-1/2} & \text{if } y > 0 \\ 0 & \text{otherwise} \end{cases}$$

which is a gamma density with $\alpha = \frac{1}{2}$ and $\beta = 2$; that is, a χ^2 density with $k = 1$ degrees of freedom. (Note that $\Gamma(\frac{1}{2}) = \sqrt{\pi}$.)

Proposition 10.4

If $X_1, X_2, \ldots, X_n$ are independent standard normal random variables, and if

$$Y = \sum_{i=1}^{n} X_i^2$$

then Y is a χ^2 random variable with n degrees of freedom.

Proof: By Proposition 10.3, X_i^2 is a χ^2 random variable with one degree of freedom, for $i = 1, 2, \ldots, n$. Consequently by Corollary 8.14,

$$M_{X_i}(t) = (1 - 2t)^{-1/2}$$

We then apply Proposition 5.6 to find that

$$M_Y(t) = [(1 - 2t)^{-1/2}]^n = (1 - 2t)^{-n/2}$$

which is the moment-generating function of the χ^2 distribution with n degrees of freedom. Consequently, by Lemma 8.21, Y is a χ^2 random variable with n degrees of freedom.

It is evident from the method of this proof that a sum of k independent χ^2 random variables with $n_1, n_2, \ldots, n_k$ degrees of freedom, respectively, is a χ^2 random variable with $\sum_{i=1}^{k} n_i$ degrees of freedom.

Proposition 10.5

Suppose that X_1 and X_2 are independent random variables such that X_1 and $X_1 + X_2$ have χ^2 distributions with k_1 and $k > k_1$ degrees of freedom, respectively. Then X_2 has a χ^2 distribution with $k - k_1$ degrees of freedom.

Proof: By Corollary 8.14,

$$M_{X_1}(t) = (1 - 2t)^{-k_1/2}$$

and

$$M_{X_1 + X_2}(t) = (1 - 2t)^{-k/2}$$

However, by Proposition 5.6,

$$M_{X_1 + X_2}(t) = M_{X_1}(t) \cdot M_{X_2}(t)$$

and so

$$
\begin{aligned}
M_{X_2}(t) &= \frac{M_{X_1 + X_2}(t)}{M_{X_1}(t)} \\
&= \frac{(1 - 2t)^{-k/2}}{(1 - 2t)^{-k_1/2}} \\
&= (1 - 2t)^{(1/2)(k_1 - k)} \\
&= (1 - 2t)^{-(1/2)(k - k_1)}
\end{aligned}
$$

which is the moment-generating function of the χ^2 distribution with $k - k_1$ degrees of freedom. Consequently, by Corollary 8.21, X_2 has χ^2 distribution with $k - k_1$ degrees of freedom.

10.3 Distribution of the Sample Variance

The methods discussed in Chapter 9 for determining confidence interval estimates for population means and for testing hypotheses concerning population means both relied on knowledge of the variance of the population. Should the population variance be unknown, we must then rely on the variance of the sample we have taken. In this case, the theory presented in Chapter 9 becomes useless, and we can no longer use the normal curve to facilitate our calculations. We shall see in Section 10.6 what we must use as a substitute.

Aside from these considerations concerning the population mean, the determination of a population variance can be of great importance in its own right. In the field of quality control, for example, it is necessary to keep check on the production performances of machines. A machine may be allowed a certain error with respect to a given measurement, but once this tolerance is exceeded, the machine must be sent in for repairs. This tolerance could be measured in terms of a variance from the expected value of the measurement, and this variance is frequently checked by means of (some-times small) samples taken from the produce of the machine. This sample variance is then tested for statistically significant difference from the allowable variance.

In Chapter 9, we introduced the sample variance $\bar{s}^2$ as an unbiased point estimate for the population variance. Actually, the sample variance, like the sample mean or the sample proportion, can be looked upon as a random variable whose domain is the set of all possible samples of the given size n, one which assigns to each such sample the value of its modified variance

$$\bar{s}^2 = \frac{\sum\limits_{i=1}^{n} (x_i - \bar{x})^2}{n - 1}$$

In this section, we will develop methods for constructing confidence interval estimates for population variances and for testing hypotheses con-cerning population variances, based on the value assumed by the sample variance. Most of our results will depend on the following:

Proposition 10.6

If $\bar{s}^2$ denotes the sample variance for samples of size n taken from a normal population with mean μ and variance σ^2, then $(n - 1)\bar{s}^2/\sigma^2$ has χ^2 distribu-tion with $n - 1$ degrees of freedom.

Proof: A complete proof of this theorem is beyond the scope of this book. The proof we shall give is almost complete; the verification of but one key step in the proof has been omitted. We assume that our sample is taken with replacement and that X_k represents the value of the kth element in the sample. By assumption, X_k is normal, $E(X_k) = \mu$, and $\text{Var}(X_k) = \sigma^2$. Consequently, $(X_k - \mu)/\sigma$ is standard normal, and so by Proposition 10.4,

$$\sum_{k=1}^{n} \left(\frac{X_k - \mu}{\sigma}\right)^2 = \frac{1}{\sigma^2} \sum_{k=1}^{n} (X_k - \mu)^2$$

has χ^2 distribution with n degrees of freedom. By definition,

$$\frac{(n - 1)\bar{s}^2}{\sigma^2} = \frac{1}{\sigma^2} \sum_{k=1}^{n} (X_k - \bar{x})^2$$

which is quite similar in appearance to the previous expression. But

$$\frac{1}{\sigma^2} \sum_{k=1}^{n} (X_k - \mu)^2 = \left(\frac{\bar{x} - \mu}{\sigma/\sqrt{n}}\right)^2 - \frac{1}{\sigma^2} \sum_{k=1}^{n} (X_k - \bar{x})^2$$

which can be verified by writing $x_k - \mu = x_k - \bar{x} + \bar{x} - \mu$. Therefore,

$$\frac{1}{\sigma^2} \sum_{k=1}^{n} (X_k - \mu)^2 = \left(\frac{\bar{x} - \mu}{\sigma/\sqrt{n}}\right)^2 - \frac{(n - 1)\bar{s}^2}{\sigma^2}$$

The left-hand side of this expression has χ^2 distribution with n degrees of freedom, and since $(\bar{x} - \mu)/(\sigma/\sqrt{n})$ is standard normal, its square has χ^2 distribution with 1 degree of freedom. Under the assumption that $[(\bar{x} - \mu)/(\sigma/\sqrt{n})]^2$ and $(n - 1)\bar{s}^2/\sigma^2$ are independent random variables, we can apply Proposition 10.5 to show that $(n - 1)\bar{s}^2/\sigma^2$ has χ^2 distribution with $n - 1$ degrees of freedom. The proof of this independence, however, is omitted.

Corollary 10.7

If $\bar{s}^2$ denotes the sample variance for samples of size n taken with replacement from a normal population with mean μ and variance σ^2, then

(a) $E(\bar{s}^2) = \sigma^2$
(b) $\text{Var}(\bar{s}^2) = 2\sigma^4/(n - 1)$
(c) $\bar{s}^2$ is a consistent point estimator for σ^2.

Proof: Since $(n - 1)\bar{s}^2/\sigma^2$ has χ^2 distribution with $n - 1$ degrees of freedom, we have

$$E\left[\frac{(n - 1)\bar{s}^2}{\sigma^2}\right] = n - 1$$

and

$$\text{Var}\left[\frac{(n - 1)\bar{s}^2}{\sigma^2}\right] = 2(n - 1)$$

The first of these equations gives

$$E\left[\frac{(n - 1)\bar{s}^2}{\sigma^2}\right] = \left(\frac{n - 1}{\sigma^2}\right)E(\bar{s}^2) = n - 1$$

which implies that $E(\bar{s}^2) = \sigma^2$. The second equation gives

$$\text{Var}\left[\frac{(n - 1)\bar{s}^2}{\sigma^2}\right] = \left[\frac{(n - 1)^2}{\sigma^4}\right]\text{Var}(\bar{s}^2) = 2(n - 1)$$

which implies that

$$\text{Var}(\bar{s}^2) = \frac{2(n - 1)\sigma^4}{(n - 1)^2} = \frac{2\sigma^4}{n - 1}$$

The fact that $\bar{s}^2$ is consistent now follows from (b) together with Propositions 9.1 and 9.4.

It follows from Proposition 10.6 that we can assert with probability $1 - \alpha$ that

$$\chi^2_{1 - (\alpha/2), n - 1} < \frac{(n - 1)\bar{s}^2}{\sigma^2} < \chi^2_{\alpha/2, n - 1}$$

Since the sample variance is known, and the population variance is the unknown to be determined, we must rewrite this confidence interval in terms of σ^2, which can be done as follows:

$$\frac{(n-1)\bar{s}^2}{\chi^2_{\alpha/2,\,n-1}} < \sigma^2 < \frac{(n-1)\bar{s}^2}{\chi^2_{1-(\alpha/2),\,n-1}}$$

Example 1   In order to investigate the effect of a new drug on the pulse rates of patients, a sample of 25 patients was studied, and it was found that the increase in their pulse rates had a standard deviation of 2.8. We wish to construct a 95 percent confidence interval for σ^2, the actual variability in the drug's effect on pulse rates. Since $\chi^2_{0.025,\,24} = 39.364$ and $\chi^2_{0.975,\,24} = 12.401$, we have

$$\frac{24 \cdot (2.8)^2}{39.364} < \sigma^2 < \frac{24 \cdot (2.8)^2}{12.401}$$

which tells us that we can be 95 percent sure that

$$4.780 < \sigma^2 < 15.173$$

or that

$$2.186 < \sigma < 3.895$$

Suppose, now, that we wish to test the hypothesis that a certain variance does not exceed a given tolerance σ_0. We would then test the hypothesis

$$H_0 : \sigma^2 = \sigma_0{}^2$$

against the one-sided alternative

$$H_1 : \sigma^2 > \sigma_0{}^2$$

Notice that we have chosen a one-sided critical region on the *right*, because it is precisely these values of $\bar{s}^2$, those higher than the allowable tolerance, which should lead us to reject the null hypothesis that the tolerance is not being exceeded. At level of significance α, we would have the one-sided critical region

$$\chi^2 > \chi^2_{\alpha,\,n-1}$$

where χ^2 is the statistic $(n-1)\bar{s}^2/\sigma_0{}^2$, $\bar{s}^2$ being the variance of the sample of size n that was taken.

Hypothesis tests for population variances with a two-sided alternative are also possible, in which case the critical region for the statistic χ^2 would be

$$\{\chi^2 > \chi^2_{\alpha/2,\,n-1}\} \cup \{\chi^2 < \chi^2_{1-(\alpha/2),\,n-1}\}$$

Example 2   A machine is deemed to be in good operating condition if the length of the items it produces have a standard deviation which does not exceed 0.02 inches from the expected measurement of 2 inches. If a sample of 10 items produced by the machine has a deviation of 0.034 inches, does the owner of the machine have cause for concern? To answer this question, we test the null hypothesis

$H_0: \sigma^2 = 0.0004$

against the one-sided alternative

$H_1: \sigma^2 > 0.0004$

At the 5 percent level of significance, this gives us the critical region

$\chi^2 > \chi^2_{0.05, 9} = 16.919$

Since the statistic

$$\chi^2 = \frac{(10 - 1)(0.034)^2}{(0.02)^2} = 26.01$$

does fall in the critical region, we can reject the null hypothesis and conclude that the machine is in need of repairs.

Example 3 Government regulations specify that the standard dosage in a certain biological preparation should be 600 activity units per cubic centimeter, with a standard deviation that does not exceed 10 activity units per cubic centimeter. A study of $n = 10$ samples of this preparation revealed an average of 592.5 activity units per cc, with a standard deviation of $\bar{s} = 11.2$ activity units per cc. Since we do not have a population deviation available, we must defer until Section 10.6 any discussion concerning the population mean. We can however test whether the sample deviation obtained indicates that the given tolerance has been exceeded. To do this, we test the null hypothesis

$H_0: \sigma^2 = 100$

against the one-sided alternative

$H_1: \sigma^2 > 100$

at level of significance, say, $\alpha = 0.05$. Since $\chi^2_{0.05, 9} = 16.919$, our critical region is

$\chi^2 > 16.919$

and since our statistic

$$\chi^2 = \frac{9 \cdot (11.2)^2}{100} = 11.29$$

does not fall in the critical region, we can either accept the null hypothesis, or subject our sample to further testing.

The method we have just discussed is usually used when the sample size n is small (say, $n \le 30$), and the population is close to being normally distributed. On the other hand, when the sample size is reasonably large, and regardless of what the distribution of the population might look like, the method that we are about to discuss can be used. We preceed our main result with two lemmas.

Lemma 10.8

If X is a random variable which assumes only positive values, then

$$P(\sqrt{2X} - \sqrt{2n} < k) = P\left(\frac{X - n}{\sqrt{2n}} < k + \frac{k^2}{2\sqrt{2n}}\right)$$

Proof: The proof is a simple exercise in arithmetic manipulation:

$$P(\sqrt{2X} - \sqrt{2n} < k) = P(\sqrt{2X} < k + \sqrt{2n})$$
$$= P[2X < (k + \sqrt{2n})^2 = k^2 + 2k\sqrt{2n} + 2n]$$
$$= P[2(X - n) < k^2 + 2k\sqrt{2n}]$$
$$= P\left[\frac{2(X - n)}{2\sqrt{2n}} < \frac{k^2 + 2k\sqrt{2n}}{2\sqrt{2n}}\right]$$
$$= P\left(\frac{X - n}{\sqrt{2n}} < k + \frac{k^2}{2\sqrt{2n}}\right)$$

As a consequence of Lemma 10.8, we have the following:

Lemma 10.9

If X has χ^2 distribution with n degrees of freedom, then for large values of n, $\sqrt{2X} - \sqrt{2n}$ is approximately standard normal.

Proof: By Corollary 8.14, $E(X) = n$ and $\mathrm{Var}(X) = 2n$, and so the random variable $(X - n)/\sqrt{2n}$ is standardized. It is also normal, for n large enough, according to Proposition 8.26, since the χ^2 distribution is, in fact, a gamma distribution. For n large enough, $k + k^2/2\sqrt{2n}$ is approximately equal to k, and therefore, according to Lemma 10.8, the probability distributions of $\sqrt{2X} - \sqrt{2n}$ and $(X - n)/\sqrt{2n}$ are quite similar. Therefore, for n large enough, $\sqrt{2X} - \sqrt{2n}$ is approximately standard normal.

Our next result is stated in terms of the sample standard deviation, which we shall denote $\bar{s}$ and which is simply the positive square root of the sample variance. It, too, is a random variable defined on the set of all samples of a given size n.

Proposition 10.10

(a) $\bar{s}$ is approximately normal;
(b) $E(\bar{s}) = \sigma$
(c) $\mathrm{Var}(\bar{s}) = \sigma^2/2(n - 1)$

where σ is the population standard deviation.

Proof: We apply Lemma 10.9 to the χ^2 distribution $(n - 1)\bar{s}^2/\sigma^2$ and find that

$$\sqrt{\frac{2(n - 1)\bar{s}^2}{\sigma^2}} - \sqrt{2(n - 1)} = \left(\frac{\bar{s}}{\sigma} - 1\right)\sqrt{2(n - 1)}$$

is approximately standard normal. Hence, $\bar{s}$, which is a linear function of $[(\bar{s}/\sigma) - 1]\sqrt{2(n-1)}$, is also normal. Since

$$E\left[\left(\frac{\bar{s}}{\sigma} - 1\right)\sqrt{2(n-1)}\right] = 0$$

we have

$$E\left[\frac{\bar{s}}{\sigma} - 1\right] = 0$$

and therefore, $E(\bar{s}/\sigma) = 1$, which implies (b). And since

$$\text{Var}\left[\left(\frac{\bar{s}}{\sigma} - 1\right)\sqrt{2(n-1)}\right] = 1$$

we have

$$2(n-1)\,\text{Var}\left(\frac{\bar{s}}{\sigma} - 1\right) = 1$$

Therefore,

$$\text{Var}\left(\frac{\bar{s}}{\sigma} - 1\right) = \frac{1}{2(n-1)}$$

and since

$$\text{Var}\left(\frac{\bar{s}}{\sigma} - 1\right) = \frac{1}{\sigma^2}\,\text{Var}(\bar{s}),$$

we have

$$\text{Var}(\bar{s}) = \frac{\sigma^2}{2(n-1)}$$

As a result of Proposition 10.10, we can form a $1 - \alpha$ percent confidence interval for σ as follows:

$$\bar{s} - z_{\alpha/2}\left(\frac{\sigma}{\sqrt{2(n-1)}}\right) < \sigma < \bar{s} + z_{\alpha/2}\left(\frac{\sigma}{\sqrt{2(n-1)}}\right)$$

However, this form of the confidence interval requires the value of σ to determine the upper and lower confidence limits. Some simple arithmetic manipulations (which we leave as an exercise), though, allow us to rewrite this confidence interval in the form:

$$\frac{\bar{s}}{1 + \dfrac{z_{\alpha/2}}{\sqrt{2(n-1)}}} < \sigma < \frac{\bar{s}}{1 - \dfrac{z_{\alpha/2}}{\sqrt{2(n-1)}}}$$

Example 4 If a sample of 100 patients using a new drug showed an increase in pulse rate having a standard deviation of 2.8, then we can form the following "large sample" confidence interval for the true population deviation:

$$\frac{2.8}{1 + \dfrac{1.96}{\sqrt{2 \cdot 99}}} < \sigma < \frac{2.8}{1 - \dfrac{1.96}{\sqrt{2 \cdot 99}}}$$

which reduces to $2.458 < \sigma < 3.253$. Compare this with the confidence interval we obtained in Example 1: $2.186 < \sigma < 3.895$ based on a small sample of 25 patients.

Proposition 10.10 also gives us a large sample method for testing a null hypothesis of the type

$$H_0 : \sigma = \sigma_0$$

against any one of the alternatives $\sigma \neq \sigma_0$, $\sigma < \sigma_0$, or $\sigma > \sigma_0$. In any of these cases, the critical region is determined exactly as it was for normal tests of population means or proportions; that is, the critical region is either $|z| > z_{\alpha/2}$, $z < -z_\alpha$, or $z > z_\alpha$, respectively. The statistic z that must be calculated is

$$z = \frac{\bar{s} - \sigma_0}{\sigma_0 / \sqrt{2(n - 1)}}$$

which is the standardized form of the normal variable $\bar{s}$.

Example 5   Referring back to Example 2, suppose that a sample of 50 items showed a deviation of 0.025 inches. Does this evidence support the hypothesis that $\sigma_0 = 0.02$ inches? At the five percent level of significance, with a one-sided alternative, we have the critical region $z > 1.64$, and the statistic

$$z = \frac{0.025 - 0.020}{0.02 / \sqrt{2 \cdot 49}}$$

$$= 2.475$$

which happens to fall in the critical region. Therefore, the evidence in this sample does not support the null hypothesis, which consequently must be rejected.

EXERCISES

1. When a certain process is operating in statistical control, it will have a standard deviation of 2.55. After a small change has been made in the process, a sample of ten items showed a standard deviation of 3.30. Has the standard deviation for the process shown a significant increase?

2. A sample of 100 measurements of the lifetime of a certain type of TV picture tube had a standard deviation of 6.0 hours. Find a 95 percent confidence interval for the true deviation.

3. The records of 26 workers in a factory showed a variance of four days absent for the previous year. Find a 95 percent confidence interval for the variance of all the workers at the factory.

4. A manufacturer of car mufflers claims that his brand has a standard deviation of 0.64 years in life span. A random sample of 100 of his mufflers showed a deviation of 1 year. Should his claim be accepted?
5. Verify that the large sample confidence interval for a population standard deviation can be written in the final form presented in the text.

10.4 Differences Among Several Population Proportions

An important and frequently occurring problem in statistics is to decide whether observed differences among k sample proportions reflect a significant difference among the true proportions for the k populations involved. We have already seen in Section 10.1 that a normal test can be used in the case $k = 2$. In this section, we will study how a χ^2 test can be used for arbitrary k.

We will work in the following general situation: Suppose that we have k different populations, each of which contains a certain proportion of items having a certain characteristic or set of characteristics. For the ith population, we shall denote this proportion by p_i. We shall be interested in testing one of the null hypotheses

$$H_0: p_1 = p_2 = p_3 = \cdots = p_k$$

or

$$H_0: p_1 = p_2 = p_3 = \cdots = p_k = p_0$$

for some specific value p_0, against the alternative hypothesis that there is a significant difference between at least two of these population proportions.

To test either of these hypotheses, we will choose k samples, one from each of the k populations, with the sample from the ith population being of size n_i. We then define k binomial random variables $X_1, X_2, \ldots, X_k$ counting the number of items with the given characteristic(s) in each of these samples. We assume, of course, that these random variables are independent; that is, the sampling from the different populations is independent. We can then prove, under the assumption that all of the sample sizes n_i are sufficiently large ($n_i \geq 5$ is usually large enough), that

Proposition 10.11

The random variable

$$\sum_{i=1}^{k} \frac{(X_i - n_i p_i)^2}{n_i p_i (1 - p_i)}$$

has χ^2 distribution with k degrees of freedom.

Proof: The random variable

$$\frac{X_i - n_i p_i}{\sqrt{n_i p_i (1 - p_i)}}$$

is a standardized binomial random variable, and consequently, if n_i is sufficiently large, is approximately standard normal. By Proposition 10.3, then, the random variable

$$\left[\frac{X_i - n_i p_i}{\sqrt{n_i p_i(1 - p_i)}}\right]^2 = \frac{(X_i - n_i p_i)^2}{n_i p_i(1 - p_i)}$$

has a χ^2 distribution with one degree of freedom. It then follows from Proposition 10.4 that

$$\sum_{i=1}^{k} \frac{(X_i - n_i p_i)^2}{n_i p_i(1 - p_i)}$$

has a χ^2 distribution with k degrees of freedom.

Consequently, should we wish to test the null hypothesis

$$H_0: p_1 = p_2 = \cdots = p_k = p_0$$

that the k population proportions all equal the specific value p_0, against the alternative that there is a significant difference between at least two of these proportions, or between all of these proportions and the value p_0, then we would evaluate the statistic

$$\chi^2 = \sum_{i=1}^{k} \frac{(X_i - n_i p_0)^2}{n_i p_0(1 - p_0)}$$

(recall that we assume that all p_i equal p_0), and determine if it falls in the critical region $\chi^2 > \chi^2_{\alpha, k}$, where α is the level of significance of the test.

On the other hand, if p_0 is unknown, and we just wish to test the hypothesis

$$H_0: p_1 = p_2 = p_3 = \cdots = p_k$$

then we must first form the pooled proportion

$$\bar{p} = \frac{X_1 + X_2 + \cdots + X_k}{n_1 + n_2 + \cdots + n_k}$$

as an approximation for p_0 (that is, as an approximation for each of the p_i). Recall that $\bar{p}$ indicates what proportion of the items in the combination of all k samples have the given characteristic(s). We then evaluate the statistic

$$\chi^2 = \sum_{i=1}^{k} \frac{(X_i - n_i \bar{p})^2}{n_i \bar{p}(1 - \bar{p})}$$

which has a χ^2 distribution, but with only $k - 1$ degrees of freedom. Although it is beyond the scope of this book to explain why, one degree of freedom is lost because we substitute the estimate $\bar{p}$ for the unknown parameter p_0. Finally, we must determine if this statistic falls in the critical region $\chi^2 > \chi^2_{\alpha, k-1}$

We should explain why the critical region for this test is one-sided. The answer lies in the term $X_i - n_i p_0$, or $X_i - n_i \bar{p}$, as the case may be. The X_i

represents the actual frequency with which items having the desired characteristic(s) occurred in the ith sample, while the $n_i p_0$, or the $n_i \bar{p}$, represent how frequently these items should have occurred, under the assumption of the null hypothesis that the k population proportions are identical. Hence, the term $X_i - n_i p_0$, or $X_i - n_i \bar{p}$, is measuring the difference between the observed frequencies and the expected frequencies. Consequently, large values of the statistic χ^2 reflect large differences between the observed and expected frequencies, which, in turn, cast doubt on the validity of the null hypothesis. As a result, it is the large values of the statistic χ^2 which are critical.

It is often helpful in problems of this type to form what are called *contingency tables*. There are two such tables, one containing the observed data, and the other containing the expected data. They are formed as shown in Table 10.1. The category Yes refers to the members of the sample that have the desired characteristic(s), and the category No refers to those that do not.

Upon contemplating these contingency tables for a moment, the concept of "degrees of freedom" may become more meaningful. Notice that the column sums are fixed—column j must add up to n_j, the number of elements in the jth sample. Consequently, only one value in each column is free to take on whatever value it chooses (within certain limits, of course), while the second value is then determined by the first value and the predetermined sum of the column. Therefore, k of the $2k$ elements in the contingency table are "free," and the other k are constrained.

Although it is beyond the scope of this book to explain why, it is a statistical fact of life that the number of degrees of freedom must be decreased by 1

Table 10.1
Observed and Expected Contingency Tables

Observed table

	Sample 1	Sample 2	Sample 3	$\cdots$	Sample k
Yes	X_1	X_2	X_3	$\cdots$	X_k
No	$n_1 - X_1$	$n_2 - X_2$	$n_3 - X_3$	$\cdots$	$n_k - X_k$
Sum	n_1	n_2	n_3	$\cdots$	n_k

Expected table

	Sample 1	Sample 2	Sample 3	$\cdots$	Sample k
Yes	$n_1 p_0$	$n_2 p_0$	$n_3 p_0$	$\cdots$	$n_k p_0$
No	$n_1(1 - p_0)$	$n_2(1 - p_0)$	$n_3(1 - p_0)$	$\cdots$	$n_k(1 - p_0)$
Sum	n_1	n_2	n_3	$\cdots$	n_k

for every parameter needed to form the expected table that must be approximated from the information contained in the observed table. The pooled proportion $\bar{p}$ defined above is such a statistic.

Example 1 In an effort to determine the effectiveness of a new drug, 200 patients were tested, of whom only 100 were treated with the drug. The results of the experiment are contained in the following table:

	With drug	Without drug
Recovered	75	65
Did not recover	25	35
	100	100

The hypothesis that we wish to test is the following: that the proportion p_1 of people suffering from the disease who were treated with this drug and recovered is the same as the proportion p_2 of people who were not treated with the drug yet recovered. Of course, the alternative is the one-sided $p_1 > p_2$, and it is hoped by the medical researchers involved that the evidence obtained from the experiment will lead to the rejection of the null hypothesis. Now the numbers contained in the table above are the observed frequencies; in order to calculate the expected frequencies, we must first calculate the pooled proportion

$$\bar{p} = \frac{75 + 65}{100 + 100} = \frac{140}{200} = 0.70$$

Notice that $\bar{p}$ is actually just the percentage of the 200 patients participating in the test who recovered, regardless of whether or not they were treated with the drug. Using the value $\bar{p} = 0.70$, we can compute the expected table as follows: of the 100 patients in either category, exactly $100 \cdot \bar{p} = 70$ should have recovered, and exactly $100 \cdot (1 - \bar{p}) = 30$ should have failed to recover. Consequently, we have the expected table

	With drug	Without drug
Recovered	70	70
Did not recover	30	30
	100	100

We can then calculate the statistic χ^2 as follows:

$$\chi^2 = \frac{(75 - 70)^2}{100 \cdot 0.7 \cdot 0.3} + \frac{(65 - 70)^2}{100 \cdot 0.7 \cdot 0.3}$$

$$= \frac{25}{21} + \frac{25}{21} = \frac{50}{21} = 2.30$$

At a five percent level of significance, the critical region would be defined by the value $\chi^2_{0.05, 1} = 3.841$. Clearly, the statistic $\chi^2 = 2.30$ does not fall in the critical region $\chi^2 > 3.841$. Thus, on the basis of the evidence from this sample, we cannot reject the null hypothesis, and, unfortunately, cannot conclude that the new drug is especially effective.

The method we have just described requires that, in the calculation of the statistic χ^2, we evaluate one term for each column in the contingency table. There is another way to calculate χ^2, which we shall now explain. We adopt the following notation: for $i = 1, 2, \ldots, k$, we let

$$f_{i1} = X_i$$

$$f_{i2} = n_i - X_i$$

$$e_{i1} = n_i \cdot \bar{p}$$

$$e_{i2} = n_i \cdot (1 - \bar{p}) = n_i - e_{i1}$$

In other words, the f_{ij} ($j = 1, 2$) represent the observed frequencies (of success and failure), while the e_{ij} ($j = 1, 2$) represent the corresponding expected frequencies. We can then prove the following:

Proposition 10.12

$$\chi^2 = \sum_{i=1}^{k} \frac{(X_i - n_i\bar{p})^2}{n_i\bar{p}(1 - \bar{p})}$$

$$= \sum_{j=1}^{2} \sum_{i=1}^{k} \frac{(f_{ij} - e_{ij})^2}{e_{ij}}$$

Proof: Working with the right-hand side of this equation, we find

$$\sum_{j=1}^{2} \sum_{i=1}^{k} \frac{(f_{ij} - e_{ij})^2}{e_{ij}} = \sum_{i=1}^{k} \frac{(f_{i1} - e_{i1})^2}{e_{i1}} + \frac{(f_{i2} - e_{i2})^2}{e_{i2}}$$

$$= \sum_{i=1}^{k} \frac{(X_i - n_i\bar{p})^2}{n_i\bar{p}} + \frac{[(n_i - X_i) - n_i(1 - \bar{p})]^2}{n_i(1 - \bar{p})}$$

$$= \sum_{i=1}^{k} \frac{(X_i - n_i\bar{p})^2}{n_i\bar{p}} + \frac{(n_i\bar{p} - X_i)^2}{n_i(1 - \bar{p})}$$

$$= \sum_{i=1}^{k} \frac{[(X_i - n_i\bar{p})^2 n_i(1 - \bar{p}) + (X_i - n_i\bar{p})^2 n_i\bar{p}]}{n_i^2\bar{p}(1 - \bar{p})}$$

$$= \sum_{i=1}^{k} \frac{[(X_i - n_i\bar{p})^2(1 - \bar{p}) + (X_i - n_i\bar{p})^2\bar{p}]}{n_i\bar{p}(1 - \bar{p})}$$

$$= \sum_{i=1}^{k} \frac{(X_i - n_i\bar{p})^2}{n_i\bar{p}(1 - \bar{p})}$$

as required.

If χ^2 is to be calculated by this method, it would be necessary to evaluate one term for each position in the contingency table. But the form of these terms is very easy to remember—simply the squared differences between the observed and expected frequencies divided by the expected frequencies.

Example 2   A teacher uses three different teaching methods in three different sections of the same course, with the following results:

	Method 1	Method 2	Method 3
Pass	50	47	56
Fail	5	14	8

The question the teacher wants to answer, of course, is whether any one method is more effective than the others. To do this, he starts by testing the hypothesis that all three methods are equally effective; that is, the same proportion of students will pass regardless of which teaching method is used. The first step in answering this question is to form the expected table, and to do this, we must first calculate the pooled proportion

$$\bar{p} = \frac{50 + 47 + 56}{55 + 61 + 64} = \frac{153}{180} = 0.85$$

Then, since

$$55 \cdot 0.85 = 46.75$$
$$61 \cdot 0.85 = 51.85$$
$$64 \cdot 0.85 = 54.40$$

we have the expected table

	Method 1	Method 2	Method 3
Pass	46.75	51.85	54.40
Fail	8.25	9.15	9.60

from which we calculate

$$\chi^2 = \frac{(50 - 46.75)^2}{46.75} + \frac{(47 - 51.85)^2}{51.85} + \frac{(56 - 54.40)^2}{54.40}$$
$$+ \frac{(5 - 8.25)^2}{8.25} + \frac{(14 - 9.15)^2}{9.15} + \frac{(8 - 9.60)^2}{9.60}$$
$$= 4.856$$

Therefore, at a 5 percent level of significance, our value of χ^2 does not exceed the critical value $\chi^2_{0.05, 2} = 5.991$. Therefore, the teacher cannot conclude that there is any significant difference between the three teaching methods, at least not on the basis of his present data.

Example 3 The following table indicates how the 100 Democratic and Republican senators voted on a recent piece of legislation:

	Democrats	Republicans
For	38	22
Against	22	18
	60	40

We would like to determine if, as the data seem to suggest, the Democrats were more in favor of this legislation than were the Republicans. From the pooled proportion

$$\bar{p} = \frac{38 + 22}{60 + 40} = \frac{60}{100} = 0.60$$

we calculate

$$60 \cdot 0.60 = 36$$
$$40 \cdot 0.60 = 24$$

which gives us the expected table

	Democrats	Republicans
For	36	24
Against	24	16
	60	40

Therefore,

$$\chi^2 = \frac{(38 - 36)^2}{36} + \frac{(22 - 24)^2}{24} + \frac{(22 - 24)^2}{24} + \frac{(18 - 16)^2}{16}$$

$$= 0.694$$

This exceptionally small value of χ^2 indicates very strongly that there was no "party-line" vote on this particular piece of legislation.

Suppose now that the elements in each of our k populations can be categorized as having any one of a set of r characteristics. For example, we may have two populations, those patients suffering from a particular disease who were treated with a certain drug and those who were treated with a placebo, and we may categorize these patients as either having recovered or having failed to recover after treatment. Or we may partition the population of all automobile drivers in a certain city into several subpopulations, by age perhaps, and classify the drivers in each subpopulation as having had none, one, two, or more traffic accidents during the past three years.

The object of our experiment is to determine whether the different populations are distributed independently of one another with respect to the r characteristics, or whether the same pattern of distribution emerges in each of the k populations. In order to determine which of the above alternatives is true, we set up the latter as our null hypothesis. That is, we assume that there is no difference in the distributions of the k populations. We then take samples of sizes $n_1, n_2, \ldots, n_k$ from the k populations, and form an $r \times k$ contingency table of observed values (Table 10.2).

Notice, once again, that the sums of the columns must equal the sizes of the corresponding samples. Consequently, $r - 1$ of the elements in each column appear to be "free," within certain limits, to assume whatever value they choose, while the rth, or last, element in the column is then constrained to make up the difference between the predetermined column sum and the sum of the other $r - 1$ elements in the column. Therefore, our first guess at the number of degrees of freedom involved in this situation would be $k(r - 1)$.

In order to make a decision concerning our null hypothesis, we would have to construct a contingency table of expected values (Table 10.3). Recall that when the observed table contained just two rows, we were able to construct the expected table using just the one pooled proportion $\bar{p}$. In our more

Table 10.2
$r \times k$ **Contingency Table of Observed Values**

	Population 1	Population 2	$\cdots$	Population k
Characteristic 1	f_{11}	f_{12}	$\cdots$	f_{1k}
Characteristic 2	f_{21}	f_{22}	$\cdots$	f_{2k}
$\vdots$				
Characteristic r	f_{r1}	f_{r2}	$\cdots$	f_{rk}

Table 10.3
$r \times k$ **Contingency Table of Expected Values**

	Population 1	Population 2	$\cdots$	Population k
Characteristic 1	e_{11}	e_{12}	$\cdots$	e_{1k}
Characteristic 2	e_{21}	e_{22}	$\cdots$	e_{2k}
$\vdots$				
Characteristic r	e_{r1}	e_{r2}	$\cdots$	e_{rk}

general case, we must calculate $r - 1$ pooled proportions

$$\bar{p}_1 = \frac{f_{11} + f_{12} + \cdots + f_{1k}}{n_1 + n_2 + \cdots + n_k}$$

$$\bar{p}_2 = \frac{f_{21} + f_{22} + \cdots + f_{2k}}{n_1 + n_2 + \cdots + n_k}$$

$$\vdots$$

$$\bar{p}_{r-1} = \frac{f_{r-1,1} + f_{r-1,2} + \cdots + f_{r-1,k}}{n_1 + n_2 + \cdots + n_k}$$

If we were to consider our k samples as merged together into one large sample, then these pooled proportions $\bar{p}_1 \, \bar{p}_2, \ldots, \bar{p}_{r-1}$ represent the percentage of the elements in the merged sample which possess each of the first $r - 1$ characteristics.

The values in the first $r - 1$ rows of the expected table can then be calculated by the formula

$$e_{ij} = n_j \cdot \bar{p}_i \qquad \text{for } \begin{Bmatrix} i = 1, 2, \ldots, r - 1 \\ j = 1, 2, \ldots, k \end{Bmatrix}$$

And the values in the last (rth) row of the expected table can be calculated by the formula

$$e_{rj} = n_j - \sum_{i=1}^{r-1} e_{ij}$$

That is, for any characteristic (for any row in the table), we assume the pooled proportion $\bar{p}_i$ to hold true for each of the k populations, and then calculate the expected frequency to be that proportion $\bar{p}_i$ of the corresponding sample size. We do this for the first $r - 1$ rows of the table, and then use the fact that the remaining row is constrained by the values in the preceding rows and the column sums (sample sizes).

Since we had to calculate $r - 1$ parameters from the data in the observed table to construct the expected table, we should expect a reduction of $r - 1$ in the number of degrees of freedom. Therefore, we should expect to have

$$k(r - 1) - (r - 1) = (k - 1)(r - 1)$$

degrees of freedom. That is, the number of degrees of freedom is the product of one less than the number of rows (characteristics) multiplied by one less than the number of columns (populations).

Although we shall omit the details, it turns out to be true that the statistic

$$\chi^2 = \sum_{j=1}^{r} \sum_{i=1}^{k} \frac{(f_{ji} - e_{ji})^2}{e_{ji}}$$

has approximately a χ^2 distribution with $(r - 1)(k - 1)$ degrees of freedom, and it is this statistic that we use to test our null hypothesis.

Example 4 Suppose that we wish to determine if a person's income has any effect on his thinking concerning a certain political issue. The results of a survey of 600 people are contained in the table below:

	Low income	Middle income	High income
For	96	115	29
Undecided	59	128	43
Against	45	57	28
	200	300	100

In order to calculate the table of expected values, we must first calculate the pooled proportions

$$\bar{p}_1 = \frac{96 + 115 + 29}{600} = \frac{240}{600} = 0.40$$

$$\bar{p}_2 = \frac{59 + 128 + 43}{600} = \frac{230}{600} = 0.383$$

In order to determine the first row of the expected table, we would calculate

$e_{11} = n_1 \cdot \bar{p}_1 = 200 \cdot 0.40 = 80$

$e_{12} = n_2 \cdot \bar{p}_1 = 300 \cdot 0.40 = 120$

$e_{13} = n_3 \cdot \bar{p}_1 = 100 \cdot 0.40 = 40$

and in order to determine the second row of the expected table, we would calculate

$e_{21} = n_1 \cdot \bar{p}_2 = 200 \cdot 0.383 = 76.67$

$e_{22} = n_2 \cdot \bar{p}_2 = 300 \cdot 0.383 = 115.0$

$e_{23} = n_3 \cdot \bar{p}_2 = 100 \cdot 0.383 = 38.33$

From these values, we find that

$e_{31} = n_1 - e_{11} - e_{21} = 200 - 80 - 76.67 = 43.33$

$e_{32} = n_2 - e_{12} - e_{22} = 300 - 120 - 115 = 65$

$e_{33} = n_3 - e_{13} - e_{23} = 100 - 40 - 38.33 = 21.67$

Thus, we have the following table of expected values:

	Low income	Middle income	High income
For	80.00	120.00	40.00
Undecided	76.67	115.00	38.33
Against	43.33	65.00	21.67

From these two tables, we can calculate

$$\chi^2 = \frac{(96 - 80)^2}{80} + \frac{(115 - 120)^2}{120} + \frac{(29 - 40)^2}{40} + \frac{(59 - 76.67)^2}{76.67} + \frac{(128 - 115)^2}{115}$$

$$+ \frac{(43 - 38.33)^2}{38.33} + \frac{(45 - 43.33)^2}{43.33} + \frac{(57 - 65)^2}{65} + \frac{(28 - 21.67)^2}{21.67}$$

$$= 15.44$$

Since we have $k = 3$ populations and $r = 3$ characteristics, our statistic has approximately the χ^2 distribution with $(3 - 1)(3 - 1) = 4$ degrees of freedom. Consequently, the critical region for our test is $\chi^2 > \chi^2_{0.05, 4} = 9.488$, and since our statistic χ^2 does fall in this critical region, we must reject the null hypothesis and conclude that income does have some effect on a person's thinking regarding this issue.

Example 5 A survey of 1000 men and 1000 women revealed the following preferences regarding television shows:

	Men	*Women*
Variety	140	260
Situation comedy	40	140
Drama	210	320
Movie	200	200
Sports	310	30
News	100	50

Based on this sample, we wish to determine if men and women have the same televiewing preferences. To do this, we must test the hypothesis that the two groups do have the same preferences. Our first step is to calculate the pooled proportions

$$\bar{p}_1 = \frac{140 + 260}{2000} = 0.200$$

$$\bar{p}_2 = \frac{40 + 140}{2000} = 0.090$$

$$\bar{p}_3 = \frac{210 + 320}{2000} = 0.265$$

$$\bar{p}_4 = \frac{200 + 200}{2000} = 0.200$$

$$\bar{p}_5 = \frac{310 + 30}{2000} = 0.170$$

from which we can construct the expected values table:

	Men	Women
Variety	200	200
Situation comedy	90	90
Drama	265	265
Movie	200	200
Sports	170	170
News	75	75
	1000	1000

We then calculate our χ^2 statistic to be

$$\chi^2 = \frac{(140 - 200)^2}{200} + \frac{(260 - 200)^2}{200} + \frac{(40 - 90)^2}{90} + \frac{(140 - 90)^2}{90}$$

$$+ \frac{(210 - 265)^2}{265} + \frac{(320 - 265)^2}{265} + \frac{(200 - 200)^2}{200} + \frac{(200 - 200)^2}{200}$$

$$+ \frac{(310 - 170)^2}{170} + \frac{(30 - 170)^2}{170} + \frac{(100 - 75)^2}{75} + \frac{(50 - 75)^2}{75}$$

$$= 361.64$$

Since we have $k = 2$ populations and $r = 6$ characteristics, the critical region is $\chi^2 > \chi_{0.01, 5} = 15.086$. Since our statistic is quite clearly in this critical region, we must reject the null hypothesis of "no difference," and conclude that men and women have different tastes as far as television viewing is concerned.

EXERCISES

1. Refer back to Example 5 above. Eliminate those men (310) and women (30) whose top preference was sports shows. Use the remaining 690 men and 970 women to determine if men and women have the same preferences, other than sports shows.
2. Do Exercise 1 in Section 10.1 using the χ^2 test.
3. Do Exercise 2 in Section 10.1 using the χ^2 test.
4. Do Exercise 5 in Section 10.1 using the χ^2 test.
5. Do Exercise 6 in Section 10.1 using the χ^2 test.
6. Do Exercise 7 in Section 10.1 using the χ^2 test.
7. Do Exercise 8 in Section 10.1 using the χ^2 test.
8. Use the data in the following table to test the hypothesis that there is no relationship between politics and religion:

	Catholic	Protestant	Jewish	Other
Democrats	220	280	130	70
Republicans	195	320	90	65
Independents	60	225	40	35
Others	25	175	40	30

9. When asked which was the more important issue, inflation or political corruption, a group of politicians responded as follows:

	Republicans	Democrats	Independents
Inflation	187	316	191
Corruption	154	187	97

Do we have evidence of a nonpartisan response to the question?

10. Use the data in the table below to determine if there is any relationship between a student's grade point average (GPA) and his income five years after graduation:

	Under $5,000	$5,000– $9,999	$10,000– $14,999	$15,000– $19,999	$20,000 up
3.5–4.0	1	18	22	9	2
3.0–3.5	5	63	39	16	11
2.5–3.0	10	152	77	38	14
2.0–2.5	11	84	46	34	2

10.5 Goodness of Fit

The methods described in the previous section can be used to determine how well observed data can be explained by a particular probability distribution. Suppose that our observed data falls into k categories with frequencies $f_1, f_2, \ldots, f_k$, respectively, while the expected frequencies should be $e_1, e_2, \ldots, e_k$, respectively, according to the probability distribution in question. In order to test the null hypothesis that the given probability distribution does provide a good fit to the observed data, we calculate the statistic

$$\chi^2 = \sum_{i=1}^{k} \frac{(f_i - e_i)^2}{e_i}$$

which turns out to have a χ^2 distribution with $k - 1$ degrees of freedom. Should the statistic χ^2 fall in the appropriate critical region $\chi^2 > \chi^2_{\alpha, k-1}$, then we must conclude that the probability distribution does not provide a good fit.

Since

$$\sum_{i=1}^{k} f_i = \sum_{i=1}^{k} e_i = n$$

where n is the sample size, we see that $k - 1$ of the k frequencies are "free," but that the kth (or last) frequency is constrained by the sum n and the values of the other frequencies. Hence, it is reasonable that there be $k - 1$ degrees of freedom in this situation.

Example 1   We wish to test whether a coin is biased, so we toss the coin 200 times, with 115 heads resulting. If the coin were fair, we would expect 100 heads in 200 tosses. We have the observed frequencies $f_1 = 115$ and $f_2 = 85$, and the expected frequencies $e_1 = e_2 = 100$. Our sample statistic is therefore

$$\chi^2 = \frac{(115 - 100)^2}{100} + \frac{(85 - 100)^2}{100}$$

$$= 4.5$$

At the five percent level of significance, our sample value $\chi^2 = 4.5$ exceeds the critical value $\chi^2_{0.05, 1} = 3.841$. Hence, the evidence leads us to believe that the coin is biased in favor of heads.

Example 2   In order to decide if the percentage of males in a certain society is $\frac{1}{2}$, a study of 80 four-child families was made, with the following results:

Number of boys	0	1	2	3	4
Number of families	3	15	33	24	6

If there were 50 percent males in this society, we would expect the following probability distribution for the number of boys in four-child families:

Number of boys	0	1	2	3	4
% of families	$\frac{1}{16}$	$\frac{4}{16}$	$\frac{6}{16}$	$\frac{4}{16}$	$\frac{1}{16}$

and consequently, we have the following expected values for the 80 families studied:

Number of boys	0	1	2	3	4
Number of families	5	20	30	20	5

Evaluating our sample statistic:

$$\chi^2 = \frac{(3 - 5)^2}{5} + \frac{(15 - 20)^2}{20} + \frac{(33 - 30)^2}{30} + \frac{(24 - 20)^2}{20} + \frac{(6 - 5)^2}{5}$$

$$= 3.35$$

we find, at level of significance 0.05, that the critical value of $\chi^2_{0.05, 4} = 9.488$ has not been exceeded. Therefore, we cannot, on the basis of this study, reject the hypothesis that the percentage of males in this society is 50 percent.

Example 3   Suppose that we wish to test the outcome of a dihybrid cross in genetics in which the expected ratio of phenotypes is $9:3:3:1$. The table below gives the observed frequencies in the first column, the expected frequencies in the second column, and the calculations leading to the evaluation of χ^2 in the remaining two columns:

	f_i	e_i	$f_i - e_i$	$(f_i - e_i)^2/e_i$
9	926	909	17	0.318
3	290	303	−13	0.558
3	295	303	−8	0.211
1	105	101	4	0.158
	1616	1616		1.245

Notice that the values of e_1, e_2, e_3, and e_4 are, respectively, 9/16, 3/16, 3/16, and 1/16 of 1616. The sum of the values in the fourth column gives the value of the sample statistic χ^2. At the five percent level of significance, the critical value for this statistic is $\chi^2_{0.05,\,3} = 7.815$. Since the value of our sample statistic is far from being critical, we would tend to lean towards believing that the Mendelian theory explains the observed data remarkably well.

Example 4 Suppose that statistics from ten years ago indicate that 20 percent of all women over 21 are single, 65 percent are married, 11 percent are widowed, and 4 percent are divorced. A recent sample of 100 women over 21 contained 26 who were single, 51 who were married, 10 who were widowed, and 13 who were divorced. This sample would seem to suggest that fewer women are married today than 10 years ago, while more are either single or divorced. Is this a statistically justifiable conclusion? To test whether or not the percentages are the same today as they were ten years ago, we compare the observed frequencies with the expected frequencies based on the ten-year old percentages. Our work is done in the following table:

	f_i	e_i	$f_i - e_i$	$(f_i - e_i)^2/e^i$
Single	26	20	6	1.800
Married	51	65	−14	3.015
Widowed	10	11	−1	0.091
Divorced	13	4	9	20.250
	100	100		25.156

Since our sample statistic $\chi^2 = 25.156$ exceeds the critical value $\chi^2_{0.05,\,3} = 7.815$, we must conclude that fewer women are married today than were married ten years ago.

We have seen on several occasions since introducing the normal distribution in Chapter 8 that the work involved in statistical calculations can be reduced quite dramatically when we are able to use a normal-curve approximation. Given an arbitrary set of data, therefore, it would be helpful to know if it could reasonably be approximated by a normal curve. In the following example, we will demonstrate how this can be done:

Example 5 A sample of the heights of 100 people yielded a mean of 67 inches with a standard deviation of 3 inches. If we group our data into intervals, we obtain the following distribution:

Height (in.)	Frequency
60–62	8
63–65	20
66–68	41
69–71	25
72–74	6
	100

The normal curve which figures to give the best approximation to this data is the normal curve with the same mean and standard deviation as the data itself; that is, the normal curve with mean $\mu = 67$ and $\sigma = 3$ (Figure 10.3). In order to calculate the expected frequencies for each interval (bar), we simply calculate the area under this normal curve between the left and right endpoints of the intervals (these endpoints are 59.5, 62.5, 65.5, 68.5, 71.5, and 74.5), and multiply by 100. The areas so calculated indicate the percentage of the total area (1.0) corresponding to the individual bar (interval), and this percentage must then be multiplied by the size of the population, which in our example is 100. As a first step in this process, we have indicated in the

Figure 10.3

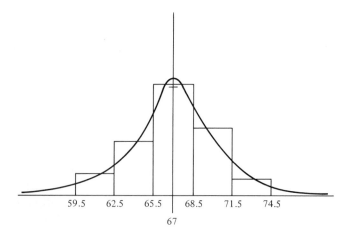

following table the area under this normal curve to the left of each of these left endpoints:

Endpoints	Standardized endpoints	Area
59.5	−2.5	0.0062
62.5	−1.5	0.0668
65.5	−0.5	0.3085
68.5	0.5	0.6915
71.5	1.5	0.9332
74.5	2.5	0.9938

The second column above was obtained from the first column by means of the transformation $z = (x - 67)/3$. In the next table, we calculate the expected area corresponding to each interval, and from this, the expected frequency for each interval. Note that we have included the area from the two tails with the areas of the left and right endmost intervals.

Interval	Area	Expected frequency
60–62	0.0668	6.68
63–65	0.2417	24.17
66–68	0.3830	38.30
69–71	0.2417	24.17
72–74	0.0668	6.68

We therefore have the following table of observed and expected frequencies, from which we can calculate a χ^2 statistic which will indicate how well the normal curve with parameters $\mu = 67$ and $\sigma = 3$ fits the observed data:

	f_i	e_i	$f_i - e_i$	$(f_i - e_i)^2/e_i$
60–62	8	6.68	1.32	0.261
63–65	20	24.17	−4.17	0.719
66–68	41	38.30	2.70	0.190
69–71	25	24.17	0.83	0.029
72–74	6	6.68	−0.68	0.069
	100	100.00		1.270

At the five percent level of significance, the critical region for our test is $\chi^2 > \chi^2_{0.05, 4} = 9.488$, and since our sample statistic is far from being critical, we may assume that this normal curve does a fairly good job of approximating our observed data.

Example 6 Suppose that we wish to determine if the Poisson distribution with parameter $\lambda = 2.0$ is an appropriate distribution for the number of accidents per hour on the Long Island Expressway during the rush hour. A sample of 198 such hours revealed the following statistics:

Number of accidents	Number of hours
0	22
1	59
2	55
3	40
4	20
5	2

(Note that the average number of accidents per hour was 1.91 for this sample of 198 hours.) We can calculate the expected number of hours with zero, one, two, three, four, or five (or more) accidents per hour using the Poisson probabilities corresponding to $\lambda = 2.0$. The expected values, and the calculations necessary to determine the sample value of χ^2, are included in the table below:

	f_i	e_i	$f_i - e_i$	$(f_i - e_i)^2/e_i$
0	22	27	-5	1.136
1	59	54	5	0.424
2	55	54	1	0.018
3	40	36	4	0.400
4	20	18	2	0.200
5+	2	9	-7	24.500
	198	198		26.678

At a five percent level of significance, the critical region for our test is $\chi^2 > \chi^2_{0.05,\, 5} = 11.070$. Since our sample statistic does fall in this critical region, we must conclude that the Poisson distribution with parameter $\lambda = 2.0$ does not provide a good fit for our observed data, and probably is not the correct probability distribution to explain the number of accidents per hour on the Long Island Expressway during the rush hour. Notice for a minute *why* our sample statistic fell in the critical region. The last category (5+ accidents per hour) was totally responsible for the high value of χ^2, to the extent that, if we had combined the last two categories into one category (4+ accidents per hour), the total value of χ^2 would have been extremely low instead, and would have led us to the exact opposite conclusion. Had we done so, we would have had $f_4 = 22$, $e_4 = 27$, and $(f_4 - e_4)^2/e_4 = (22 - 27)^2/27 = 0.926$, and a total χ^2 value of 2.90, well below the critical value of 9.488.

Another test of the agreement between observed frequencies and frequencies expected on the basis of a hypothesis defining a probability distribution is called the *likelihood ratio test* (or the *G test*). Suppose once again that we have a population that can be partitioned into k categories, and that a sample of size n taken from this population yielded frequencies of f_1, $f_2, \ldots, f_k$ for these categories. Suppose that the expected frequencies based on the hypothesized probability distribution are $e_1, e_2, \ldots, e_k$. If we define

$$p_i = \frac{e_i}{n} \qquad \text{for } i = 1, 2, \ldots, k$$

then the probability that the given sample will occur, on the assumption that the hypothesized probability distribution is valid, equals

$$L_h = \binom{n}{f_1, f_2, \ldots, f_k} p_1^{f_1} p_2^{f_2} \cdots p_k^{f_k}$$

On the other hand, if we define

$$q_i = \frac{f_i}{n} \qquad \text{for } i = 1, 2, \ldots, k$$

then the probability that the observed sample will occur, on the assumption that the probability distribution of the population is exactly as indicated in the sample, equals

$$L_s = \binom{n}{f_1, f_2, \ldots, f_k} q_1^{f_1} q_2^{f_2} \cdots q_k^{f_k}$$

The ratio L_s/L_h is called the *likelihood ratio* and is denoted L. Notice that the likelihood ratio will always be at least 1, since the hypothesis governing the probability L_s is based on the observed data. In other words, the observed sample is the most likely sample to be drawn from a population distributed exactly as is the observed sample. On the other hand, if the observed distribution of the sample closely approximates the distribution hypothesized for the population, then the probability L_h will be just slightly less than the probability L_s, and so the likelihood ratio will be just slightly larger than 1. Consequently, it is the large values of L which are critical in that they would cast doubt that the hypothesized distribution gives an accurate description of the population.

Unfortunately, the distribution of the likelihood ratio statistic L is rather complex. However, the following is true (although the proof is beyond the scope of this book):

Proposition 10.13

The statistic $G = 2 \log_e L$, where L is the likelihood ratio statistic, can be closely approximated by a χ^2 distribution with $k - 1$ degrees of freedom.

The following will give us a computational formula for calculating G:

Proposition 10.14

$$G = 2 \sum_{i=1}^{k} f_i \cdot \log\left(\frac{f_i}{e_i}\right)$$

Proof: Since $G = 2 \log L$, and since

$$L = \frac{L_s}{L_h} = \frac{\left(\dfrac{n}{f_1, f_2, \ldots, f_k}\right) q_1^{f_1} q_2^{f_2} \cdots q_k^{f_k}}{\left(\dfrac{n}{f_1, f_2, \ldots, f_k}\right) p_1^{f_1} p_2^{f_2} \cdots p_k^{f_k}}$$

we have the simple formula

$$L = \prod_{i=1}^{k} \left(\frac{q_i}{p_i}\right)^{f_i}$$

Consequently,

$$G = 2 \log \prod_{i=1}^{k} \left(\frac{q_i}{p_i}\right)^{f_i}$$

$$= 2 \sum_{i=1}^{k} f_i \cdot \log\left(\frac{q_i}{p_i}\right)$$

Since $q_i = f_i/n$ and $p_i = e_i/n$, so that

$$\frac{q_i}{p_i} = \frac{f_i}{e_i}$$

therefore,

$$G = 2 \sum_{i=1}^{k} f_i \cdot \log\left(\frac{f_i}{e_i}\right)$$

Example 3 (*revisited*) The following table includes all the calculations necessary to evaluate the G statistic:

f_i	e_i	f_i/e_i	$\log(f_i/e_i)$	$f_i \log(f_i/e_i)$
926	909	1.019	0.019	17.158
290	303	0.957	−0.044	−12.717
295	303	0.974	−0.027	−7.893
105	101	1.039	0.039	4.078
				0.626

Consequently, $G = 2 \cdot 0.626 = 1.251$ (as compared with the value $\chi^2 = 1.245$ obtained using the χ^2 test), which does not fall in the critical region $\chi^2 > \chi^2_{0.05, 3} = 7.815$.

Example 5 (*revisited*) The following table includes all the calculations needed to evaluate the G statistic:

f_i	e_i	f_i/e_i	$\log(f_i/e_i)$	$f_i \log(f_i/e_i)$
8	6.68	1.198	0.180	1.443
20	24.17	0.827	−0.189	−3.788
41	38.30	1.070	0.068	2.793
25	24.17	1.034	0.034	0.844
6	6.68	0.898	−0.107	−0.644
				0.648

Consequently, $G = 2 \cdot 0.648 = 1.296$, which also is quite close to the value $\chi^2 = 1.27$ we obtained using the χ^2 test. Neither value, G nor χ^2, falls in the critical region $\chi^2 > \chi^2_{0.05,\,4} = 9.488$.

EXERCISES

1. Prove the following calculation formula for use with the χ^2 test:

$$\chi^2 = \sum_{i=1}^{k} \frac{(f_i - e_i)^2}{e_i} = \sum_{i=1}^{k} \frac{f_i^2}{e_i} - k$$

2. Determine if the number of books borrowed from a library depends on the day of the week, based on the following data for one week:

	Number of books
Monday	153
Tuesday	108
Wednesday	120
Thursday	114
Friday	145

3. Many students believe that teachers grade according to the normal curve. Does the following distribution of grades suggest that the teacher responsible used such a practice?

	Number of students
A	21
B	36
C	50
D	27
F	11

4. Mendel's experiment of crossing two different types of peas resulted in the following frequencies:

Round and yellow peas: 315
Round and green peas: 108

Wrinkled and yellow peas: 101
Wrinkled and green peas: 32

Do these frequencies support Mendel's theory that these four categories should occur in a 9:3:3:1 ratio?

5. The number of gamma rays per second emitted by a radioactive substance are thought to have a Poisson distribution with parameter $\lambda = 2.4$. Does the following data support this hypothesis?

gamma rays/second	Frequency
0	320
1	636
2	704
3	601
4	350
5	175
6+	120

6. Suppose that General Motors produces 54 percent, Ford 27 percent, Chrysler 15 percent, and American Motors 4 percent of the automobiles produced in this country. Suppose that, of the 1000 cars purchased in a certain city during the last year, 520 were from General Motors, 250 were from Ford, 200 were from Chrysler, and 30 were from American Motors. Are the number of cars purchased from these companies in this city out of proportion with the national percentages?

7. A pair of dice were rolled 1000 times with the following distribution of sums:

Sum	Frequency
2	28
3	56
4	98
5	125
6	162
7	180
8	133
9	97
10	72
11	40
12	9

Should one suspect these dice are loaded?

10.6 The F Distribution

The F distribution is still another continuous probability distribution which is of great importance in the field of statistical inference. To obtain its density function, we start with the beta density with parameters $\alpha = k_1/2$ and

$\beta = k_2/2$, which is

$$f(x) = \begin{cases} \dfrac{\Gamma[(k_1 + k_2)/2]}{\Gamma(k_1/2)\Gamma(k_2/2)} x^{(k_1/2)-1}(1-x)^{(k_2/2)-1} & 0 < x < 1 \\ 0 & \text{otherwise} \end{cases}$$

We then apply the transformation

$$y = \frac{k_2 x}{k_1(1-x)}$$

to obtain the F density. As x ranges over the interval $0 < x < 1$, y assumes values between 0 and $+\infty$, and

$$\frac{dy}{dx} = \frac{k_1 k_2}{k_1^2(1-x)^2} = \frac{(k_2 + k_1 y)^2}{k_1 k_2}$$

and so

$$\frac{dx}{dy} = \frac{k_1 k_2}{(k_2 + k_1 y)^2}$$

Applying Proposition 8.3 (the function $y = k_2 x/k_1(1-x)$ is an increasing function), we obtain the following density function for the F distribution:

$$g(y) = \begin{cases} \dfrac{\Gamma((k_1 + k_2)/2)}{\Gamma(k_1/2)\Gamma(k_2/2)} \left(\dfrac{k_1 y}{k_2 + k_1 y} \right)^{(k_1/2)-1} \left(\dfrac{k_2}{k_2 + k_1 y} \right)^{(k_2/2)-1} \dfrac{k_1 k_2}{(k_2 + k_1 y)^2} & \text{if } y > 0 \\ 0 & \text{otherwise} \end{cases}$$

This is equivalent to

$$g(y) = \begin{cases} \dfrac{\Gamma((k_1 + k_2)/2)}{\Gamma(k_1/2)\Gamma(k_2/2)} k_1^{k_1/2} k_2^{k_2/2} y^{(k_1/2)-1}(k_2 + k_1 y)^{-[(k_1 + k_2)/2]} & \text{if } y > 0 \\ 0 & \text{otherwise} \end{cases}$$

The F distribution whose density is defined above is said to have k_1 and k_2 degrees of freedom. The graphs of the F density functions are quite similar to those of the χ^2 density functions.

In the way of notation, we will let F_{α, k_1, k_2} denote that point on the x-axis to the right of which lies α percent of the area under the F density with k_1 and k_2 degrees of freedom, as indicated in Figure 10.4. A table of values for $F_{0.05, k_1, k_2}$ is given in Appendix E. Since this table must reflect two different degrees of freedom, the value of α is held constant (at 0.05) throughout this table.

The F distribution is important in statistics mainly because of the following result and its applications.

Proposition 10.15

If X_1 and X_2 are independent χ^2 distributions with k_1 and k_2 degrees of freedom, respectively, then the random variable

$$Y = \frac{X_1/k_1}{X_2/k_2}$$

has an F distribution with k_1 and k_2 degrees of freedom.

Proof: Suppose that

$$f_i(x) = \left[\left(\frac{1}{2}\right)^{k_i/2} \Gamma\left(\frac{k_i}{2}\right)\right] x^{(k_i/2)-1} e^{-x/2}$$

for $i = 1, 2$ and $x > 0$, are the density functions of X_1 and X_2, respectively. Since we assume that X_1 and X_2 are independent, their joint density function is

$$f(x, y) = \left[\left(\frac{1}{2}\right)^{(k_1+k_2)/2} \Gamma\left(\frac{k_1}{2}\right)\Gamma\left(\frac{k_2}{2}\right)\right] x^{(k_1/2)-1} y^{(k_2/2)-1} \exp\left(-\frac{x+y}{2}\right)$$

Letting

$$U = \frac{X_1/k_1}{X_2/k_2} \quad \text{and} \quad V = X_2$$

we have

$$X_1 = \frac{k_1 U V}{k_2} \quad \text{and} \quad X_2 = V$$

and so

$$\frac{\partial X_1}{\partial U} = \frac{k_1 V}{k_2}$$

Figure 10.4

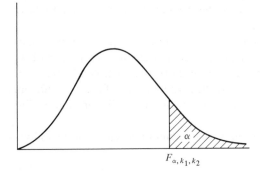

$$F_{\alpha, k_1, k_2}$$

Applying Proposition 8.10, we find that the density function $h(u)$ of U equals

$$h(u) = \frac{1}{2^{(k_1+k_2)/2}\Gamma(k_1/2)\Gamma(k_2/2)} \int_0^\infty f\left(\frac{k_1 uv}{k_2}, v\right)\left|\frac{k_1 v}{k_2}\right| dv$$

Both k_1 and k_2 are positive, as is v, and so

$$h(u) = \frac{1}{2^{(k_1+k_2)/2}\Gamma(k_1/2)\Gamma(k_2/2)} \frac{k_1}{k_2} \int_0^\infty \left(\frac{k_1 uv}{k_2}\right)^{(k_1/2)-1} v^{(k_2/2)-1}$$

$$\times \exp\left(-\frac{k_1 uv}{k_2} + v\right) v\, dv$$

$$= \frac{(k_1/k_2)^{k_1/2}}{2^{(k_1+k_2)/2}\Gamma(k_1/2)\Gamma(k_2/2)} \int_0^\infty u^{(k_1/2)-1} v^{[(k_1+k_2)/2]-1}$$

$$\times \exp\left[-\left(\frac{k_1 uv + k_2 v}{2k_2}\right)\right] dv$$

$$= \frac{(k_1/k_2)^{k_1/2} u^{(k_1/2)-1}}{2^{(k_1+k_2)/2}\Gamma(k_1/2)\Gamma(k_2/2)} \int_0^\infty v^{[(k_1+k_2)/2]-1} \exp\left[-\left(\frac{k_1 uv + k_2 v}{2k_2}\right)\right] dv$$

(since the variable u is being held constant.) Making the substitution

$$z = \frac{k_1 uv + k_2 v}{2k_2}$$

we find

$$v = \frac{2k_2 z}{k_1 u + k_2} \quad \text{and} \quad dv = \frac{2k_2}{k_1 u + k_2}\, dz$$

This transforms the integral into the form

$$\int_0^\infty \left(\frac{2k_2 z}{k_1 u + k_2}\right)^{[(k_1+k_2)/2]-1} \exp\left[-z\left(\frac{2k_2}{k_1 u + k_2}\right)\right] dz$$

$$= \int_0^\infty \left(\frac{2k_2}{k_1 u + k_2}\right)^{(k_1+k_2)/2} z^{[(k_1+k_2)/2]-1} e^{-z}\, dz$$

$$= \left(\frac{2k_2}{k_1 u + k_2}\right)^{(k_1+k_2)/2} \int_0^\infty z^{[(k_1+k_2)/2]-1} e^{-z}\, dz$$

$$= \left(\frac{2k_2}{k_1 u + k_2}\right)^{(k_1+k_2)/2} \Gamma\left(\frac{k_1 + k_2}{2}\right)$$

Therefore,

$$h(u) = \frac{(k_1/k_2)^{k_1/2} u^{(k_1/2)-1}}{2^{(k_1+k_2)/2}\Gamma(k_1/2)\Gamma(k_2/2)}\left(\frac{2k_2}{k_1 u + k_2}\right)^{(k_1+k_2)/2} \Gamma\left(\frac{k_1 + k_2}{2}\right)$$

$$= \frac{k_1^{k_1/2} u^{(k_1/2)-1} 2^{(k_1+k_2)/2} k_2^{k_1/2} k_2^{k_2/2} \Gamma\left(\dfrac{k_1 + k_2}{2}\right)}{2^{(k_1+k_2)/2}\Gamma(k_1/2)\Gamma(k_2/2) k_2^{k_1/2}(k_1 u + k_2)^{(k_1+k_2)/2}}$$

which is equivalent to

$$\frac{\Gamma\!\left(\dfrac{k_1 + k_2}{2}\right)}{\Gamma(k_1/2)\Gamma(k_2/2)}\, k_1^{\,k_1/2} k_2^{\,k_2/2} u^{(k_1/2)-1} (k_1 u + k_2)^{-(k_1+k_2)/2}$$

the density function of the F distribution with k_1 and k_2 degrees of freedom.
As an application of Proposition 10.15, we have the following:

Proposition 10.16

If $\bar{s}_1^{\,2}$ and $\bar{s}_2^{\,2}$ are the sample variances for samples of sizes n_1 and n_2 taken from two normal populations with equal variances, then $\bar{s}_1^{\,2}/\bar{s}_2^{\,2}$ has an F distribution with $n_1 - 1$ and $n_2 - 1$ degrees of freedom.

Proof: By Proposition 10.6,

$$(n_1 - 1)\bar{s}_1^{\,2}/\sigma_1^{\,2} \quad \text{and} \quad (n_2 - 1)\bar{s}_2^{\,2}/\sigma_2^{\,2}$$

have χ^2 distributions with $n_1 - 1$ and $n_2 - 1$ degrees of freedom, respectively. By Proposition 10.15,

$$\frac{\dfrac{[(n_1 - 1)\bar{s}_1^{\,2}/\sigma_1^{\,2}]}{n_1 - 1}}{\dfrac{[(n_2 - 1)\bar{s}_2^{\,2}/\sigma_2^{\,2}]}{n_2 - 1}}$$

has an F distribution with $n_1 - 1$ and $n_2 - 1$ degrees of freedom. Under the assumption that $\sigma_1^{\,2} = \sigma_2^{\,2}$, this reduces to $\bar{s}_1^{\,2}/\bar{s}_2^{\,2}$, and so, the stated result is proved.

It is because of Proposition 10.16 that we can use the F distribution to test the equality of the variances of two (normal) populations. For if we wish to test the null hypothesis

$$H_0\!: \sigma_1^{\,2} = \sigma_2^{\,2}$$

against a one-sided alternative (say, $\sigma_1^{\,2} > \sigma_2^{\,2}$), we simply check if the quotient of the sample variances lies in the critical region; that is, we check if

$$\frac{\bar{s}_1^{\,2}}{\bar{s}_2^{\,2}} > F_{\alpha,\,n_1-1,\,n_2-1}$$

Should we wish to use a two-sided alternative $\sigma_1^{\,2} \neq \sigma_2^{\,2}$, then we use $\bar{s}_1^{\,2}/\bar{s}_2^{\,2}$ if $s_1 \geq s_2$, or we use $\bar{s}_2^{\,2}/\bar{s}_1^{\,2}$ if $s_1 \leq s_2$, and compare to the appropriate critical value.

Note that since our table of F values contains only the values $F_{0.05,\,k_1,\,k_2}$ which delineate the right-hand tail of the F density, our critical values must turn out to be extremely positive. Consequently, when we form the quotient of our two sample variances, we must place the larger variance in the numerator.

Example 1 Measurements performed on two random samples of ten cigarettes each gave the following results: The first sample had a mean of 26.2 mg of nicotine with a standard deviation of 1.2 mg of nicotine, while the second sample had a mean of 26.6 mg of nicotine with a standard deviation of 1.6 mg of nicotine. We wish to determine if it is likely that these two samples were taken from the same brand of cigarettes. What we would like to know, more specifically, is whether it is reasonable to believe that these two samples came from populations with equal means and deviations. Since the test on means relies on an accurate estimate of deviations, we must first perform a test on the deviations. This can be accomplished by means of the *F* distribution, provided that the population of cigarette milligram contents is (approximately) normal. If this is true, then all we must calculate is

$$F = \frac{(1.6)^2}{(1.2)^2} = 1.778$$

which must be compared to the critical value $F_{0.05, 9, 9} = 3.18$. Since our sample statistic does not fall in the critical region (values greater than 3.18), we can accept the hypothesis that the two populations have the same standard deviation. We can then proceed to one of our tests for the equality of the two means.

Example 2 A certain disease is characterized by wildly fluctuating temperatures. To help determine if a particular patient, who is suspected of having the disease, does, in fact, have the disease, the temperatures of the patient and another patient, who is known *not* to have the disease, are recorded each hour over a 12-hour period. Due to the carelessness of a nurse, one of the twelve temperatures for the patient believed to be suffering from the disease has been lost, leaving a sample of $n_1 = 11$ temperatures which had a standard deviation of $\bar{s}_1 = 1.4°F$. The sample of $n_2 = 12$ temperatures for the other patient had a standard deviation of $\bar{s}_2 = 0.7°F$. We wish to test the null hypothesis

$$H_0: \sigma_1 = \sigma_2$$

that the two sets of temperatures came from the same population (and consequently, that the first patient is really not suffering from a disease characterized by wildly fluctuating temperatures), against the alternative

$$H_1: \sigma_1 > \sigma_2$$

which would indicate that the first patient actually does have the disease. According to Proposition 10.16, we must test the statistic

$$F = \frac{(1.4)^2}{(0.7)^2} = 4$$

against the critical value $F_{0.05, 10, 11} = 2.85$. Since our sample statistic does fall in the critical region, we must reject our null hypothesis and conclude that the two patients come from two different populations, the second patient from the population of normal, healthy people, and the first patient from the population of people suffering from this particular disease. (Note that we must also make the assumption that the distribution of body temperatures is normal, or approximately normal, an assumption which may not be quite valid. This population may be slightly skewed to the right.)

Example 3 A psychology professor wishes to give a psychological test to a group of students attending an urban university and to another group of students who attend a suburban university. The professor wants his two samples to be similar in that the students who comprise the two samples have approximately the same mean IQ with approximately the same variance in IQ. Suppose that a sample of $n_1 = 20$ students from the urban university have a standard deviation in IQ of $\bar{s}_1 = 10.2$, while a sample of $n_2 = 20$ students from the suburban university have a standard deviation in IQ of $\bar{s}_2 = 7.8$. Are these two samples acceptable? To answer this question, we simply test the statistic

$$F = \frac{(10.2)^2}{(7.8)^2} = 1.71$$

against the critical value $F_{0.05,19,19} = 2.17$. Since our sample statistic does not fall in this critical region, the two samples may be acceptable for the professor's purposes.

EXERCISES

1. Suppose the X has an F distribution with k_1 and k_2 degrees of freedom. Prove that $1/X$ has an F distribution with k_2 and k_1 degrees of freedom.
2. Suppose that X has an F distribution with k_1 and k_2 degrees of freedom. Prove that, as $k_2 \to \infty$, then $k_1 X$ has a χ^2 distribution with k_1 degrees of freedom.
3. The ages of a sample of 20 men attending a college showed a variance of 3.6 years, while the ages of a sample of 12 women attending the same college showed a variance of 3.2 years. Does this indicate a significant difference in the age distributions of the men and women attending the college?
4. A sample of 26 brand A tires had a standard deviation in lifetime of 182 miles, while a sample of 21 brand B tires had a standard deviation of 167 miles. Test the hypothesis that the two brands have the same standard deviation in lifetimes.
5. Two different strains of mice showed standard deviations of 0.36 grams and 0.87 grams in weight at birth for samples of size 20. Test the hypothesis that the two strains have the same standard deviation in weight at birth.
6. Suppose that the blood cholesterol level for 31 patients on a low-fat diet averaged 174 with a variance of 184, while the blood cholesterol level for 41 patients on a normal diet averaged 194 with a variance of 278. Test the hypothesis that patients on either diet have the same variance in blood cholesterol level.

10.7 The *t* Distribution

Another continuous probability distribution which is of great importance in statistics is known as the Student *t* distribution. This distribution is attributed to W. S. Gosset, who published his mathematics under the pen name "Student." (Gosset himself was a worker in the Guiness Brewery in Dublin, and this has often caused the author to wonder if "t" might not be short for "tipsy"!)

Like the F distribution, the t distribution is defined as a transformation of known distributions. Let us suppose that X and Y are independent, continuous random variables, with X having standard normal distribution and Y having χ^2 distribution with k degrees of freedom. The joint density of X and Y is therefore equal to

$$f(x, y) = \frac{1}{\sqrt{2\pi}} e^{-x^2/2} \frac{1}{2^{k/2}\Gamma(k/2)} y^{(k-2)/2} e^{-y/2} \qquad \begin{matrix} x > 0 \\ y > 0 \end{matrix}$$

DEFINITION

Under the assumptions made above, the random variable

$$U = \frac{X}{\sqrt{Y/k}}$$

is said to have a *t distribution* with k degrees of freedom.

In order to determine the density function h of U, we define

$$V = Y$$

and find that

$$X = \frac{U}{\sqrt{V/k}} \quad \text{and} \quad Y = V$$

Consequently,

$$\frac{\partial X}{\partial U} = \sqrt{\frac{V}{k}}$$

so, by Proposition 8.10, we have

$$h(u) = \int_{-\infty}^{\infty} f\left(u \sqrt{\frac{v}{k}}, v\right) \cdot \sqrt{\frac{v}{k}} \, dv$$

$$= \int_{-\infty}^{\infty} \frac{1}{\sqrt{2\pi} \, 2^{k/2}\Gamma(k/2)} e^{-u^2 v/2k} v^{(k-2)/2} e^{-v/2} \sqrt{\frac{v}{k}} \, dv \qquad (v > 0)$$

$$= \frac{1}{\sqrt{2k\pi} \, 2^{k/2}\Gamma(k/2)} \int_0^{\infty} v^{(k-1)/2} \exp\left[-\frac{v}{2}\left(1 + \frac{u^2}{k}\right)\right] dv$$

If we make the substitution

$$z = v \frac{(1 + u^2/k)}{2}$$

we have

$$v = \frac{2z}{1 + u^2/k} \quad \text{and} \quad dv = 2\frac{dz}{1 + u^2/k}$$

which transforms the integral into

$$
\begin{aligned}
h(u) &= \int_0^\infty \left(\frac{2z}{1 + u^2/k}\right)^{(k-1)/2} e^{-z}\left(\frac{2}{1 + u^2/k}\right) dz \\
&= \int_0^\infty \frac{2^{(k-1)/2} z^{(k-1)/2}}{(1 + u^2/k)^{(k-1)/2}} \frac{2e^{-z}}{1 + u^2/k} dz \\
&= \frac{2^{(k+1)/2}}{(1 + u^2/k)^{(k+1)/2}} \int_0^\infty z^{(k-1)/2} e^{-z} dz \\
&= \frac{2^{(k+1)/2}\Gamma\left(\dfrac{k+1}{2}\right)}{(1 + u^2/k)^{(k+1)/2}}
\end{aligned}
$$

We therefore have, for $-\infty < u < \infty$, that

$$
\begin{aligned}
h(u) &= \frac{1}{\sqrt{2k\pi}\ 2^{k/2}\Gamma(k/2)} \cdot \frac{2^{(k+1)/2}\Gamma\left(\dfrac{k+1}{2}\right)}{(1 + u^2/k)^{(k+1)/2}} \\
&= \frac{\Gamma\left(\dfrac{k+1}{2}\right)}{\sqrt{k\pi}\ \Gamma(k/2)} (1 + u^2/k)^{-((k+1)/2)}
\end{aligned}
$$

which is our final form for the t density.

The t densities are quite similar in appearance to the standard normal density. They are symmetric about the y-axis, and therefore have mean 0, and can be shown (Exercise 4) to have variance $k/(k - 2)$ provided that

Figure 10.5

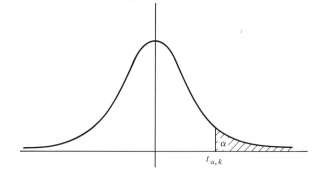

$k > 2$. For values of $k > 30$, the t densities are so closely approximated by the standard normal density that tables of functional values and areas for the t distributions are seldom calculated beyond $k = 30$.

In Appendix F, we have included a table of values $t_{\alpha, k}$, for $k = 1, 2, 3, \ldots, 29$, and for $\alpha = 0.05, 0.025, 0.01, 0.005$, where $t_{\alpha, k}$ is that point on the x-axis to the right of which lies α percent of the area (1.0) under the t density with k degrees of freedom, as indicated in Figure 10.5.

As we learned in Section 8.5, the standardized sampling distribution of the mean

$$\frac{\bar{x} - \mu}{\sigma/\sqrt{n}}$$

is approximately standard normal (precisely standard normal, if the population from which the sample was taken is normal) for large enough values of n. However, as we have mentioned in Chapter 9, this is no longer true if the population variance σ^2 is unknown, and we are forced to use the sample variance $\bar{s}^2$ in its place. The major justification for having a t distribution, and the reason for which it was invented, is that it describes the distribution of the statistic

$$\frac{\bar{x} - \mu}{\bar{s}/\sqrt{n}}$$

provided that the population is known to have a normal distribution.

Proposition 10.17

If $\bar{x}$ and $\bar{s}^2$ are the sample mean and sample variance for samples of size n taken from a normal population with mean μ and variance σ^2, then

$$\frac{\bar{x} - \mu}{\bar{s}/\sqrt{n}}$$

has a t distribution with $n - 1$ degrees of freedom.

Proof: Since $(\bar{x} - \mu)/(\sigma/\sqrt{n})$ has standard normal distribution and $(n - 1)\bar{s}^2/\sigma^2$ has χ^2 distribution with $n - 1$ degrees of freedom (the latter following from Proposition 10.6, since the population is normal), we find, by definition, that

$$\frac{(\bar{x} - \mu)/(\sigma/\sqrt{n})}{\sqrt{((n - 1)\bar{s}^2/\sigma^2)/(n - 1)}} = \frac{\bar{x} - \mu}{\bar{s}/\sqrt{n}}$$

has a t distribution with $n - 1$ degrees of freedom.

This result can be used to form the $(1 - \alpha)$ percent confidence interval

$$\bar{x} - t_{\alpha/2, n-1}\left(\frac{\bar{s}}{\sqrt{n}}\right) \leq \mu \leq \bar{x} + t_{\alpha/2, n-1}\left(\frac{\bar{s}}{\sqrt{n}}\right)$$

for the population mean μ. Note that the value $t_{\alpha/2, n-1}$ is used instead of the value $z_{\alpha/2}$ associated with the standard normal curve.

This result also can be used to test the hypothesis

$$H_0: \mu = \mu_0$$

against either a one-sided or a two-sided alternative at level of significance α. We would consider the sample statistic

$$t = \frac{\bar{x} - \mu}{\bar{s}/\sqrt{n}}$$

and determine if it falls in the critical region, which will be one of $|t| > t_{\alpha/2, n-1}$, $t > t_{\alpha, n-1}$, or $t < -t_{\alpha, n-1}$.

In actual practice, the value $t_{\alpha/2, n-1}$ is replaced by $z_{\alpha/2}$ (which it had been defined to replace) should the sample size n exceed 30. As we mentioned above, the t distribution with n degrees of freedom is very closely approximated by the standard normal distribution when $n \geq 30$.

Example 1 A random sample of $n = 25$ brand A cigarettes yields an average nicotine content of $\bar{x} = 26$ mg with a standard deviation of $\bar{s} = 6$ mg of nicotine. We wish to construct a 95 percent confidence interval for μ, the true nicotine content of this brand of cigarettes. Under the assumption that the population under question is normal, we can use the value $t_{0.025, 24} = 2.064$ ($\alpha = 0.05$ and $df = 25 - 1 = 24$) to form the confidence interval

$$26 - 2.064 \cdot \frac{6}{\sqrt{25}} \leq \mu \leq 26 + 2.064 \cdot \frac{6}{\sqrt{25}}$$

which works out to $23.52 \leq \mu \leq 28.48$. Note that if our sample size had been $n = 121$, say, we could have used *either* $t_{0.025, 120} = 1.98$ *or* $z_{0.025} = 1.96$ to form the confidence interval.

Example 2 A manufacturer claims that the average life span of his product is 1000 hours. A random sample of ten of these items showed an average life of $\bar{x} = 950$ hours with a standard deviation of $\bar{s} = 100$ hours. Is this conclusive evidence that the manufacturer's claim is false? To decide, we must test the hypothesis

$$H_0: \mu = 1000$$

against the one-sided alternative

$$H_1: \mu < 1000$$

at, say, a five percent level of significance. Under the assumption that the population under question is normal, our test statistic is

$$t = \frac{950 - 1000}{100/\sqrt{10}} = -1.58$$

which does not fall in the one-sided critical region

$$t < -t_{0.05, 9} = -1.833$$

Therefore, the evidence contained in this small sample is not sufficient to refute the manufacturer's claim.

Example 3 In order to support its aquatic life, the oxygen content of a river must be a minimum of 5 parts per million of dissolved oxygen. Suppose that 15 samples taken from the river revealed an average of 4.6 parts per million with a standard deviation of 0.2 parts per million. Is this strong evidence that the river's waters are polluted? To answer this question, we must test the hypothesis

$$H_0: \mu = 5$$

that the river's waters are not polluted, against the one-sided alternative

$$H_1: \mu < 5$$

at level of significance $\alpha = 0.01$. Our sample statistic is

$$t = \frac{4.6 - 5.0}{0.2/\sqrt{15}} = -7.75$$

which is well within the critical region

$$t < -t_{0.01, 14} = -2.624$$

Hence, this rather small sample of 15 would seem to provide very strong evidence that the river's waters are polluted.

In Section 10.1 we saw that our test of the hypothesis

$$H_0: \mu_1 = \mu_2$$

that the means of two populations are equal also relies upon knowledge of the population variances. If the variances of the two populations are known to be σ_1^2 and σ_2^2, respectively, then we know that the statistic

$$\frac{\bar{x}_1 - \bar{x}_2}{\sqrt{\sigma_1^2/n_1 + \sigma_2^2/n_2}}$$

has standard normal distribution. However, we have no information about what kind of distribution the statistic

$$\frac{\bar{x}_1 - \bar{x}_2}{\sqrt{\bar{s}_1^2/n_1 + \bar{s}_2^2/n_2}}$$

may have, in the event that the population variances are unknown, and the sample variances $\bar{s}_1^2$ and $\bar{s}_2^2$ must be used in their place. As a matter of fact, we do not even know if this is the proper statistic to consider, should this be the case.

As it turns out, if we wish to use a t test, it is necessary to make two assumptions about our two populations. Should either of these assumptions prove untenable, we would most likely have to use one of the techniques described in the next section.

Proposition 10.18

If $\bar{x}_i$ and $\bar{s}_i{}^2$ are the sampling distributions of the mean and variance for samples of size n_i taken from normally distributed populations, for $i = 1, 2$, and should these populations have equal, although unknown, variance σ^2, then the statistic

$$\frac{\bar{x}_1 - \bar{x}_2}{\sqrt{\dfrac{(n_1 - 1)\bar{s}_1{}^2 + (n_2 - 1)\bar{s}_2{}^2}{n_1 + n_2 - 2}} \sqrt{\dfrac{1}{n_1} + \dfrac{1}{n_2}}}$$

has a t distribution with $n_1 + n_2 - 2$ degrees of freedom.

Proof: By Proposition 10.1,

$$\frac{\bar{x}_1 - \bar{x}_2}{\sqrt{(\sigma^2/n_1 + \sigma^2/n_2)}}$$

has standard normal distribution. By Proposition 10.6,

$$\frac{(n_1 - 1)\bar{s}_1{}^2}{\sigma^2}$$

has χ^2 distribution with $n_1 - 1$ degrees of freedom, and

$$\frac{(n_2 - 1)\bar{s}_2{}^2}{\sigma^2}$$

has χ^2 distribution with $n_2 - 1$ degrees of freedom. By the remark following Proposition 10.4,

$$\frac{(n_1 - 1)\bar{s}_1{}^2 + (n_2 - 1)\bar{s}_2{}^2}{\sigma^2}$$

has χ^2 distribution with $(n_1 - 1) + (n_2 - 1) = n_1 + n_2 - 2$ degrees of freedom. By definition, then

$$\frac{(\bar{x}_1 - \bar{x}_2)/\sqrt{(\sigma^2/n_1) + (\sigma^2/n_2)}}{\sqrt{\dfrac{[(n_1 - 1)\bar{s}_1{}^2 + (n_2 - 1)\bar{s}_2{}^2]/\sigma^2}{n_1 + n_2 - 2}}}$$

has t distribution with $n_1 + n_2 - 2$ degrees of freedom.

Example 4   A company claims that its light bulbs are better than those of its closest competitor. A sample of $n_1 = 16$ of this company's light bulbs showed an average life of 600 hours with a standard deviation of 30 hours, while a sample of $n_2 = 12$ of the competitor's light bulbs showed an average life of 560 hours with a standard deviation of 25 hours. We wish to test the hypothesis

$$H_0: \mu_1 = \mu_2$$

that there is no difference in the two brands of light bulbs, against the one-sided alternative

$$H_1: \mu_1 > \mu_2$$

that the company's brand is indeed the better, at say, a five percent level of significance. To do so, we test the statistic

$$t = \frac{600 - 560}{\sqrt{\dfrac{15 \cdot 30^2 + 11 \cdot 25^2}{16 + 12 - 2}} \sqrt{\dfrac{1}{16} + \dfrac{1}{12}}} = 3.742$$

against the critical value $t_{0.05,\,26} = 1.706$. Since the value of our sample statistic does fall in this critical region, we must reject the null hypothesis and accept the company's claim.

Example 5 An industrial plant is suspected of polluting the river. If 15 samples taken above the plant revealed an average of 5.2 parts per million of dissolved oxygen, while 15 samples taken below the plant revealed an average of 4.8 parts per million of dissolved oxygen, with both samples having a standard deviation of 0.2 ppm, do we have solid evidence that the plant is a polluter? To decide, we must test the hypothesis

$$H_0: \mu_1 = \mu_2$$

that the oxygen content of the river is the same above and below the plant (before and after the suspected polluting), against the alternative

$$H_1: \mu_1 > \mu_2$$

that the oxygen content of the river is worse below (after) the plant, at, say, a one percent level of significance. Our sample statistic is

$$t = \frac{5.2 - 4.8}{\sqrt{\dfrac{14 \cdot 0.2^2 + 14 \cdot 0.2^2}{28}} \sqrt{\dfrac{1}{15} + \dfrac{1}{15}}} = 5.47$$

which clearly falls in the critical region $t > t_{0.01,\,28} = 2.467$. Therefore, we must reject the null hypothesis and conclude that the plant is polluting the river.

EXERCISES

1. Suppose that 15 brand A flashlight batteries had an average lifetime of 32.4 hours with a standard deviation of 2.2 hours, while 16 brand B flashlight batteries had an average lifetime of 30.8 hours with a standard deviation of 2.2 hours. Can we conclude that brand A batteries last longer than brand B batteries?

2. A sample of 10 brand A cigarettes showed an average nicotine content of 22.6 mg with a standard deviation of 2.0 mg, while a sample of 12 brand B cigarettes averaged 19.8 mg of nicotine with a standard deviation of 1.6 mg. Can we conclude that the nicotine content of brand B cigarettes is lower than the nicotine content of brand A cigarettes?

3. Two different brand golf balls were tested for distance, with the following results: 50 brand A balls averaged 234 yards with a standard deviation of 24 yards, while 50 brand B balls averaged 225 yards with a standard deviation of 21 yards. Can we conclude that brand A balls go farther?

4. Prove that the variance of the t distribution with $k > 2$ degrees of freedom is $k/(k - 2)$.

5. Suppose that X has t distribution with k degrees of freedom. Prove that X^2 has F distribution with one and k degrees of freedom.

6. Suppose that a sample of $n = 16$ brand A car batteries had an average lifetime of 2.2 years with a standard deviation of 0.5 years. Form a five percent confidence interval for the true lifetime of these batteries.

7. Suppose that 25 students of a speed-reading course read a certain amount of material in an average of 10.8 minutes with a standard deviation of 2.4 minutes. Does this support the claim that graduates of this course will be able to read the assigned material in at most 10 minutes?

8. Suppose that a group of 40 underweight young children is split in half, with the two groups observing two different diets for the next month. Suppose that the 20 children in the first group showed an average weight gain of 5.6 pounds with a standard deviation of 1.6 pounds, while the 20 children in the second group showed an average gain of 9.8 pounds with a standard deviation of 3.9 pounds. Does this imply that the second diet was the more effective?

10.8 Nonparametric Tests

If tests such as the normal test, the t test, or the χ^2 test cannot be used because the basic assumptions upon which they are based are not satisfied, then the statistician may resort to one of the nonparametric tests, a few of which are discussed in this section. These tests are called nonparametric because they do not rely on knowledge about the distribution of the underlying population, and consequently, do not rely on knowledge of the value of a parameter (or parameters) which helps shape such a distribution.

Perhaps the word "resort" should not have been used above, because, although these tests may appear to be of the "short-cut" variety, and although they make fewer assumptions about the nature of the underlying population than do some of the tests we have already discussed and therefore can be applied more generally, they have been shown in recent years to be quite effective and respectable statistical tests.

We will, in this section, study three of these nonparametric tests, the sign test, the median test, and the rank-sum test.

The first of these, the *sign test*, has many applications, although we shall be content with studying just the following three:

1. to test the null hypothesis $H_0: \mu = \mu_0$ for a symmetric population;

2. to test the null hypothesis H_0: $m = m_0$ (where m is the median) for any population;
3. to test the null hypothesis H_0: $\mu_1 = \mu_2$ for two populations.

The sign test proceeds in basically the same manner, regardless of which of these cases we are discussing. In cases 1 and 2, a sample $\{x_1\ x_2, \ldots, x_n\}$ is taken from the population, and each sample element is looked upon as either a plus($+$) or a minus($-$), depending on whether it falls above ($+$) or below ($-$) the mean or median being tested. In case 3, a paired sample $\{(x_1, y_1), (x_2, y_2), \ldots, (x_n, y_n)\}$ is taken, with the first coordinate x_i coming from the first population and the second coordinate y_i coming from the second population. Each sample element is then looked upon as a plus or a minus, depending on whether the element from the first population is larger ($+$) or smaller ($-$) than the corresponding element from the second population. (We will ignore ties so as to avoid the theoretical complications which would then arise.) In any of these three cases, if the null hypothesis to be tested were true, we would expect any sample to yield about as many plusses as minuses. Our statistical test then reduces to whether or not the distribution of plusses and minuses in our sample can reasonably be considered to be a random sample taken from a Bernoulli population with parameter $p = \frac{1}{2}$. That is, our test reduces to a test of the null hypothesis H_0: $p = \frac{1}{2}$ in a Bernoulli population.

Example 1 Suppose that a manufacturer claims that the average life span of his product is $\mu = 1000$ hours. We wish to test his claim against the alternative that $\mu < 1000$ at a five percent level of significance, and to do this, we take a sample of ten of his items, and find that their life spans are 985, 890, 1020, 980, 990, 1030, 975, 980, 920, and 1010 hours. In this sample, there are three items above the conjectured average of 1000 hours, and seven items below this average. This gives three plusses and seven minuses in our sample of ten, giving a sample proportion of $\bar{p} = 0.30$. Using the normal test of the Bernoulli hypothesis $p = \frac{1}{2}$, we find that our sample statistic

$$z = \frac{0.30 - 0.50}{\sqrt{0.50 \cdot 0.50/10}} = -1.265$$

does not fall in the critical region $z < -1.64$. Therefore, the evidence of this small sample does not allow us to reject the manufacturer's claim. Note that since

$$\sum_{i=0}^{2} b(i; 10, \tfrac{1}{2}) = 0.0547$$

it would take a sample of size 10 containing 0, 1, or 2 plusses to lead us to reject the null hypothesis, at approximately the five percent level of significance.

Example 2 We wish to determine whether a particular diet is effective in reducing cholesterol. A study of ten people who followed the diet for one month revealed the following information concerning before and after cholesterol level:

Subject	Cholesterol in mg %		Plus/Minus
	Before	After	
1	350	300	—
2	325	310	—
3	250	220	—
4	275	270	—
5	220	210	—
6	240	222	—
7	180	190	+
8	200	195	—
9	315	30u	—
10	300	272	—

Thus, our sample of ten contains just one plus and nine minuses. Such a distribution of plusses and minuses is critical at a 1.1 percent level of significance, since $b(0; 10, \frac{1}{2}) + b(1; 10, \frac{1}{2}) = 0.0108$. So the information contained in our sample leads us to reject the hypothesis that $p = \frac{1}{2}$, and therefore, to also reject the hypothesis $\mu_1 = \mu_2$ that the before and after cholesterol levels are the same. Our conclusion must be that the diet has a positive effect in reducing cholesterol.

Whereas the sign test reduces to a test of a binomial parameter, the *median test*, which we are about to describe, reduces to a χ^2 test. The median test can be used to test whether or not two populations have identical means or medians. In order to test the null hypothesis

$$H_0: \mu_1 = \mu_2$$

or the null hypothesis

$$H_1: m_1 = m_2$$

where m_1 and m_2 are the population medians, against a suitable alternative, we take samples of sizes n_1 and n_2 from the two populations, where n_1 and n_2 are chosen so that their sum $n_1 + n_2$ is even (say, $n_1 + n_2 = 2k$). We then merge the two samples into one sample and count the number of items in the first sample which fall below the median of the combined sample. (Since $n_1 + n_2$ is even, there will be no ties here.) If we denote this number by x, we can form the following table of observed values:

	Sample 1	Sample 2	
Below combined median	x	$k - x$	k
Above combined median	$n_1 - x$	$n_2 - (k - x)$	k
	n_1	n_2	

Under the assumption that the null hypothesis is true, we would expect this table to look like the following table of expected values:

	Sample 1	Sample 2	
Below combined median	$n_1/2$	$n_2/2$	k
Above combined median	$n_1/2$	$n_2/2$	k
	n_1	n_2	

We can then perform a simple χ^2 test to determine if the difference between the observed and expected tables is statistically significant. Our χ^2 statistic has one degree of freedom, since the contingency tables have $k = 2$ rows and $r = 2$ columns.

Example 3 The effect of a particular drug on pulse rate is being studied. Two groups of patients suffering from the disease in question are studied. One group is treated with the drug, while the other group is treated with a placebo. Three days after beginning the test, the two groups had the following pulse rates:

Group treated with the drug: $n_1 = 20$
73, 80, 99, 86, 68, 70, 85, 93, 98, 96, 75, 81, 82, 92, 96, 64, 79, 97, 76, 80

Group treated with the placebo: $n_2 = 24$
76, 65, 73, 95, 77, 99, 65, 75, 72, 83, 70, 65, 86, 60, 62, 90, 71, 65, 93, 89, 71, 80, 76, 94

Combining these two samples into one (merged) sample, we find that the median lies between 79 and 80, so that there are seven from the first group and 15 from the second group below the median. This gives the observed table

	Drug	Placebo	
Above median	13	9	22
Below median	7	15	22
	20	24	

which must be compared with the expected table

	Drug	Placebo	
Above median	10	12	22
Below median	10	12	22
	20	24	

Our χ^2 statistic is therefore equal to

$$\chi^2 = \frac{(13 - 10)^2}{10} + \frac{(9 - 12)^2}{12} + \frac{(7 - 10)^2}{10} + \frac{(15 - 12)^2}{12}$$

$$= 3.30$$

which, at the one percent level of significance, fails to fall in the critical region $\chi^2 > \chi^2_{0.01,\,1} = 6.635$. Hence, the null hypothesis that the two groups (populations) have the same mean cannot be rejected. That is, we cannot conclude that this drug has the effect of increasing pulse rates.

Example 4 In order to determine the effect of speed on gas mileage, ten cars had their tanks filled, and then were driven at 50 miles per hour the full length of the New Jersey Turnpike during late night hours. They were then turned around, had their tanks refilled, and were driven at 60 miles per hour back the full length of the Turnpike. The following indicates the miles per gallon achieved for each car:

At 50 mph: $n_1 = 10$
16.2, 15.8, 16.3, 17.2, 16.5, 14.9, 15.9, 17.1, 16.6, 16.3

At 60 mph: $n_2 = 10$
14.3, 14.9, 16.6, 17.2, 15.2, 13.9, 14.7, 15.3, 15.9, 16.2

The median of the combined sample of 20 lies between 15.9 and 16.2, with 7 from the second sample below the combined median. This gives us the observed table

	50 mpg	60 mpg	
Below median	3	7	10
Above median	7	3	10
	10	10	

which must be compared with the expected table

	50 mpg	60 mpg	
Below median	5	5	10
Above median	5	5	10
	10	10	

Our χ^2 statistic is

$$\chi^2 = \frac{(3 - 5)^2}{5} + \frac{(7 - 5)^2}{5} + \frac{(7 - 5)^2}{5} + \frac{(3 - 5)^2}{5}$$

$$= 3.20$$

which does not exceed the critical value $\chi^2_{0.05,\,1} = 3.841$. Therefore, we cannot conclude that gas mileage is better at 50 mph than it is at 60 mph, or vice versa.

The *rank-sum test* (which is also known as the *Mann-Whitney test*, the *Wilcoxen test*, or the *U test*) can be used to test the null hypothesis that two populations have identical means, usually against a suitable one-sided alternative. To test our hypothesis, we take a sample of size n_1 from the first population and a sample of size n_2 from the second population. As we did in the median test, we merge these two samples into one larger sample of size $N = n_1 + n_2$, and we then arrange the elements of the merged sample into ascending order. Each element of the merged sample is then assigned a rank. We will assume, at this stage, that the two populations from which we are sampling are continuous, so that ties in rank will be virtually impossible. We will explain later what adjustments can be made in the case of ties.

The basic statistic upon which the rank-sum test is based is the sum R of the ranks of the elements in one of the two samples. If the null hypothesis were true, and the two populations did have identical means, one would expect the rank-sums from the two samples to be approximately equal. However, if the rank-sum for one of the two samples were much the larger of the two, this would seem to indicate that the mean of the corresponding population might very well be larger than the mean of the other population.

The question that must be answered, then, is when does the difference between the two sample rank-sums become so pronounced, so statistically significant, that the null hypothesis should be rejected. Although we will not pursue this particular line of reasoning, we will mention that it is possible to study the exact distribution of the statistic R. Tables have been tabulated that facilitate the calculation of the probability of obtaining, by chance alone, a rank-sum less than or equal to, or greater than or equal to, the rank-sum obtained from one of our samples. If this probability should turn out to be less than the level of significance of our test, we would then reject the null hypothesis.

The tables just mentioned are restricted for use with samples of size $n_i \leq 8$ ($i = 1, 2$), and not only for reasons of economy. If both n_1 and n_2 exceed 8, a statistic known as the U statistic, and very closely related to the R statistic, is approximately normal. Although we will not be able to prove this assertion, we will proceed to define the U statistic and determine its mean and variance. We will begin with the R statistic.

Let us suppose that we have taken samples of size n_1 from the first population and n_2 from the second population, with a test of the null hypothesis

$$H_0: \mu_1 = \mu_2$$

against the one-sided alternative

$$H_1: \mu_1 > \mu_2$$

in mind. We let R represent the sum of the ranks of the elements in the first sample. Under the assumption that the null hypothesis is true, the ranks of the elements in the first sample can be looked upon as a random sample of size n_1 taken without replacement from the finite uniform population $\{1, 2, 3, \ldots, N\}$, where $N = n_1 + n_2$. The mean and variance of this population are $(N + 1)/2$ and $(N^2 - 1)/12$, respectively (see Exercise 1), and so, according to Proposition 5.22, the mean and variance of the sample mean, for samples of size n_1 taken from this population, are

$$E(\bar{x}) = \frac{N + 1}{2}$$

and

$$Var(\bar{x}) = \left(\frac{N^2 - 1}{12n_1}\right)\left(\frac{N - n_1}{N - 1}\right)$$

$$= \frac{(N + 1)(N - n_1)}{12n_1}$$

respectively. Since the sum of the ranks of the sample elements is simply $n_1 \cdot \bar{x}$, we have

$$E(R) = E(n_1 \cdot \bar{x}) = n_1 \cdot E(\bar{x})$$

and

$$Var(R) = Var(n_1 \cdot \bar{x}) = n_1{}^2 \cdot Var(\bar{x})$$

Therefore,

$$E(R) = \frac{n_1(N + 1)}{2} = \frac{n_1(n_1 + n_2 + 1)}{2}$$

and

$$Var(R) = \frac{n_1{}^2(N + 1)(N - n_1)}{12n_1}$$

$$= \frac{n_1 \cdot n_2 \cdot (n_1 + n_2 + 1)}{12}$$

As we mentioned above, the distribution of R has been tabulated for values of n_1 and n_2 not exceeding 8. The distribution of R, however, does not follow one of the well-known distributions, so instead, for values of n_1 and n_2 exceeding 8, we consider the following statistic:

$$U = n_1 \cdot n_2 + \frac{n_1 \cdot (n_1 + 1)}{2} - R$$

It is easy to see that

$$E(U) = \frac{n_1 \cdot n_2}{2}$$

and that

$$Var(U) = Var(R) = \frac{n_1 n_2 (n_1 + n_2 + 1)}{12}$$

As we mentioned above, the distribution of U is approximately normal, and so the distribution of

$$z = \frac{U - E(U)}{\sqrt{\text{Var}(U)}}$$

is approximately standard normal.

Now, if our alternative hypothesis were false, the value of R would be comparatively large, and therefore, the value of U would be comparatively small. Therefore, the critical region for our test would be $z < -z_\alpha$.

Example 5 To determine the importance of weight gained by women during pregnancy, the weights of the babies of ten women who were on a particular diet during their pregnancies were recorded along with the weights of the babies of nine women who were not on this diet. The results are as follows:

Mother on the diet: $n_1 = 10$
6.9, 6.8, 8.4, 10.2, 9.1, 6.1, 7.4, 8.9, 9.6, 7.2

Mother not on the diet: $n_2 = 9$
6.7, 7.0, 8.5, 8.1, 9.2, 6.5, 7.3, 6.6, 5.9

We wish to test the null hypothesis

$$H_0: \mu_1 = \mu_2$$

that the babies in both groups will have the same average weight at birth, against the alternative

$$H_1: \mu_1 > \mu_2$$

that babies produced by mothers who were on this particular diet during their pregnancy weigh more at birth, and therefore are healthier and stronger at birth. The first step in this process is to merge the two samples, and then place the merged sample into ascending order. The results of these two steps are indicated in Table 10.4. Consequently,

$$R = 2 + 6 + 7 + 9 + 11 + 13 + 15 + 16 + 18 + 19 = 116$$

the sum of the ranks of the ten elements in the first sample. Since $n_1 = 10$ and $n_2 = 9$, we have

$$U = 10 \cdot 9 + \frac{10 \cdot 11}{2} - 116 = 29$$

$$E(U) = \frac{10 \cdot 9}{2} = 45$$

$$\text{Var}(U) = \frac{[10 \cdot 9 \cdot (10 + 9 + 1)]}{12} = 150$$

$$\sigma_U = 12.247$$

Our test statistic is therefore

$$z = \frac{29 - 45}{12.247} = -1.306$$

which does not fall in the critical region $z < -1.64$ at the five percent level of significance. So the rank-sum test (U test) does not allow us to reject the null hypothesis at the five percent level of significance. (The reader might wish to perform a t test on this data, and compare results.)

It is possible that two of the elements in the combined sample might tie in rank. The weights of two of the babies in the example above might very well have been the same, after rounding off to one decimal place. The problem of ties can be resolved by assigning to each of the tied sample elements the average of the ranks involved. For example, if there were a tie for the seventh and eighth positions, then the two elements involved in the tie would each be assigned rank 7.5. This process will be illustrated in Example 6 below.

The rank-sum test can also be used to test whether two populations have equal variances. The only difference in the procedure is the manner in which the elements of the merged sample are ranked. This will be done by assigning rank 1 to the smallest element in the merged sample, ranks 2 and 3 to the largest and second largest elements, ranks 4 and 5 to the second and third

Table 10.4
Ranks of Merged Sample

Weight	Rank	Sample
5.9	1	2
6.1	2	1
6.5	3	2
6.6	4	2
6.7	5	2
6.8	6	1
6.9	7	1
7.0	8	2
7.2	9	1
7.3	10	2
7.4	11	1
8.1	12	2
8.4	13	1
8.5	14	2
8.9	15	1
9.1	16	1
9.2	17	2
9.6	18	1
10.2	19	1

smallest elements, and so forth. Naturally, the smaller rank-sum (larger U) would be associated with the sample with the larger variance, so if we wish to test the null hypothesis

$$H_0: \sigma_1^2 = \sigma_2^2$$

against the alternative

$$H_1: \sigma_1^2 > \sigma_2^2$$

the critical region would be $z > z_\alpha$.

Example 6 We wish to determine if a particular drug used for treating a certain disease has the dangerous effect of causing sharp increases or decreases in pulse rates. A sample of 20 patients who were suffering from the disease and who all had approximately the same pulse rate were tested, with only ten of them being administered the drug. Their pulse rates, 15 minutes after being administered the drug (or the placebo), are given below:

With drug: 70, 88, 82, 96, 72, 85, 73, 95, 90, 76
Without drug: 84, 82, 80, 84, 84, 78, 76, 90, 85, 88

If we arrange the elements in the merged sample of 20 into ascending order and assign the ranks as described above, we obtain Table 10.5. Notice how

Table 10.5
Ranks of Merged Sample

Pulse rate	Rank	Sample
70	1	1
72	4	1
73	5	1
76	8.5	1
76	8.5	2
78	12	2
80	13	2
82	16.5	1
82	16.5	2
84	19	2
84	19	2
84	19	2
85	14.5	1
85	14.5	2
88	10.5	1
88	10.5	2
90	6.5	1
90	6.5	2
95	3	1
96	2	1

the many ties have been resolved, particularly the tie for the last three positions. For sample 1, we have

$$R = 1 + 4 + 5 + 8.5 + 16.5 + 14.5 + 10.5 + 6.5 + 3 + 2$$
$$= 71.5$$

and so

$$U = 10 \cdot 10 + \frac{10 \cdot 11}{2} - 71.5 = 83.5$$

$$E(U) = \frac{10 \cdot 10}{2} = 50$$

$$\text{Var}(U) = \frac{10 \cdot 10 \cdot (10 + 10 + 1)}{12} = 175$$

and

$$\sigma_U = 13.229$$

Therefore, our sample statistic is

$$z = \frac{83.5 - 50}{13.229} = 2.53$$

which does fall in the critical region $z > 2.33$ at one percent level of significance. Hence, our test leads us to reject the null hypothesis, and conclude that the drug may be dangerous.

EXERCISES

1. Prove that the mean of the discrete uniform distribution on $\{1, 2, 3, \ldots, n\}$ is $(n + 1)/2$, and that the variance is $(n^2 - 1)/12$.

2. Use a nonparametric test to determine, on the basis of the grades given below, if the two teachers are equally effective teaching the same subject:

   Class 1: 75, 86, 65, 97, 94, 83, 62, 96, 84, 76
   Class 2: 73, 62, 73, 84, 75, 85, 90, 47, 64

3. Use a nonparametric test to determine, on the basis of the information given below, if the members of the first nationality are taller than the members of the second nationality:

   Nationality 1: 69, 74, 66, 72, 70, 71, 70, 68 in.
   Nationality 2: 66, 68, 72, 66, 67, 65, 71, 68 in.

4. Use nonparametric tests on the information below to test the hypotheses that the two brands of gasoline give the same average mileage with the same standard deviation:

   Gasoline 1: 16, 18, 14, 21, 16, 17, 24, 15, 18, 14, 16
   Gasoline 2: 18, 17, 18, 15, 14, 16, 19, 20, 15, 14, 17

5. Based on the information below, determine if it can be stated that the first born (second born) of twins weighs more at birth:

    First born: 7.6, 6.5, 6.5, 7.3, 8.1, 5.6, 6.7, 7.3
    Second born: 7.4, 6.7, 6.2, 6.8, 7.4, 5.4, 5.9, 6.5

6. Based on the information below, determine if a change in the stop-light pattern has improved travel time across New York City on Forty Second Street:

    Before: 10.2, 11.5, 15.3, 14.2, 12.5, 13.7, 9.9, 11.1, 12.0 min
    After: 11.3, 11.1, 9.7, 10.4, 12.1, 10.1, 9.4, 9.8, 10.3, 13.1 min

7. Based on the following sample, determine if it is reasonable to believe that the average IQ for convicted murderers is as high as 110:

    118, 121, 111, 99, 103, 92, 125, 118, 94, 112, 114, 120, 99, 102

10.9 Analysis of Variance

We have already, in this chapter, discussed how to test for equality between two population means or two population proportions and among k population proportions. In this section, we shall discuss the remaining case, equality among k population means. The method we shall discuss is known as *analysis of variance*, although what we shall discuss is merely analysis of variance in its most elementary form. Analysis of variance, or ANOVA as it is frequently called, was first developed by R. A. Fisher, after whom the F distribution was named, and has now grown into one of the most important branches of statistical analysis.

The problem we shall be faced with is to decide whether observed differences among two or more sample means reflect actual differences among the means of the populations involved, or whether these observed differences can reasonably be attributed to chance. In other words, we wish to test the null hypothesis

$$H_0: \mu_1 = \mu_2 = \mu_3 \cdots = \mu_k$$

where μ_i is the mean of the ith population, $i = 1, 2, \ldots, k$, against the alternative that there is at least one pair of populations whose means differ. As a first step, we take independent samples of size n from each of the k populations (we shall explain later what alterations would have to be made in our procedure should we wish to allow the samples from the different populations to be of unequal sizes), and let x_{ij} denote the jth element in the sample taken from the ith population. We then let

$$\bar{x}_i = \sum_{j=1}^{n} x_{ij}$$

denote the sample mean for the sample taken from the ith population.

To test our null hypothesis, we consider the term

$$v = \sum_{i=1}^{k} \sum_{j=1}^{n} (x_{ij} - \bar{x})^2$$

where

$$\bar{x} = \frac{\sum_{i=1}^{k} \sum_{j=1}^{n} x_{ij}}{kn}$$

the mean of *all* the elements sampled. This term v is called the *total variability* of the data from all k samples combined, and is sometimes written SS_{total}, where SS is an abbreviation for "sum of squares." Under the assumption that the null hypothesis is true, all of v is due to chance variation, since v is essentially the variance of the combined sample about the maximum likelihood estimate of the common mean of the populations involved. On the other hand, should the null hypothesis be false, part of v is due to chance, and part is due to the differences among the true means of the k populations. To make this last statement more concrete, we prove the following:

Proposition 10.19

$$v = \sum_{i=1}^{k} \sum_{j=1}^{n} (x_{ij} - \bar{x}_i)^2 + n \sum_{i=1}^{k} (\bar{x}_i - \bar{x})^2$$

Proof: We take the term $(x_{ij} - \bar{x})^2$ in the definition of v, write it as

$$[(x_{ij} - \bar{x}_i) + (\bar{x}_i - \bar{x})]^2$$

and then expand the square to obtain

$$(x_{ij} - \bar{x}_i)^2 + 2(x_{ij} - \bar{x}_i)(\bar{x}_i - \bar{x}) + (\bar{x}_i - \bar{x})^2$$

We therefore have

$$v = \sum_{i=1}^{k} \sum_{j=1}^{n} (x_{ij} - \bar{x}_i)^2 + \sum_{i=1}^{k} \sum_{j=1}^{n} 2(x_{ij} - \bar{x}_i)(\bar{x}_i - \bar{x}) + \sum_{i=1}^{k} \sum_{j=1}^{n} (\bar{x}_i - \bar{x})^2$$

We can rewrite the middle term as

$$2 \sum_{i=1}^{k} (\bar{x}_i - \bar{x}) \cdot \sum_{j=1}^{n} (x_{ij} - \bar{x}_i)$$

which equals zero since

$$\sum_{j=1}^{n} (x_{ij} - \bar{x}_i) = 0$$

this, for any i, being the sum of the differences between the n terms x_{ij} and their mean $\bar{x}_i$. Hence,

$$v = \sum_{i=1}^{k} \sum_{j=1}^{n} (x_{ij} - \bar{x}_i)^2 + \sum_{i=1}^{k} \sum_{j=1}^{n} (\bar{x}_i - \bar{x})^2$$

In the second term here, the summands are independent of j, and so this term equals

$$\sum_{i=1}^{k} n(\bar{x}_i - \bar{x})^2 = n \sum_{i=1}^{k} (\bar{x}_i - \bar{x})^2$$

Therefore, v can be written in the required form.

Before we proceed any further, it is important to understand what these two summands represent. Each component

$$\sum_{j=1}^{n} (x_{ij} - \bar{x}_i)^2$$

of the left-hand summand is measuring the variability within one of the k samples. This variability within a given sample, as usual, is due completely to chance, and has no relationship whatsoever to whether or not the null hypothesis is true. Consequently, the left-hand summand of v, which is frequently called SS_{within}, can be entirely attributed to chance variation. On the other hand, the right-hand summand is measuring the variability among the means of the k samples. For this reason, it is frequently called SS_{among}. Should the null hypothesis be true and the k population means be equal, then SS_{among} is also measuring only chance variation; but should the null hypothesis be false, then SS_{among} would also be measuring variation caused by the differences among the population means. In either case, we can always partition SS_{total} into the sum of two components, one measuring the variation existing within the individual samples, and the other measuring the variation existing among the sample means.

We now look at these two summands separately, and determine what kind of probability distribution each follows. We will work under the assumption that the null hypothesis is true; that is, that the k populations have identical means. We must also add another assumption, namely that the k populations have normal distributions with common variance σ^2.

Proposition 10.20

Under the above assumptions,

$$\left(\frac{1}{\sigma^2}\right) \sum_{i=1}^{k} \sum_{j=1}^{n} (x_{ij} - \bar{x}_i)^2$$

has a χ^2 distribution with $k(n-1)$ degrees of freedom.

Proof: Should the null hypothesis be true, each of the k populations is normal with common mean and variance. If we let $\bar{s}_i^2$ be the sample variance for the ith population, then, by Proposition 10.6, $(n-1)\bar{s}_i^2/\sigma^2$ has a χ^2 distribution with $n-1$ degrees of freedom. We have

$$\bar{s}_i^2 = \frac{\sum_{j=1}^{n}(x_{ij} - \bar{x}_i)^2}{n-1}$$

and so

$$\frac{(n-1)\bar{s}_i^2}{\sigma^2} = \left(\frac{1}{\sigma^2}\right)\sum_{j=1}^{n}(x_{ij} - \bar{x}_i)^2$$

Adding these k χ^2 distributions, we find that

$$\sum_{i=1}^{k}\left(\frac{1}{\sigma^2}\right)\sum_{j=1}^{n}(x_{ij} - \bar{x}_i)^2 = \left(\frac{1}{\sigma^2}\right)\sum_{i=1}^{k}\sum_{j=1}^{n}(x_{ij} - \bar{x}_i)^2$$

has a χ^2 distribution with $\sum_{i=1}^{k}(n-1) = k(n-1)$ degrees of freedom.

Proposition 10.21

Under the above assumptions,

$$\left(\frac{n}{\sigma^2}\right)\sum_{i=1}^{k}(\bar{x}_i - \bar{x})^2$$

has a χ^2 distribution with $k-1$ degrees of freedom.

Proof: Should the null hypothesis be true, each of the sample means $\bar{x}_i$ has a normal distribution with common mean μ and common variance σ^2/n. Again applying Proposition 10.6, we find that

$$\frac{(k-1)\bar{s}^2}{\sigma^2/n}$$

has a χ^2 distribution with $k-1$ degrees of freedom, where

$$\bar{s}^2 = \frac{\sum_{i=1}^{k}(\bar{x}_i - \bar{x})^2}{k-1}$$

Since

$$\frac{(k-1)\bar{s}^2}{\sigma^2/n} = \left(\frac{n}{\sigma^2}\right)\sum_{i=1}^{k}(\bar{x}_i - \bar{x})^2$$

the desired result follows.

We are almost in a position now to apply Proposition 10.15. We have thus far determined that the random variables $(1/\sigma^2)\,SS_{\text{within}}$ and $(1/\sigma^2)\,SS_{\text{among}}$ have χ^2 distributions, and under the additional assumption that these two random variables are independent (the proof of which is beyond the scope of this book), we can prove the following:

Proposition 10.22

Adding the above assumption to our previous assumptions, we can conclude that

$$\frac{nk(n-1) \sum_{i=1}^{k} (\bar{x}_i - \bar{x})^2}{(k-1) \sum_{i=1}^{k} \sum_{j=1}^{n} (x_{ij} - \bar{x}_i)^2}$$

has an F distribution with $k-1$ and $k(n-1)$ degrees of freedom.

Proof: By Propositions 10.15, 10.20, and 10.21,

$$\frac{\left[\frac{n}{\sigma^2} \sum_{i=1}^{k} (\bar{x}_i - \bar{x})^2 \right] \Big/ (k-1)}{\left[\frac{1}{\sigma^2} \sum_{i=1}^{k} \sum_{j=1}^{n} (x_{ij} - \bar{x}_i)^2 \right] \Big/ k(n-1)}$$

has an F distribution with $k-1$ and $k(n-1)$ degrees of freedom. The result stated is immediate from this.

How is Proposition 10.22 to be interpreted? To answer this, we first take a closer look at the term

$$\frac{\sum_{i=1}^{k} \sum_{j=1}^{n} (x_{ij} - \bar{x}_i)^2}{k(n-1)}$$

which is simply the χ^2 random variable SS_{within} divided by its number of degrees of freedom. We can rewrite this term as

$$\sum_{i=1}^{k} \left[\frac{\sum_{j=1}^{n} (x_{ij} - \bar{x}_i)^2}{n-1} \right] \Big/ k$$

the inner summands being the sample variances for our k samples. As we have seen in Sections 9.2 and 10.3, the sample variance is a consistent, unbiased estimator of the population variance. In our case, we have k populations which are assumed to have the same variance, and in the term

$$\frac{SS_{\text{within}}}{k(n-1)}$$

we are taking the average of k sample variances, or k estimations of the common population variance. Consequently, $SS_{\text{within}}/k(n-1)$ would appear to be an excellent estimator of σ^2. Recall also that the variation that occurs in the term SS_{within} is due entirely to chance variation within the individual samples, and has no relationship whatsoever to whether or not the null hypothesis is true.

It also turns out to be true that the term

$$\frac{n \sum_{i=1}^{k} (\bar{x}_i - \bar{x})^2}{k - 1}$$

which is simply the random variable SS_{among} divided by its number of degrees of freedom, gives an approximation to the population variance σ^2. For the term

$$\frac{\sum_{i=1}^{k} (\bar{x}_i - \bar{x})^2}{k - 1}$$

gives a consistent, unbiased estimate of the variance $\sigma_{\bar{x}}^2$ of the sample mean, which happens to equal σ^2/n. Consequently,

$$n \left[\frac{\sum_{i=1}^{k} (\bar{x}_i - \bar{x})^2}{k - 1} \right]$$

gives a consistent, unbiased estimate of $n(\sigma^2/n) = \sigma^2$. You will recall, however, that the variation that occurs in term SS_{among} is very much related to the null hypothesis. Consequently, should the null hypothesis be false, our second estimation of σ^2 would be large in comparison with our first estimation of σ^2. The value of Proposition 10.22 is that it tells us that we can use the F distribution to determine when this difference between these two estimations of σ^2 is large enough to cast doubts on the validity of the null hypothesis.

Example 1 A tutor uses three different teaching methods with three groups of four students each, with the following results on their final examination:

First group (method): 74, 78, 68, 72
Second group (method): 93, 83, 89, 87
Third group (method): 75, 80, 79, 82

The 12 grades given are the x_{ij}, for $i = 1$, 2, and 3 and $j = 1$, 2, 3, and 4. We first calculate the three sample means

$$\bar{x}_1 = \frac{74 + 78 + 68 + 72}{4} = 73$$

$$\bar{x}_2 = \frac{93 + 83 + 89 + 87}{4} = 88$$

$$\bar{x}_3 = \frac{75 + 80 + 79 + 82}{4} = 79$$

Therefore,

$$\bar{x} = \frac{73 + 88 + 79}{3} = 80$$

which could also have been calculated by taking the average of the 12 grades x_{ij}. Next, we calculate the variability in the three samples:

$$\sum_{j=1}^{4} (x_{1j} - \bar{x}_1)^2 = (74 - 73)^2 + (78 - 73)^2 + (68 - 73)^2 + (72 - 73)^2 = 52$$

$$\sum_{j=1}^{4} (x_{2j} - \bar{x}_2)^2 = (93 - 88)^2 + (83 - 88)^2 + (89 - 88)^2 + (87 - 88)^2 = 52$$

$$\sum_{j=1}^{4} (x_{3j} - \bar{x}_3)^2 = (75 - 79)^2 + (80 - 79)^2 + (79 - 79)^2 + (82 - 79)^2 = 26$$

Consequently, $SS_{within} = 52 + 52 + 26 = 130$. On the other hand,

$$\sum_{i=1}^{3} (\bar{x}_i - \bar{x})^2 = (73 - 80)^2 + (88 - 80)^2 + (79 - 80)^2 = 114$$

so that $SS_{among} = 4 \cdot 114 = 456$. As a result,

$$SS = 130 + 456 = 586$$

Our sample statistic

$$F = \frac{k(n - 1)SS_{among}}{(k - 1)SS_{within}} = \frac{3 \cdot (4 - 1) \cdot 456}{2 \cdot 130} = 15.78$$

which falls well within the critical region

$$F > F_{0.05,\, 2,\, 9} = 4.26$$

We therefore must reject the null hypothesis that the three teaching methods are equally effective, and conclude that there is some difference among them. Further tests of these three samples, two at a time, might reveal which is the most effective or the least effective.

Example 2 We wish to determine if the oxygen content of a lake is lower in some areas than in other areas, so we take samples of four readings each from four different sections of the lake, obtaining the following information:

Area 1: 5.0, 4.6, 5.3, 5.1 ppm oxygen
Area 2: 4.8, 5.3, 4.9, 5.4 ppm oxygen
Area 3: 4.7, 4.9, 5.0, 5.0 ppm oxygen
Area 4: 4.6, 4.8, 4.9, 4.9 ppm oxygen

Proceeding as in the preceding example, we find the four sample means to be $\bar{x}_1 = 5.0$, $\bar{x}_2 = 5.1$, $\bar{x}_3 = 4.9$, and $\bar{x}_4 = 4.8$, and we then find the overall four-sample mean to be

$$\bar{x} = \frac{5.0 + 5.1 + 4.9 + 4.8}{4} = 4.95$$

We next calculate the variability in the four samples:

$$\sum_{j=1}^{4} (x_{1j} - \bar{x}_1)^2 = (5.0 - 5.0)^2 + (4.6 - 5.0)^2 + (5.3 - 5.0)^2 + (5.1 - 5.0)^2$$
$$= 0.26$$

$$\sum_{j=1}^{4} (x_{2j} - \bar{x}_2)^2 = (4.8 - 5.1)^2 + (5.3 - 5.1)^2 + (4.9 - 5.1)^2 + (5.4 - 5.1)^2$$
$$= 0.26$$

$$\sum_{j=1}^{4} (x_{3j} - \bar{x}_3)^2 = (4.7 - 4.9)^2 + (4.9 - 4.9)^2 + (5.0 - 4.9)^2 + (5.0 - 4.9)^2$$
$$= 0.06$$

$$\sum_{j=1}^{4} (x_{4j} - \bar{x}_4)^2 = (4.6 - 4.8)^2 + (4.8 - 4.8)^2 + (4.9 - 4.8)^2 + (4.9 - 4.8)^2$$
$$= 0.06$$

so that $SS_{within} = 0.26 + 0.26 + 0.06 + 0.06 = 0.64$. Also,

$$\sum_{i=1}^{4} (\bar{x}_i - \bar{x})^2 = (5.0 - 4.95)^2 + (5.1 - 4.95)^2 + (4.9 - 4.95)^2 + (4.8 - 4.95)^2$$
$$= 0.05$$

so that $SS_{among} = 4 \cdot 0.05 = 0.20$. Notice that, in this example, SS_{among} is much smaller than SS_{within}, indicating that the chance variability of SS_{within} is the more significant part of the total variability SS_{total}. This, in turn, suggests that the null hypothesis is most likely a reasonable hypothesis. To verify this, we compare the sample statistic

$$F = \frac{k(n-1)SS_{among}}{(k-1)SS_{within}} = \frac{4 \cdot (4-1) \cdot 0.20}{(4-1) \cdot 0.64} = 1.25$$

to the critical value $F_{0.05,\,3,\,12} = 3.49$. Since our sample statistic fails to fall in the critical region beyond 3.49, we cannot reject the null hypothesis that the oxygen content of the lake is uniform throughout the lake.

The calculation of SS_{within} and SS_{among} can be simplified, especially for use with an electronic hand calculator, by the following formulas:

Proposition 10.23

(a) $SS_{total} = \displaystyle\sum_{i=1}^{k} \sum_{j=1}^{n} x_{ij}^2 - \frac{\left(\displaystyle\sum_{i=1}^{k} \sum_{j=1}^{n} x_{ij} \right)^2}{kn}$

(b) $SS_{among} = \dfrac{\displaystyle\sum_{i=1}^{k} \left(\sum_{j=1}^{n} x_{ij} \right)^2}{n} - \dfrac{\left(\displaystyle\sum_{i=1}^{k} \sum_{j=1}^{n} x_{ij} \right)^2}{kn}$

Proof: The proof consists of two applications of the result contained in Exercise 1, which states that if μ is the mean of the N values $\{x_1, x_2, \ldots, x_N\}$, then

$$\sum_{i=1}^{N} (x_i - \mu)^2 = \sum_{i=1}^{N} x_i^2 - \frac{\left(\sum_{i=1}^{N} x_i\right)^2}{N}$$

If we apply this result to the kn values x_{ij}, we obtain (a), since

$$SS_{\text{total}} = \sum_{i=1}^{k} \sum_{j=1}^{n} (x_{ij} - \bar{x})^2$$

where $\bar{x}$ is the mean of the x_{ij}. If we apply this result to the k values $\bar{x}_i$, we obtain

$$SS_{\text{among}} = n \sum_{i=1}^{k} (\bar{x}_i - \bar{x})^2$$

$$= n \sum_{i=1}^{k} \bar{x}_i^2 - n \frac{\left(\sum_{i=1}^{k} \bar{x}_i\right)^2}{k}$$

But since $\bar{x}_i = (\sum_{j=1}^{n} x_{ij})/n$, we have

$$SS_{\text{among}} = n \sum_{i=1}^{k} \left(\frac{\sum_{j=1}^{n} x_{ij}}{n}\right)^2 - \frac{n}{k}\left[\sum_{i=1}^{k} \left(\frac{\sum_{j=1}^{n} x_{ij}}{n}\right)^2\right]$$

$$= \frac{1}{n} \sum_{i=1}^{k} \left(\sum_{j=1}^{n} x_{ij}\right)^2 - \frac{1}{kn}\left(\sum_{i=1}^{k} \sum_{j=1}^{n} x_{ij}\right)^2$$

as claimed.

Both terms above involve calculating the sum of the kn terms x_{ij} and the sum of the kn terms x_{ij}^2, in addition to the k partial sums of the n terms x_{ij} in each individual sample. Once both SS_{total} and SS_{among} have been calculated, SS_{within} can be obtained from the equation

$$SS_{\text{within}} = SS_{\text{total}} - SS_{\text{among}}$$

Example 1 (continued) According to our formulas, we must calculate

$$\sum_{i=1}^{3} \sum_{j=1}^{4} x_{ij} = 960$$

$$\left(\sum_{i=1}^{3} \sum_{j=1}^{4} x_{ij}\right)^2 = 921{,}600$$

$$\sum_{i=1}^{3} \sum_{j=1}^{4} x_{ij}^2 = 77{,}386$$

$$\left(\sum_{j=1}^{4} x_{1j} \right)^2 = (292)^2 = 85{,}264$$

$$\left(\sum_{j=1}^{4} x_{2j} \right)^2 = (352)^2 = 123{,}904$$

$$\left(\sum_{j=1}^{4} x_{3j} \right)^2 = (316)^2 = 99{,}856$$

Therefore,

$$SS_{total} = 77{,}386 - \frac{921{,}600}{12} = 586$$

and

$$SS_{among} = \frac{85{,}264 + 123{,}904 + 99{,}856}{4} - 76{,}800 = 456$$

so that

$$SS_{within} = 586 - 456 = 130$$

as above.

Should we wish to allow the sample sizes to be unequal, we would have to make several slight modifications, although the overall procedure described above will not be altered. If we let n_i denote the size of the sample taken from the ith population, then we define

$$\bar{x} = \frac{\displaystyle\sum_{i=1}^{k} \sum_{j=1}^{n_i} x_{ij}}{\displaystyle\sum_{i=1}^{k} n_i}$$

and

$$SS_{total} = \sum_{i=1}^{k} \sum_{j=1}^{n_i} (x_{ij} - \bar{x})^2$$

From the definition

$$\bar{x}_i = \frac{\displaystyle\sum_{j=1}^{n_i} x_{ij}}{n_i}$$

follow the definitions

$$SS_{within} = \sum_{i=1}^{k} \sum_{j=1}^{n_i} (x_{ij} - \bar{x}_i)^2$$

and

$$SS_{among} = \sum_{i=1}^{k} n_i (\bar{x}_i - \bar{x})^2$$

The counterpart to Proposition 10.19

$$SS_{\text{total}} = SS_{\text{within}} + SS_{\text{among}}$$

can be proved in much the same manner as before. It can also be shown, as in Proposition 10.20, that

$$\frac{1}{\sigma^2} \sum_{j=1}^{n} (x_{ij} - \bar{x}_i)^2$$

has a χ^2 distribution with $n_i - 1$ degrees of freedom, and consequently, that $(1/\sigma^2)SS_{\text{within}}$ has a χ^2 distribution with

$$\sum_{i=1}^{k} (n_i - 1) = \sum_{i=1}^{k} n_i - k$$

degrees of freedom.

Proposition 10.21 can be restated and reproved in the following manner:

Proposition 10.21A

$$\frac{1}{\sigma^2} \sum_{i=1}^{k} n_i(\bar{x}_i - \bar{x})^2$$

has a χ^2 distribution with $k - 1$ degrees of freedom.

Proof: Should the null hypothesis be true, each of the sample means $\bar{x}_i$ has a normal distribution with common mean μ and variance σ^2/n_i. Consequently, $(\bar{x}_i - \mu)/(\sigma/\sqrt{n_i})$ has standard normal distribution, so by Proposition 10.3,

$$\left(\frac{\bar{x}_i - \mu}{\sigma/\sqrt{n_i}}\right)^2 = \frac{n_i(\bar{x}_i - \mu)^2}{\sigma^2}$$

has χ^2 distribution with one degree of freedom. Therefore,

$$\sum_{i=1}^{k} n_i \frac{(\bar{x}_i - \mu)^2}{\sigma^2}$$

has a χ^2 distribution with k degrees of freedom. Replacing the unknown parameter μ with the estimate $\bar{x}$, we find that

$$\frac{1}{\sigma^2} \sum_{i=1}^{k} n_i(\bar{x}_i - \bar{x})^2 = \frac{1}{\sigma^2} SS_{\text{among}}$$

has a χ^2 distribution with $k - 1$ degrees of freedom.

Our test statistic then becomes

$$\frac{SS_{\text{among}}/(k - 1)}{SS_{\text{within}} \bigg/ \left(\sum_{i=1}^{k} n_i - k\right)}$$

which has an F distribution with $k - 1$ and $\sum_{i=1}^{k} n_i - k$ degrees of freedom.

The calculation formulas for SS_{total} and SS_{among} presented in Proposition 10.23 must also be altered slightly. The revised formulas are

$$SS_{total} = \sum_{i=1}^{k} \sum_{j=1}^{n_i} x_{ij}^2 - \frac{\left(\sum_{i=1}^{k} \sum_{j=1}^{n_i} x_{ij}\right)^2}{\sum_{i=1}^{k} n_i}$$

and

$$SS_{among} = \sum_{i=1}^{k} \left[\frac{\left(\sum_{j=1}^{n_i} x_{ij}\right)^2}{n_i}\right] - \frac{\left(\sum_{i=1}^{k} \sum_{j=1}^{n_i} x_{ij}\right)^2}{\sum_{i=1}^{k} n_i}$$

Example 3 The number of months survival after initiation of four different treatments on four different groups of patients suffering from a terminal disease are given below:

A	B	C	D
3	4	7	7
4	5	9	6
2	3	8	5
4	6		4
2	2		3
3			

We wish to determine if these four treatments are equally effective in prolonging life for patients suffering from this disease. We have $n_1 = 6$, $n_2 = n_4 = 5$, and $n_3 = 3$. Using the calculation formulas, we find

$$\sum_{i=1}^{4} \sum_{j=1}^{n_i} x_{ij} = 87$$

$$\left(\sum_{i=1}^{4} \sum_{j=1}^{n_i} x_{ij}\right)^2 = 7569$$

$$\sum_{i=1}^{4} \sum_{j=1}^{n_i} x_{ij}^2 = 477$$

$$\sum_{j=1}^{6} x_{1j} = 18 \qquad \frac{\left(\sum_{j=1}^{6} x_{1j}\right)^2}{6} = 54$$

$$\sum_{j=1}^{5} x_{2j} = 20 \qquad \frac{\left(\sum_{j=1}^{5} x_{2j}\right)^2}{5} = 80$$

$$\sum_{j=1}^{3} x_{3j} = 24 \qquad \frac{\left(\sum_{j=1}^{3} x_{3j}\right)^2}{3} = 192$$

$$\sum_{j=1}^{5} x_{4j} = 25 \qquad \frac{\left(\sum_{j=1}^{5} x_{4j}\right)^2}{5} = 125$$

Therefore,

$$SS_{total} = 477 - \frac{7569}{19} = 78.63$$

$$SS_{among} = (54 + 80 + 192 + 125) - \frac{7569}{19} = 52.63$$

$$SS_{within} = 78.63 - 52.63 = 26.00$$

So our sample statistic equals

$$F = \frac{SS_{among}/(k-1)}{SS_{within}/\left(\sum_{i=1}^{k} n_i - k\right)} = \frac{52.63/3}{26.0/(19-4)} = 10.12$$

which falls well within the critical region $F > F_{0.05, 3, 15} = 3.29$. We can therefore reject the null hypothesis that the four treatments are equally effective in prolonging life for these patients and conclude instead that at least one of these treatments is more (or less) effective than at least one of the others.

EXERCISES

1. There are four different routes that may be taken from city A to city B. In order to determine if the times it takes to travel each of these routes are approximately the same, each route was traversed four times, with the following results:

   Route #1: 20, 22, 19, 21 minutes
   Route #2: 23, 18, 19, 21 minutes
   Route #3: 17, 19, 21, 24 minutes
   Route #4: 18, 23, 19, 22 minutes

   What can be concluded from this data?

2. Based on the data given below, determine if the five different brands of gasoline deliver approximately the same mileage per gallon:

   Brand A: 15.6, 16.3, 12.3, 15.3, 15.7 mpg
   Brand B: 14.8, 15.9, 14.1, 16.0, 13.9 mpg
   Brand C: 17.1, 16.5, 14.9, 15.6, 15.2 mpg
   Brand D: 12.9, 14.5, 15.3, 13.9, 15.0 mpg
   Brand E: 15.8, 17.0, 15.6, 14.9, 16.4 mpg

3. Based on the samples below, determine if a student's grade point average is independent of year:

Freshmen: 3.5, 2.2, 4.0, 3.6, 2.9, 3.1, 3.4
Sophomores: 3.8, 2.4, 3.1, 3.5, 2.8, 3.9, 3.5
Juniors: 3.5, 3.2, 3.3, 2.9, 3.9, 3.5, 3.4
Seniors: 3.6, 3.2, 2.9, 2.4, 3.6, 3.7, 2.7, 2.9, 3.4

4. Based on the information given below, determine if the three different diets are equally effective weight reducers:

Diet 1: 5.6, 4.5, 3.7, 4.7, 5.1, 8.2, 4.6 pounds
Diet 2: 3.2, 3.3, 4.3, 4.5, 3.4, 5.6, 3.2 pounds
Diet 3: 4.5, 6.7, 2.8, 4.7 pounds

ADDITIONAL PROBLEMS

1. In a true-false test, a student answered 60 of the 100 questions correctly. Is it likely that the student was guessing?
2. Test the hypothesis that two students did equally well in all subjects, if their final grades in ten courses were:

Student 1: 80, 60, 85, 90, 85, 75, 80, 65, 70, 80
Student 2: 75, 80, 70, 86, 95, 60, 60, 90, 70, 75

3. In 600 throws of a pair of dice, sum 7 occurred 148 times, and sum 11 occurred 37 times. Should we suspect that the dice are loaded?
4. A company manufactures large numbers of washing machines which have an average lifetime of 4.5 years with a standard deviation of 1.2 years. How long a guarantee should the company offer if it plans to replace no more than five percent of its sales?
5. In order to answer the question of whether people who do not attend college marry earlier than people who do attend college, samples of 100 from each group were taken, with the following results: the average age at the time of marriage for the 100 people who did not attend college was 22.5 years with a standard deviation of 1.4 years, while the average age for the 100 who attended college was 23.0 years with a standard deviation of 1.8 years. What does this data imply?

Regression
and Correlation

11.1 Introduction to Regression

One of the more interesting problems in statistics is to determine if some form of relationship might exist among two or more variables. The objective in determining such a relationship, of course, would be the ability to predict the values of one (dependent) variable based on the value(s) assumed by the other (independent) variable(s). For example, one might want to be able to predict a person's weight in terms of his height, a product's potential sales in terms of the amount of money spent for advertising, a student's success in college based on his high school ranking and his score on a college placement exam, or a thoroughbred's success in today's race based on the number of days since the horse last raced, the weight the horse must carry, and the post position.

The first two example above are cases of what is called *simple regression*, where one variable is predicted in terms of one other variable. The other two examples are cases of what is called *multiple regression*, where one variable is predicted in terms of several other variables. In this book, we will restrict ourselves to the study of simple regression.

If the functional relationship relating the dependent variable Y to the independent variable X is of the straight line variety

$$Y = \alpha + \beta X$$

then we have what is called a *linear regression* of Y on X. We shall restrict our study, for the most part, to this simple case, although we shall at times explain how one might proceed should the relationship between X and Y turn out to be, say, of the polynomial or exponential variety.

It would be ideal, of course, if we could predict the exact value of Y corresponding to any given value of X. Unfortunately, this will not be possible very often, and we will usually have to be satisfied with predicting the average value of Y corresponding to any given value of X. If the joint density of X and Y is completely known, this can usually be done fairly easily. If $f(x, y)$ is the joint density of X and Y, then we can find (see Section 8.2) the conditional density $h_2(y|x_0)$ of Y, given $X = x_0$, and, in turn, the regression equation

$$E(Y|x_0) = \int_{-\infty}^{\infty} y h_2(y|x_0)\, dy$$

On the other hand, if the joint density of X and Y is not completely known, then we are faced with the problem of estimating the parameters α and β in the regression equation. How this might be done in the case of linear regression will be the topic of the next section.

But, for now, let us suppose that the joint density $f(x, y)$ is completely known, and let us also suppose that we have a linear regression equation

$$E(Y|x) = \alpha + \beta x$$

The constants α and β are called the *regression coefficients*, and we are now going to show how they can be expressed in terms of the means, standard deviations, and convariance of the random variables X and Y. As our starting point, we have

$$\int_{-\infty}^{\infty} y h_2(y|x)\, dy = \alpha + \beta x$$

If $f_X(x)$ denotes the (marginal) density of X, then upon multiplying both sides of the above equation by $f_X(x)$, and integrating with respect to x, we have

$$\int_{-\infty}^{\infty} \int_{-\infty}^{\infty} y h_2(y|x) f_X(x)\, dy\, dx = \int_{-\infty}^{\infty} [\alpha f_X(x) + \beta x f_X(x)]\, dx$$

But $f_X(x) h_2(y|x) = f(x, y)$, so we have

$$\int_{-\infty}^{\infty} \int_{-\infty}^{\infty} y f(x, y)\, dy\, dx = \alpha \int_{-\infty}^{\infty} f_X(x)\, dx + \beta \int_{-\infty}^{\infty} x f_X(x)\, dx$$

which is equivalent to

$$\int_{-\infty}^{\infty} y \left(\int_{-\infty}^{\infty} f(x, y)\, dx \right) dy = \alpha \cdot 1 + \beta \cdot E(X)$$

or

$$\int_{-\infty}^{\infty} y f_Y(y)\, dy = \alpha + \beta \cdot E(X)$$

which implies that $E(Y) = \alpha + \beta \cdot E(X)$.

If we had multiplied our original equation by $x f_X(x)$ instead, we would have obtained

$$\int_{-\infty}^{\infty} \int_{-\infty}^{\infty} xy f(x, y)\, dy\, dx = \alpha \int_{-\infty}^{\infty} x f_X(x)\, dx + \beta \int_{-\infty}^{\infty} x^2 f_X(x)\, dx$$

which is equivalent to

$$E(XY) = \alpha \cdot E(X) + \beta \cdot E(X^2)$$

We therefore have the two equations

$$\left\{ \begin{array}{l} E(Y) = \alpha + \beta \cdot E(X) \\ E(XY) = \alpha \cdot E(X) + \beta \cdot E(X^2) \end{array} \right\}$$

in the two unknowns α and β. Solving simultaneously, we find

$$\beta = \frac{1 \cdot E(XY) - E(X) \cdot E(Y)}{\mathrm{Var}(X)} = \frac{\mathrm{Cov}(X, Y)}{\mathrm{Var}(X)}$$

and

$$\alpha = \frac{E(Y) \cdot E(X^2) - E(X) \cdot E(XY)}{\mathrm{Var}(X)}$$

Using the equations $E(X^2) = \mathrm{Var}(X) + E(X)^2$ and $E(XY) = \mathrm{Cov}(X, Y) + E(X) \cdot E(Y)$, we can transform the solution for α into

$$\alpha = E(Y) - \frac{\mathrm{Cov}(X, Y)}{\mathrm{Var}(X)} \cdot E(X)$$

Our regression equation then becomes

$$E(Y|x) = E(Y) - \frac{\text{Cov}(X, Y)}{\text{Var}(X)} \cdot E(X) + \frac{\text{Cov}(X, Y)}{\text{Var}(X)} \cdot x$$

or

$$E(Y|x) = E(Y) + \frac{\text{Cov}(X, Y)}{\text{Var}(X)} \cdot [x - E(X)]$$

Since the correlation $\rho(X, Y) = \text{Cov}(X, Y)/\sigma_X \sigma_Y$, we can rewrite this regression equation as

$$E(Y|x) = E(Y) + \rho(X, Y)\left(\frac{\sigma_Y}{\sigma_X}\right)[x - E(X)]$$

One particularly interesting situation arises when the joint density $f(x, y)$ of the two random variables X and Y has what is called the bivariate normal distribution. In this case, the regression curves of Y on X, and of X on Y, will always be linear, and the theory of normal regression and correlation that results is quite beautiful. We begin our discussion with the following definition:

DEFINITION

The joint probability function defined by

$$f(x, y) = \frac{\exp\left\{\dfrac{-1}{2(1 - \rho^2)}\left[\left(\dfrac{x - \mu_1}{\sigma_1}\right)^2 - 2\rho\left(\dfrac{x - \mu_1}{\sigma_1}\right)\left(\dfrac{y - \mu_2}{\sigma_2}\right) + \left(\dfrac{y - \mu_2}{\sigma_2}\right)^2\right]\right\}}{2\pi\sigma_1\sigma_2\sqrt{1 - \rho^2}}$$

where $-\infty < x, y < +\infty, \sigma_1, \sigma_2 > 0$, and $-1 < \rho < 1$, is called the *bivariate normal distribution.*

Obviously, $f(x, y) > 0$ for all possible choices of x and y. It is a straightforward, though tedious, task to verify that

$$\int_{-\infty}^{\infty} \int_{-\infty}^{\infty} f(x, y)\, dx\, dy = 1$$

using two separate applications of Proposition 8.9. First, there is the change of variables

$$U = \frac{X - \mu_1}{\sigma_1} \quad \text{and} \quad V = \frac{Y - \mu_2}{\sigma_2}$$

and then, the second change of variables

$$T = \frac{U - \rho \cdot V}{\sqrt{1 - \rho^2}} \quad \text{and} \quad S = V$$

We will leave the details as an exercise for the interested reader.

We are particularly interested in the marginal densities of X and Y and the regression curves of Y on X and X on Y. Now the marginal density of

X is defined to equal

$$f_X(x) = \int_{-\infty}^{\infty} f(x, y)\, dy$$

$$= \frac{1}{2\pi\sigma_1\sigma_2\sqrt{1 - \rho^2}} \int_{-\infty}^{\infty}$$

$$\times \exp\left\{\frac{-1}{2(1 - \rho^2)}\left[\left(\frac{x - \mu_1}{\sigma_1}\right)^2 - 2\rho\left(\frac{x - \mu_1}{\sigma_1}\right)\left(\frac{y - \mu_2}{\sigma_2}\right) + \left(\frac{y - \mu_2}{\sigma_2}\right)^2\right]\right\} dy$$

Under the change of variables

$$v = \frac{y - \mu_2}{\sigma_2} \quad \text{and} \quad dv = \frac{dy}{\sigma_2}$$

this integral becomes

$$\frac{1}{2\pi\sigma_1\sqrt{1 - \rho^2}} \int_{-\infty}^{\infty} \exp\left\{\frac{-1}{2(1 - \rho^2)}\left[\left(\frac{x - \mu_1}{\sigma_1}\right)^2 - 2\rho v\left(\frac{x - \mu_1}{\sigma_1}\right) + v^2\right]\right\} dv$$

Completing the square in the exponent, we have

$$v^2 - 2\rho\left(\frac{x - \mu_1}{\sigma_1}\right)v + \rho^2\left(\frac{x - \mu_1}{\sigma_1}\right)^2 - \rho^2\left(\frac{x - \mu_1}{\sigma_1}\right)^2 + \left(\frac{x - \mu_1}{\sigma_1}\right)^2$$

which equals

$$\left[v - \rho\left(\frac{x - \mu_1}{\sigma_1}\right)\right]^2 + (1 - \rho^2)\left(\frac{x - \mu_1}{\sigma_1}\right)^2$$

This transforms the integral into

$$\frac{1}{2\pi\sigma_1\sqrt{1 - \rho^2}} \int_{-\infty}^{\infty} \exp\left\{\frac{1}{2(1 - \rho^2)}\left[v - \rho\left(\frac{x - \mu_1}{\sigma_1}\right)^2 + (1 - \rho^2)\left(\frac{x - \mu_1}{\sigma_1}\right)^2\right]\right\} dv$$

which, upon rearranging terms, equals

$$\frac{\exp\left[-\frac{1}{2}\left(\frac{x - \mu_1}{\sigma_1}\right)^2\right]}{\sigma_1\sqrt{2\pi}} \int_{-\infty}^{\infty} \frac{1}{\sqrt{2\pi}\sqrt{1 - \rho^2}} \exp\left\{-\frac{1}{2}\left[\frac{v - \rho\left(\frac{x - \mu_1}{\sigma_1}\right)}{\sqrt{1 - \rho^2}}\right]^2\right\} dv$$

Under the further change of variables

$$t = \frac{v - \rho\left(\frac{x - \mu_1}{\sigma_1}\right)}{\sqrt{1 - \rho^2}}, \quad dt = \frac{dv}{\sqrt{1 - \rho^2}}$$

this integral becomes

$$\frac{\exp\left[-\frac{1}{2}\left(\frac{x - \mu_1}{\sigma_1}\right)^2\right]}{\sigma_1\sqrt{2\pi}} \int_{-\infty}^{\infty} \frac{1}{\sqrt{2\pi}} e^{-t^2/2}\, dt$$

But this last integral equals 1, being the total area under the standard normal curve, so that

$$f_X(x) = \frac{\exp\left[-\frac{1}{2}\left(\frac{x-\mu_1}{\sigma_1}\right)^2\right]}{\sigma_1\sqrt{2\pi}}$$

So the marginal density of X is normal with mean μ_1 and standard deviation σ_1. Likewise, the marginal density of Y will be

$$f_Y(y) = \frac{\exp\left[-\frac{1}{2}\left(\frac{y-\mu_2}{\sigma_2}\right)^2\right]}{\sigma_2\sqrt{2\pi}}$$

which is normal with mean μ_2 and standard deviation σ_2.

The conditional density of Y, given X, is defined to equal

$$h_2(y\,|\,x) = \frac{f(x, y)}{f_X(x)}$$

and so we have

$$h_2(y\,|\,x) = \frac{\dfrac{\exp\left\{\dfrac{-1}{2(1-\rho^2)}\left[\left(\dfrac{x-\mu_1}{\sigma_1}\right)^2 - 2\rho\left(\dfrac{x-\mu_1}{\sigma_1}\right)\left(\dfrac{y-\mu_2}{\sigma_2}\right) + \left(\dfrac{y-\mu_2}{\sigma_2}\right)^2\right]\right\}}{2\pi\sigma_1\sigma_2\sqrt{1-\rho^2}}}{\dfrac{\exp\left[-\dfrac{1}{2}\left(\dfrac{x-\mu_1}{\sigma_1}\right)^2\right]}{\sigma_1\sqrt{2\pi}}}$$

which equals

$$h_2(y\,|\,x) = \frac{\exp\left[\dfrac{-1}{2(1-\rho^2)}\left\{\rho^2\left(\dfrac{x-\mu_1}{\sigma_1}\right)^2 - 2\rho\left(\dfrac{x-\mu_1}{\sigma_1}\right)\left(\dfrac{y-\mu_2}{\sigma_2}\right) + \left(\dfrac{y-\mu_2}{\sigma_2}\right)^2\right\}\right]}{\sigma_2\sqrt{2\pi}\sqrt{1-\rho^2}}$$

This is equivalent to

$$h_2(y\,|\,x) = \frac{\exp\left\{-\dfrac{1}{2}\left[\dfrac{\left(\dfrac{y-\mu_2}{\sigma_2}\right) - \rho\left(\dfrac{x-\mu_1}{\sigma_1}\right)}{\sqrt{1-\rho^2}}\right]^2\right\}}{\sigma_2\sqrt{2\pi}\sqrt{1-\rho^2}}$$

which, in turn, is equal to

$$h_2(y\,|\,x) = \frac{\exp\left\{-\dfrac{1}{2}\left[\dfrac{y-(\mu_2+\rho(\sigma_2/\sigma_1)(x-\mu_1))}{\sigma_2\sqrt{1-\rho^2}}\right]^2\right\}}{\sigma_2\sqrt{2\pi}\sqrt{1-\rho^2}}$$

Consequently, the conditional expectation of Y, given $X = x_0$, equals

$$E(Y|x_0) = \int_{-\infty}^{\infty} y h_2(y|x_0)\,dy$$

$$= \frac{1}{\sigma_2\sqrt{2\pi}\sqrt{1-\rho^2}} \int_{-\infty}^{\infty} y \cdot \exp\left\{-\frac{1}{2}\left[\frac{y - (\mu_2 + \rho(\sigma_2/\sigma_1)(x_0 - \mu_1))}{\sigma_2\sqrt{1-\rho^2}}\right]^2\right\} dy$$

Under the change of variables

$$u = \frac{y - \mu_2 + \rho(\sigma_2/\sigma_1)(x_0 - \mu_1)}{\sqrt{\sigma_2^2(1-\rho^2)}}$$

$$du = \frac{dy}{\sqrt{\sigma_2^2(1-\rho^2)}}$$

we have

$$y = \sqrt{\sigma_2^2(1-\rho^2)} \cdot u + \left[\mu_2 + \rho\left(\frac{\sigma_2}{\sigma_1}\right)(x_0 - \mu_1)\right]$$

and the integral becomes

$$\frac{1}{\sqrt{2\pi}} \int_{-\infty}^{\infty} \left\{\sqrt{\sigma_2^2(1-\rho^2)} \cdot u + \left[\mu_2 + \rho\left(\frac{\sigma_2}{\sigma_1}\right)(x_0 - \mu_1)\right]\right\} e^{-u^2/2}\,du$$

Separating into two integrals, we have

$$\frac{\sqrt{\sigma_2^2(1-\rho^2)}}{\sqrt{2\pi}} \int_{-\infty}^{\infty} u e^{-u^2/2}\,du + \frac{\mu_2 + \rho(\sigma_2/\sigma_1)(x_0 - \mu_1)}{\sqrt{2\pi}} \int_{-\infty}^{\infty} e^{-u^2/2}\,du$$

The first of these integrals vanishes since the integrand is an odd function. The second integral equals 1 since the integrand is a probability density. Therefore,

$$E(Y|x_0) = \mu_2 + \rho\left(\frac{\sigma_2}{\sigma_1}\right)(x_0 - \mu_1)$$

which establishes the fact that the regression curve of Y on X is linear. In a similar manner, we can show that

$$E(X|y_0) = \mu_1 + \rho\left(\frac{\sigma_1}{\sigma_2}\right)(y_0 - \mu_2)$$

so that the regression curve of X on Y is also linear.
We will return to the bivariate normal density in Section 11.4.

11.2 Method of Least Squares

If the joint distribution of our two random variables X and Y is not completely known (one or more of its defining parameters may be unknown), then we must use the tactics of statistical estimation to determine (if possible) the form of the regression curve of Y on X (or X on Y). Our starting point, as usual, will be observed data, which in this case will take the form of a bivariate sample of size n. By this we mean a sampling of n pairs (x_i, y_i), where the first coordinate is sampled from the random variable X and the second coordinate is sampled from the random variable Y. For example, the x_i might be a person's height and the y_i the weight of the same person. We will always assume that the first coordinate corresponds to the independent variable.

Having obtained our bivariate sample and having plotted the n ordered pairs of the sample (such a graph is referred to as a *scatter diagram*), we may notice that a certain type of curve, such as a straight line, a quadratic, or an exponential describes, or "fits," our data fairly well. For the sake of explaining the techniques that follow, we will assume that a straight line appears to give a fairly good approximation to our sample data. This leads us to suspect that there is a linear relationship

$$Y = \alpha + \beta X$$

existing between our random variables. We can only hope to approximate this line, and the way we shall do this is by determining the equation of the line

$$y = a + bx$$

which "best fits" our sample data. We shall employ what is known as the

Figure 11.1

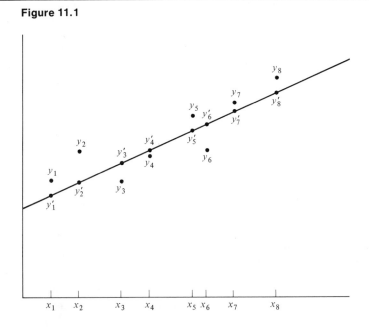

method of least squares, which dates back to the French mathematician Legendre.

Corresponding to each of our n values x_i, we have the observed y-value y_i from our bivariate sample. If Y is a (linear) function of X (as above), we can also associate with each x_i an expected y value

$$y_i' = a + bx_i$$

The difference between y_i and y_i' is illustrated in the scatter diagram in Figure 11.1. Notice that each of the y_i' values lie on the straight line $y = a + bx$, as the definition indicates they should. We note that there is nothing mathematical involved (yet) in the drawing of the line $y = a + bx$ in Figure 11.1. It is drawn simply to please the author's eye, because it appears to approximate the data as well as any other straight line might.

How do we determine the coefficients a and b that define the line which best fits our data? Obviously, we would like to find that line which reduces the error terms $y_i - y_i'$, the differences between the observed and expected y-values, to a minimum. For reasons of mathematical expedience, it turns out to be wiser to try to minimize the squared differences $(y_i - y_i')^2$ instead. This leads us to the criterion of the least-squares method:

DEFINITION

The line $y = a + bx$ which minimizes the term

$$\sum_{i=1}^{n} (y_i - y_i')^2$$

is called the *least-squares line*.

We will see in Section 11.4 a partial justification for this approach, in that the least-squares line, under the assumption that the joint density of X and Y is bivariate normal, does give the best possible fit for the observed data.

Since $y_i' = a + bx_i$, we must, according to the least-squares criterion, minimize

$$\sum_{i=1}^{n} [y_i - (a + bx_i)]^2$$

This latter form has the added attraction of bringing the unknowns a and b into play. To determine which values of a and b minimize this summation, we take the partial derivatives of the summation with respect to a and b, set them equal to 0, and solve for a and b:

$$\frac{\delta}{\delta a}\left(\sum_{i=1}^{n} [y_i - (a + bx_i)]^2\right) = \sum_{i=1}^{n} 2[y_i - (a + bx_i)](-1)$$

$$= -2\sum_{i=1}^{n} [y_i - (a + bx_i)] = 0$$

and

$$\frac{\delta}{\delta b}\left(\sum_{i=1}^{n} [y_i - (a + bx_i)]^2\right) = \sum_{i=1}^{n} 2[y_i - (a + bx_i)](-x_i)$$

$$= -2 \sum_{i=1}^{n} x_i[y_i - (a + bx_i)] = 0$$

These two equations are equivalent to

$$\sum_{i=1}^{n} y_i = \sum_{i=1}^{n} (a + bx_i)$$

and

$$\sum_{i=1}^{n} x_i y_i = \sum_{i=1}^{n} (ax_i + bx_i^2)$$

which, in turn, are equivalent to

$$\sum_{i=1}^{n} y_i = na + \left(\sum_{i=1}^{n} x_i\right) \cdot b$$

and

$$\sum_{i=1}^{n} x_i y_i = \left(\sum_{i=1}^{n} x_i\right) \cdot a + \left(\sum_{i=1}^{n} x_i^2\right) \cdot b$$

These last two equations, which both contain the unknowns a and b, are called the *normal equations*. Their exact form can easily be remembered in the following way: assuming that the regression line gives a perfect fit, we would have

$$y_i = a + bx_i \quad (i = 1, 2, \ldots, n)$$

Add these n equations, and you have the first of the normal equations. Then multiply each of these equations by the corresponding x_i to obtain the n equations

$$x_i y_i = ax_i + bx_i^2 \quad (i = 1, 2, \ldots, n)$$

Add these n equations, and you obtain the second of the normal equations.
Solving the normal equations simultaneously for a and b, we obtain

$$a = \frac{\left(\sum_{i=1}^{n} y_i\right)\left(\sum_{i=1}^{n} x_i^2\right) - \left(\sum_{i=1}^{n} x_i\right)\left(\sum_{i=1}^{n} x_i y_i\right)}{n \sum_{i=1}^{n} x_i^2 - \left(\sum_{i=1}^{n} x_i\right)^2}$$

and

$$b = \frac{n \sum_{i=1}^{n} x_i y_i - \left(\sum_{i=1}^{n} x_i\right)\left(\sum_{i=1}^{n} y_i\right)}{n \sum_{i=1}^{n} x_i^2 - \left(\sum_{i=1}^{n} x_i\right)^2}$$

Example 1 Suppose that we wish to determine if there is a linear relationship between a student's IQ and a student's final college grade point average (GPA). Suppose that a small sample of ten students is taken, with the following results:

IQ	GPA
120	3.6
105	2.2
130	3.7
117	3.1
119	3.4
122	3.6
109	2.4
121	3.2
127	3.9
115	2.9

We wish to determine the equation of the straight line $y = a + bx$ which best fits our data, and therefore is best suited for use in predicting a student's final GPA in terms of the student's IQ. We group our calculations into the following table, where x_i denotes the IQ and y_i the corresponding GPA:

x_i	y_i	$x_i y_i$	x_i^2	y_i^2
120	3.6	432.0	14,400	12.96
105	2.2	231.0	11,025	4.84
130	3.7	481.0	16,900	13.69
117	3.1	362.7	13,689	9.61
119	3.4	404.6	14,161	11.56
122	3.6	439.2	14,884	12.96
109	2.4	261.6	11,881	5.76
121	3.2	387.2	14,641	10.24
127	3.9	495.3	16,129	15.21
115	2.9	333.5	13,225	8.41
1,185	32.0	3,828.1	140,935	105.24

Consequently,

$$a = \frac{32 \cdot 140{,}935 - 1{,}185 \cdot 3{,}828.1}{10 \cdot 140{,}935 - (1{,}185)^2} = -5.147$$

and

$$b = \frac{10 \cdot 3{,}828.1 - 1{,}185 \cdot 32}{10 \cdot 140{,}935 - (1{,}185)^2} = 0.07$$

and hence, the equation of the least-squares regression line is

$$y = 0.07x - 5.147$$

Using this equation, we can predict that a student with an IQ of 130 should have a final GPA of $0.07 \cdot 130 - 5.147 = 3.953$, while a student with an IQ of 110 should have a final GPA of $0.07 \cdot 110 - 5.147 = 2.553$.

Example 2 In this example we describe what in marine biology is known as the Leslie method, used in estimating hard clam abundance in Chesapeake Bay. The hard-clam abundance is to be estimated on the basis of a limited number of observations of the number of bushels harvested per hour by means of a hydraulic escalator. The catch per unit effort C_t (the number of bushels per hour) at time t is regressed on the cumulative catch K_t prior to time t. The regression line will be in the form

$$C_t = a - cK_t$$

where the first degree coefficient $-c$, termed *catchability*, will be negative since the number of bushels harvested at any given time decreases with time as more and more of the clam population has already been harvested. Factoring the constant term a into the form cN, we have

$$C_t = cN - cK_t = c(N - K_t)$$

Therefore, at time $t = 1$, when we have $K_t = 0$, we find that $C_1 = cN$. Hence, if the catchability c remains constant throughout the process (this is the usual assumption), the constant N represents an estimation, based on the regression line, of the size of the clam population. For example, suppose that over a five-hour time period, we obtained the following results:

t	K_t	C_t
1	0.0	3.2
2	3.2	2.8
3	6.0	2.4
4	8.4	2.2
5	10.6	1.8

Our regression table of calculations looks like:

K_t	C_t	$K_t C_t$	K_t^2	C_t^2
0.00	3.20	0.00	0.00	10.24
3.20	2.80	8.96	10.24	7.84
6.00	2.40	14.40	36.00	5.76
8.40	2.20	18.48	70.56	4.84
10.60	1.80	19.08	112.36	3.24
28.20	12.40	60.92	229.16	31.92

Our regression coefficients therefore equal

$$a = \frac{12.4 \cdot 229.16 - 28.2 \cdot 60.92}{5 \cdot 229.16 - (28.2)^2} = 3.205$$

and

$$b = \frac{5 \cdot 60.92 - 28.2 \cdot 12.4}{5 \cdot 229.16 - (28.2)^2} = -0.129$$

so that the least-squares regression line is

$$C_t = 3.205 - 0.129 K_t$$

or

$$C_t = 0.129(24.845 - K_t)$$

Therefore, our estimation of the population size is 24.845 bushels.

The regression line $y = a + bx$ can be put into an equivalent form which we will find useful in the next section.

Proposition 11.1

The regression line $y = a + bx$ is equivalent to

$$y - \bar{y} = b(x - \bar{x})$$

where b can be found by the formula

$$\frac{\sum\limits_{i=1}^{n} (x_i - \bar{x})(y_i - \bar{y})}{\sum\limits_{i=1}^{n} (x_i - \bar{x})^2}$$

Proof: According to the first normal equation, we have

$$\sum_{i=1}^{n} y_i = na + b\left(\sum_{i=1}^{n} x_i\right)$$

Dividing through this equation by n, we find that

$$\bar{y} = a + b\bar{x}$$

Subtracting this from the original equation, we have

$$y - \bar{y} = b(x - \bar{x})$$

We have seen that

$$b = \frac{n \sum\limits_{i=1}^{n} x_i y_i - \left(\sum\limits_{i=1}^{n} x_i\right)\left(\sum\limits_{i=1}^{n} y_i\right)}{n \sum\limits_{i=1}^{n} x_i^2 - \left(\sum\limits_{i=1}^{n} x_i\right)^2}$$

Letting $x_i^* = x_i - \bar{x}$ and $y_i^* = y_i - \bar{y}$, we have

$$b = \frac{n \sum_{i=1}^{n} (x_i^* + \bar{x})(y_i^* + \bar{y}) - \left(\sum_{i=1}^{n} (x_i^* + \bar{x}) \right)\left(\sum_{i=1}^{n} (y_i^* + \bar{y}) \right)}{n \sum_{i=1}^{n} (x_i^* + \bar{x})^2 - \left(\sum_{i=1}^{n} (x_i^* + \bar{x}) \right)^2}$$

Expanding these products, we find that

$$b = \frac{n \sum_{i=1}^{n} x_i^* y_i^* + n\bar{y} \sum_{i=1}^{n} x_i^* + n\bar{x} \sum_{i=1}^{n} y_i^* + n^2\bar{x}\bar{y} - \left(\sum_{i=1}^{n} x_i^* + n\bar{x} \right)\left(\sum_{i=1}^{n} y_i^* + n\bar{y} \right)}{n \sum_{i=1}^{n} x_i^{*2} + 2n\bar{x} \sum_{i=1}^{n} x_i^* + n^2\bar{x}^2 - \left(\sum_{i=1}^{n} x_i^* + n\bar{x} \right)^2}$$

However,

$$\sum_{i=1}^{n} x_i^* = \sum_{i=1}^{n} (x_i - \bar{x}) = 0$$

and

$$\sum_{i=1}^{n} y_i^* = \sum_{i=1}^{n} (y_i - \bar{y}) = 0$$

Therefore, our expression for b simplifies to

$$b = \frac{n \sum_{i=1}^{n} x_i^* y_i^* + n^2\bar{x}\bar{y} - n^2\bar{x}\bar{y}}{n \sum_{i=1}^{n} x_i^{*2} + n^2\bar{x}^2 - n^2\bar{x}^2} = \frac{\sum_{i=1}^{n} x_i^* y_i^*}{\sum_{i=1}^{n} x_i^{*2}}$$

Since $x_i^* = x_i - \bar{x}$ and $y_i^* = y_i - \bar{y}$, we have

$$b = \frac{\sum_{i=1}^{n} (x_i - \bar{x})(y_i - \bar{y})}{\sum_{i=1}^{n} (x_i - \bar{x})^2}$$

as required.

As a Corollary to the first few steps of our proof, we have:

Corollary 11.2

The point $(\bar{x}, \bar{y})$ lies on the regression line $y = a + bx$.

This result allows us to check if the coefficients a and b have been calculated correctly. If $a + b\bar{x}$ does not equal $\bar{y}$, then an arithmetic error has probably been made.

The techniques we have discussed in this section can be modified slightly to handle cases where the regression curve cannot be assumed to be linear.

If, for example, the regression curve is assumed to be the quadratic

$$y = a + bx + cx^2$$

we would define

$$y_i' = a + bx_i + cx_i^2$$

to be the expected y-value corresponding to the x-value x_i. We would then attempt, once again, to minimize the summation

$$\sum_{i=1}^{n} (y_i - y_i')^2$$

Partial differentiation with respect to a, b, and c would then lead to three normal equations in the three unknowns a, b, and c:

$$\sum_{i=1}^{n} y_i = na + \left(\sum_{i=1}^{n} x_i \right) b + \left(\sum_{i=1}^{n} x_i^2 \right) c$$

$$\sum_{i=1}^{n} x_i y_i = \left(\sum_{i=1}^{n} x_i \right) a + \left(\sum_{i=1}^{n} x_i^2 \right) b + \left(\sum_{i=1}^{n} x_i^3 \right) c$$

$$\sum_{i=1}^{n} x_i^2 y_i = \left(\sum_{i=1}^{n} x_i^2 \right) a + \left(\sum_{i=1}^{n} x_i^3 \right) b + \left(\sum_{i=1}^{n} x_i^4 \right) c$$

The mnemonic device mentioned above for remembering the form of the normal equations in the linear regression case can also be used in the quadratic case (and, as a matter of fact, in the case of a polynomial of arbitrary degree n). Simply think of the normal equations as arising from the addition, in turn, of n equations of the forms

$$y_i = a + bx_i + cx_i^2$$
$$x_i y_i = ax_i + bx_i^2 + cx_i^3$$
$$x_i^2 y_i = ax_i^2 + bx_i^3 + cx_i^4$$

As you can probably guess, a regression polynomial of degree $n - 1$ would lead to a system of n equations in n unknowns (the n coefficients of the polynomial). A computer program that solves linear systems, or at least inverts $n \times n$ matrices, would almost be mandatory.

Example 3 According to U.S. Department of Health, Education, and Welfare statistics, the number of births per each 1000 people in the United States during the years 1915–1955 were as follows:

1915	1920	1925	1930	1935	1940	1945	1950	1955
25.0	23.7	21.3	18.9	16.9	17.9	19.5	23.6	24.6

We would like to use this data to predict the birth rate in 1960. Consequently, we would like to express the birth rate as a function of time, and to do this, we would like to use some form of regression. However, a quick look at the data (see Figure 11.2) suggests that linear regression would not be appropriate. Due to the effects of the two world wars and the depression, the birth rate dropped steadily from 1915 until 1940, when it started an ascent. Clearly, the graph of this data would not look like a straight line, but rather is more in the form of a parabola. So we will use the method of least squares to determine which parabola

$$y = a + bx + cx^2$$

best approximates our sample data. To simplify our calculations, we will transform the values of our independent variable (the year) as follows:

$$x' = \frac{x - 1935}{5}$$

In essence, we have placed the origin at 1935, in such a way that all of our x-values are simple, small integers, and in such a way that guarantees that

$$\sum_{i=1}^{n} x_i = 0 \qquad \text{and} \qquad \sum_{i=1}^{n} x_i^3 = 0$$

Figure 11.2

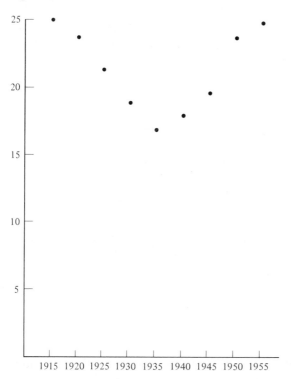

as the following tables indicates:

x_i	y_i	x_i^2	x_i^3	x_i^4	$x_i y_i$	$x_i^2 y_i$
-4	25.0	16	-64	256	-100.0	400.0
-3	23.7	9	-27	81	-71.1	213.3
-2	21.3	4	-8	16	-42.6	85.2
-1	18.9	1	-1	1	-18.9	18.9
0	16.9	0	0	0	0.0	0.0
1	17.9	1	1	1	17.9	17.9
2	19.5	4	8	16	39.0	78.0
3	23.6	9	27	81	70.8	212.4
4	24.6	16	64	256	98.4	393.6
0	191.4	60	0	708	-6.5	1419.3

Substituting these values into our three normal equations, we obtain

$$191.4 = 9a + 60c$$
$$-6.5 = 60b$$
$$1419.3 = 60a + 708c$$

The second equation gives $b = -6.5/60 = -0.1083$. Solving the first equation for a, and substituting into the third equation, we obtain

$$1419.3 = 1276 - 400c + 708c = 1276 + 308c$$

which gives $c = 143.3/308 = 0.4653$. Substituting this value into the first equation, we obtain

$$191.4 = 9a + 27.918$$

which gives $a = 163.482/9 = 18.1647$. Therefore, the equation of the least-squares parabola is

$$y = 18.1647 - 0.1083x + 0.4653x^2$$

To obtain our prediction for 1960, we then need only substitute the value $(1960 - 1935)/5 = 5$ into this equation:

$$y_{1960} = 18.1647 - 0.1083 \cdot 5 + 0.4653 \cdot 25$$
$$= 29.2556 \text{ births per 1000 people}$$

We should point out the danger of "extrapolation" here. The regression curve we obtain provides us with the best fit (by a curve of that type) over the range of x-values specified by the data. Should we attempt to use this curve to predict a y-value corresponding to an x-value outside the range of the data, we have no guarantee of the accuracy of our prediction. We will have more to say about this problem in Section 11.4.

Let us now suppose that our regression curve appears to be an exponential

$$y = ae^{bx}$$

as would be the case when linear changes in the independent variable produce proportional changes in the dependent variable. Applying a logarithmic transformation to this regression curve, we obtain

$$\log_e y = \log_e ae^{bx} = \log_e a + bx$$

Therefore, when y is an exponential function of x, $\log_e y$ is a linear function of x. We can then use the techniques of linear regression to determine the coefficients $\log_e a$ and b of the least squares line expressing $\log_e y$ as a function of x. The normal equations in this case would be

$$\sum_{i=1}^{n} \log_e y_i = n \log_e a + \left(\sum_{i=1}^{n} x_i \right) b$$

and

$$\sum_{i=1}^{n} x_i \log_e y_i = \left(\sum_{i=1}^{n} x_i \right) \log_e a + \left(\sum_{i=1}^{n} x_i^2 \right) b$$

which have solutions

$$b = \frac{n \sum_{i=1}^{n} x_i \log_e y_i - \left(\sum_{i=1}^{n} x_i \right) \left(\sum_{i=1}^{n} \log_e y_i \right)}{n \sum_{i=1}^{n} x_i^2 - \left(\sum_{i=1}^{n} x_i \right)^2}$$

and, using the first normal equation,

$$\log_e a = \frac{\sum_{i=1}^{n} \log_e y_i - b \sum_{i=1}^{n} x_i}{n}$$

where the value of b determined above is then used to determine the value of $\log_e a$. Using an exponential transformation, we find that

$$a = \exp \left[\frac{\left(\sum_{i=1}^{n} \log_e y_i - b \sum_{i=1}^{n} x_i \right)}{n} \right]$$

giving us the coefficients a and b of the least-squares exponential expressing y as a function of x.

Example 4 According to U.S. Census Bureau figures, the population of the United States grew during the twentieth century as indicated in the following table:

1900	76 million
1910	92 million
1920	106 million
1930	123 million
1940	132 million
1950	151 million
1960	179 million
1970	203 million

We would like to use this information to project population figures in 1980, 1990, and 2000. Looking at a scatter diagram of this data (Figure 11.3), one might suspect that a linear approximation would be in order. However, noting the effects of the two world wars and the depression once again, and also noting the post-World War II population boom, one could certainly justify an exponential approximation. As far as projecting population figures beyond 1970 is concerned, the exponential approximation would appear to be the proper choice, because the population growth in the years 1950–1970

Figure 11.3

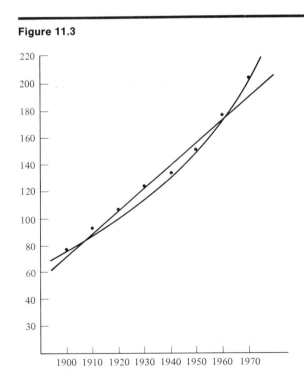

seems more exponential than linear. Placing the origin at 1900 and letting one unit represent ten years (to simplify calculations), we proceed as described above to find the least squares exponential curve $y = ae^{bx}$. All calculations are included in the table below:

x_i	x_i^2	y_i	$\log_e y_i$	$x_i \log_e y_i$
0	0	76	4.33	0.00
1	1	92	4.52	4.52
2	4	106	4.66	9.32
3	9	123	4.81	14.43
4	16	132	4.88	19.52
5	25	151	5.02	25.10
6	36	179	5.19	31.14
7	49	203	5.31	37.17
28	140		38.73	141.20

We then calculate

$$b = \frac{8 \cdot 141.20 - 28 \cdot 38.73}{8 \cdot 140 - 28 \cdot 28} = 0.134$$

and then

$$\log_e a = \frac{38.73 - 0.134 \cdot 28}{8} = 4.37$$

so that $a = e^{4.37} = 79.22$. Therefore, the least-squares exponential curve is

$$y = 79.22e^{0.134x}$$

According to this curve, the population figures during 1900–1970 should have been:

1900	79.22 million
1910	90.58 million
1920	103.57 million
1930	118.42 million
1940	135.40 million
1950	154.81 million
1960	177.01 million
1970	202.40 million

(As an exercise, the interested reader should perform a χ^2 test comparing this table with the preceding table to determine how good a fit our exponential curve gives. We shall discuss another method for determining how good the fit is in the next section.) We could then estimate the population in 1945 (which corresponds to $x = 4.5$) to have been

$$y_{1945} = 79.22e^{0.134 \cdot 4.5} = 144.78 \text{ million}$$

We can project the population in 1980, 1990, and 2000 (which correspond to $x = 8, 9,$ and 10) to be

$$y_{1980} = 79.22e^{0.134 \cdot 8} = 231.42 \text{ million}$$
$$y_{1990} = 79.22e^{0.134 \cdot 9} = 264.60 \text{ million}$$
$$y_{2000} = 79.22e^{0.134 \cdot 10} = 302.54 \text{ million}$$

EXERCISES

1. Determine the least-squares exponential curve for the data in Example 1.
2. Determine the least-squares regression line for the data in Example 3.
3. Determine the least-squares regression line for the data in Example 4.
4. Find the least-squares regression line relating weight to height, based on the following data:

   Height: 70 63 72 60 66 70 74 65 62 67
   Weight: 155 150 180 135 156 168 178 160 132 145

5. If X represents weekly advertising expenditures in units of \$1000, and Y represents weekly sales, in the same units, find the least-squares regression line that best explains sales in terms of advertising, based on the following data:

   X: 55 40 50 60 70 80 90 60 55 100
   Y: 130 125 140 155 160 200 210 150 140 200

6. Determine the least-squares regression line that best predicts a student's final grade from his midterm grade, based on the following data:

   Midterm: 98 68 100 74 88 98 45 85 64 87
   Final: 90 82 97 78 77 93 82 77 80 99

7. Determine the regression curve which best explains weight loss in terms of the number of months on a particular diet, based on the following information:

Months:	1	2	3	4	5	6	7	8	9	10	11	12
Weight loss:	4	9	14	21	30	41	?	?	?	?	?	?

   Calculate the patient's expected weight loss over his second six months on this diet.
8. A large clothing store's sales over the past ten years have been 1.2, 1.3, 1.5, 1.4, 1.7, 1.9, 1.9, 2.0, 1.9, and 1.8 million dollars. Find the least-squares regression curve which best explains the trend of sales and use this curve to predict the store's sales five years from now.
9. Find the least-squares regression curve which best explains savings as a function of income, based on the following data:

   Income: 8 9 10 12 15 16 18 24 30 50
   Savings: 0.4 0.6 0.9 1.5 2.3 2.3 2.6 3.6 5.0 15.0

   (Earnings and savings are expressed in units of \$1000.)

11.3 Introduction to Correlation

Up to this point, we have not discussed how strongly two random variables might be related. Some (random) variables may be exactly related by an equation, such as the one which relates the area of a circle to the radius of the circle. In other cases, it would be logical to expect a strong relationship to exist between two random variables. For example, we would expect a student's college grade point average to be strongly related to his high school ranking. We might expect there to be some form of relationship between a person's height and weight, but we would not expect there to be any relationship between government spending and the number of marriages per year.

What we have done thus far has all been very mechanical. We look at a scatter diagram of our sample data, decide upon a particular type of regression curve that appears to make sense, plug our observed values x_i and y_i into formulas for the coefficients of our regression curve, mix, and voila: a regression curve. We have said nothing at all about how well, if at all, this curve fits (or explains) the observed data.

Suppose that the scatter diagram of our observed data does not clearly indicate what type of regression curve should be used. If we cannot decide whether a straight line or a lazy parabola best fits our data, we can try both. Using the techniques described in the preceding section, we can find the straight line that best fits our data, and we can find the parabola that best fits our data. But how then are we to decide which does the better job?

Clearly, we need some method for determining how well a particular regression curve fits the observed data. In this section, we will discuss two statistics which can be used for this purpose (in addition to the χ^2, as suggested in the preceding section). The first, called the standard error of the estimate of Y on X, is quite similar to the standard deviation. The second, called the coefficient of correlation, is closely related to the correlation coefficient discussed in Chapter 5.

We will assume, unless specifically mentioned to the contrary, that our regression curve is linear: $Y = a + bX$. Most of the formulas we will derive, however, are valid regardless of the form of the regression curve.

DEFINITION

The *standard error of the estimate of Y on X*, which we shall denote $s_{Y, X}$, is defined by

$$s_{Y, X} = \sqrt{\frac{\sum_{i=1}^{n} (y_i - y_i')^2}{n}}$$

As the definition indicates, $s_{Y, X}$ is calculating the average amount by which the sample data points miss the regression curve. Also, the least squares curve, by its definition, is the curve of its particular species which yields the smallest value of $s_{Y, x}$.

It is helpful to be able to calculate statistics such as $s_{Y, x}$ (and the coefficient of correlation, as well) in terms of values already calculated in determining the coefficients of the regression curve. In that respect, we have the following formula (which is valid *only* for linear regression):

Proposition 11.3

$$s_{Y, x} = \sqrt{\frac{\sum\limits_{i=1}^{n} y_i^2 - a \sum\limits_{i=1}^{n} y_i - b \sum\limits_{i=1}^{n} x_i y_i}{n}}$$

Proof: Since $y_i' = a + bx_i$, we have

$$s_{Y, x}^2 = \frac{1}{n} \sum_{i=1}^{n} [y_i - (a + bx_i)]^2$$

$$= \frac{1}{n} \sum_{i=1}^{n} (y_i - a - bx_i)^2$$

$$= \frac{1}{n} \sum_{i=1}^{n} y_i(y_i - a - bx_i)$$

$$- \frac{a}{n} \sum_{i=1}^{n} (y_i - a - bx_i)$$

$$- \frac{b}{n} \sum_{i=1}^{n} x_i(y_i - a - bx_i)$$

Due to the normal equations, the second and third summations equal zero, so

$$s_{Y, x}^2 = \frac{1}{n} \sum_{i=1}^{n} (y_i^2 - ay_i - bx_iy_i)$$

from which the desired formula follows.

We point out that, as with the standard deviation, should we wish to use the *sample* standard error of the estimate to estimate the *population* standard error of the estimate, we would have an unbiased estimator only if we altered our definition slightly and divided by $n - 2$ rather than by n.

Example 1   Referring back to Example 1 of the preceding section, we find that

$$s_{Y, x} = \sqrt{\frac{105.24 + 5.147 \cdot 32.0 - 0.07 \cdot 3828.1}{10}} = 0.445$$

Hence, the average predicted GPA differs from the corresponding actual GPA by 0.445 points.

Example 2 Referring back to Example 2 in the preceding section, we find that

$$s_{Y,X} = \sqrt{\frac{31.92 - 3.205 \cdot 12.40 + 0.129 \cdot 60.92}{5}} = 0.086$$

The standard error of the estimate is nowhere near as popular or widely used as the coefficient of correlation. The coefficient of correlation concerns the total variation

$$\sum_{i=1}^{n} (y_i - \bar{y})^2$$

of the observed y-values about their mean, and how much of this variation can be attributed to the regression line. To pave the way for our definition, we first prove the following:

Proposition 11.4

$$\sum_{i=1}^{n} (y_i - \bar{y})^2 = \sum_{i=1}^{n} (y_i - y_i')^2 + \sum_{i=1}^{n} (y_i' - \bar{y})^2$$

Proof: We write

$$(y_i - \bar{y})^2 = [(y_i - y_i') + (y_i' - \bar{y})]^2$$

expand the square, to find that:

$$\sum_{i=1}^{n} (y_i - \bar{y})^2 = \sum_{i=1}^{n} [(y_i - y_i')^2 + 2(y_i - y_i')(y_i' - \bar{y}) + (y_i' - \bar{y})^2]$$

Our proof will be complete if we can show that the middle term

$$\sum_{i=1}^{n} (y_i - y_i')(y_i' - \bar{y}) = 0$$

But since $y_i' = a + bx_i$, this term equals

$$\sum_{i=1}^{n} (y_i - a - bx_i)(a + bx_i - \bar{y})$$

which, when expanded, equals

$$a \sum_{i=1}^{n} (y_i - a - bx_i) + b \sum_{i=1}^{n} x_i(y_i - a - bx_i) - \bar{y} \sum_{i=1}^{n} (y_i - a - bx_i)$$

It then follows from the normal equations that all three of these summands vanish, and so the middle term is, in fact, equal to zero.

The two summands on the right-hand side of this equation are important, and so are given names. The first,

$$\sum_{i=1}^{n} (y_i - y_i')^2$$

is called the *unexplained variation*, because it reflects the differences between the observed values of Y and the expected values of Y on the regression line. The second term,

$$\sum_{i=1}^{n} (y_i' - \bar{y})^2$$

is called the *explained variation*, because it reflects the differences between the expected y-values on the regression line and the overall average of the y-values (Exercise 1), that amount of the variation among the y-values that can be explained by the regression line. And Proposition 11.4 says that the total variation among the observed y-values can always be partitioned into these two components.

If the regression curve provides a close fit to the observed data, the unexplained variation will be fairly small in comparison with the total variation. On the other hand, if the regression curve provides a poor fit to the data, the unexplained variation will be quite significant. With this in mind, we make the following definition:

DEFINITION

The *coefficient of correlation* between the two random variables X and Y, denoted by r, is defined by

$$r = \pm \sqrt{\frac{\text{explained variation}}{\text{total variation}}} = \pm \sqrt{\frac{\sum_{i=1}^{n} (y_i' - \bar{y})^2}{\sum_{i=1}^{n} (y_i - \bar{y})^2}}$$

(where r is defined to have the same sign as the slope of the regression line, should the regression curve be linear).

It is an immediate consequence of Proposition 11.4 that $|r| \leq 1$. A value of $|r|$ close to 1 means that the unexplained variation is relatively insignificant, and consequently, that the regression curve provides a good fit to the observed data. On the other hand, a value of $|r|$ close to zero means that the explained variation is the insignificant component, and that the unexplained variation is relatively large. In this case, the regression curve gives a poor approximation to the observed data.

In the case of linear regression, we emphasize that if $|r|$ is close to 1, there is a strong linear relationship between the two variables. But if $|r|$ is close to zero, then there is no linear relationship between the two variables, although there may well be some other type of relationship existing between the two variables.

A quick look at the way the coefficient of correlation is defined indicates what an intermediate value of r means. For example, a value of $r = 0.50$, and therefore, a value of $r^2 = 0.25$, indicates that 25 percent of the variation among the y-values in our sample can be explained by the regression line.

The following is an immediate consequence of the definition:

Proposition 11.5

$r = \sqrt{1 - (s_{Y,X}^2/s_Y^2)}$, where s_Y^2 is the variation among the observed y-values. Equivalently, $s_{Y,X} = s_Y\sqrt{(1 - r^2)}$.

Proof: By definition,

$$r^2 = \frac{\sum\limits_{i=1}^{n} (y_i' - \bar{y})^2}{\sum\limits_{i=1}^{n} (y_i - \bar{y})^2}$$

which, by Proposition 11.4 can be written as

$$r^2 = \frac{\sum\limits_{i=1}^{n} (y_i - \bar{y})^2 - \sum\limits_{i=1}^{n} (y_i - y_i')^2}{\sum\limits_{i=1}^{n} (y_i - \bar{y})^2}$$

$$= 1 - \frac{\sum\limits_{i=1}^{n} (y_i - y_i')^2}{\sum\limits_{i=1}^{n} (y_i - \bar{y})^2}$$

$$= 1 - \frac{n \cdot s_{Y,X}^2}{n \cdot s_Y^2}$$

$$= 1 - \frac{s_{Y,X}^2}{s_Y^2}$$

Therefore,

$$r^2 \cdot s_Y^2 = s_Y^2 - s_{Y,X}^2$$

or

$$s_{Y,X}^2 = (1 - r^2) \cdot s_Y^2$$

That is,

$$s_{Y,X} = s_Y\sqrt{1 - r^2}$$

We can apply the formula in Proposition 11.1 to the definition of r to obtain what is called the *product moment* formula for r:

Proposition 11.6

$$r = \frac{\sum\limits_{i=1}^{n} (x_i - \bar{x})(y_i - \bar{y})}{\sqrt{\sum\limits_{i=1}^{n} (x_i - \bar{x})^2 \sum\limits_{i=1}^{n} (y_i - \bar{y})^2}}$$

Proof: By definition,

$$r^2 = \frac{\sum\limits_{i=1}^{n} (y_i' - \bar{y})}{\sum\limits_{i=1}^{n} (y_i - \bar{y})^2}$$

which, according to Proposition 11.1, equals

$$\frac{\sum\limits_{i=1}^{n} b^2 (x_i - \bar{x})^2}{\sum\limits_{i=1}^{n} (y_i - \bar{y})^2}$$

where

$$b^2 = \left(\frac{\sum\limits_{i=1}^{n} (x_i - \bar{x})(y_i - \bar{y})}{\sum\limits_{i=1}^{n} (x_i - \bar{x})^2} \right)^2$$

Therefore,

$$r^2 = \left(\frac{\sum\limits_{i=1}^{n} (x_i - \bar{x})(y_i - \bar{y})}{\sum\limits_{i=1}^{n} (x_i - \bar{x})^2} \right)^2 \frac{\sum\limits_{i=1}^{n} (x_i - \bar{x})^2}{\sum\limits_{i=1}^{n} (y_i - \bar{y})^2}$$

which gives

$$\frac{\left[\sum\limits_{i=1}^{n} (x_i - \bar{x})(y_i - \bar{y}) \right]^2}{\left(\sum\limits_{i=1}^{n} (x_i - \bar{x})^2 \right)\left(\sum\limits_{i=1}^{n} (y_i - \bar{y})^2 \right)}$$

which is equivalent to the desired result.

If we let s_{XY} denote the sample covariance

$$s_{XY} = \frac{\sum\limits_{i=1}^{n} (x_i - \bar{x})(y_i - \bar{y})}{n}$$

then we find, as an immediate corollary of the above, that

Corollary 11.7

$$r = \frac{s_{XY}}{s_X \cdot s_Y}$$

Proof: Simply divide the numerator and denominator of the result in Proposition 11.6 by n.

Therefore, the sample coefficient of correlation can be looked upon as an estimate, based on data in a sample, of the correlation coefficient $\rho(X, Y)$, which was defined in Chapter 5 to equal

$$\rho(X, Y) = \frac{\text{Cov}(X, Y)}{\sigma_X \sigma_Y}$$

Another immediate consequence of Proposition 11.6 is the following computational formula for r which makes use of the terms already calculated in determining the coefficients of the regression curve.

Proposition 11.8

$$r = \frac{n \sum\limits_{i=1}^{n} x_i y_i - \left(\sum\limits_{i=1}^{n} x_i\right)\left(\sum\limits_{i=1}^{n} y_i\right)}{\sqrt{n \sum\limits_{i=1}^{n} x_i^2 - \left(\sum\limits_{i=1}^{n} x_i\right)^2} \sqrt{n \sum\limits_{i=1}^{n} y_i^2 - \left(\sum\limits_{i=1}^{n} y_i\right)^2}}$$

Proof: By Proposition 11.6,

$$r = \frac{\sum\limits_{i=1}^{n} (x_i - \bar{x})(y_i - \bar{y})}{\sqrt{\sum\limits_{i=1}^{n} (x_i - \bar{x})^2 \sum\limits_{i=1}^{n} (y_i - \bar{y})^2}}$$

But it is straightforward to show that

$$\sum\limits_{i=1}^{n} (x_i - \bar{x})(y_i - \bar{y}) = \sum\limits_{i=1}^{n} x_i y_i - \frac{1}{n}\left(\sum\limits_{i=1}^{n} x_i\right)\left(\sum\limits_{i=1}^{n} y_i\right)$$

$$\sum\limits_{i=1}^{n} (x_i - \bar{x})^2 = \sum\limits_{i=1}^{n} x_i^2 - \frac{1}{n}\left(\sum\limits_{i=1}^{n} x_i\right)^2$$

and

$$\sum_{i=1}^{n} (y_i - \bar{y})^2 = \sum_{i=1}^{n} y_i^2 - \frac{1}{n}\left(\sum_{i=1}^{n} y_i\right)^2$$

Therefore,

$$r = \frac{\displaystyle\sum_{i=1}^{n} x_i y_i - \frac{1}{n}\left(\sum_{i=1}^{n} x_i\right)\left(\sum_{i=1}^{n} y_i\right)}{\sqrt{\displaystyle\sum_{i=1}^{n} x_i^2 - \frac{1}{n}\left(\sum_{i=1}^{n} x_i\right)^2}\sqrt{\displaystyle\sum_{i=1}^{n} y_i^2 - \frac{1}{n}\left(\sum_{i=1}^{n} y_i\right)^2}}$$

The required formula for r then follows upon multiplying both numerator and denominator by n.

Example 1 (*continued*) Using Proportion 11.8, we find that

$$r = \frac{(10 \cdot 3828.1 - 1185 \cdot 32)}{\sqrt{10 \cdot 140935 - (1185)^2}\,\sqrt{10 \cdot 105.24 - (32)^2}} = 0.946$$

indicating that the linear regression line $y = 0.07x - 5.147$ provides an extremely good fit to the observed data.

Example 2 (*continued*) Using Proposition 11.8 once again, we find that

$$r = \frac{(5 \cdot 60.92 - 28.20 \cdot 12.40)}{\sqrt{5 \cdot 229.16 - (28.20)^2}\,\sqrt{5 \cdot 31.92 - (12.40)^2}} = -0.996$$

again indicating that our regression line provides an extremely good fit to the observed data.

Proposition 11.9

The regression line of Y on X can be expressed as

$$y - \bar{y} = r\left(\frac{s_Y}{s_X}\right)(x - \bar{x})$$

where

$$s_X = \sqrt{\frac{\displaystyle\sum_{i=1}^{n} (x_i - \bar{x})^2}{n}} \quad \text{and} \quad s_Y = \sqrt{\frac{\displaystyle\sum_{i=1}^{n} (y_i - \bar{y})^2}{n}}$$

Proof: By Proposition 11.1, we have

$$y - \bar{y} = \frac{\displaystyle\sum_{i=1}^{n} (x_i - \bar{x})(y_i - \bar{y})}{\displaystyle\sum_{i=1}^{n} (x_i - \bar{x})^2}(x - \bar{x})$$

but, by Proposition 11.6,

$$r = \frac{\sum_{i=1}^{n} (x_i - \bar{x})(y_i - \bar{y})}{\sqrt{\sum_{i=1}^{n} (x_i - \bar{x})^2} \sqrt{\sum_{i=1}^{n} (y_i - \bar{y})^2}}$$

so that

$$\sum_{i=1}^{n} (x_i - \bar{x})(y_i - \bar{y}) = r \sqrt{\sum_{i=1}^{n} (x_i - \bar{x})^2 \sum_{i=1}^{n} (y_i - \bar{y})^2}$$

Substituting into the first equation above, we have

$$y - \bar{y} = \frac{r \sqrt{\sum_{i=1}^{n} (x_i - \bar{x})^2 \sum_{i=1}^{n} (y_i - \bar{y})^2}}{\sum_{i=1}^{n} (x_i - \bar{x})^2}$$

$$= r \cdot \frac{\sqrt{\sum_{i=1}^{n} (y_i - \bar{y})^2}}{\sqrt{\sum_{i=1}^{n} (x_i - \bar{x})^2}}$$

and then, upon dividing numerator and denominator by n, the desired result follows.

Example 3 In Example 4 of the preceeding section, we fit an exponential curve to our observed data. Unfortunately, since our regression curve is not linear, the formulas in Propositions 11.3 and 11.8 cannot be used to calculate the standard error of the estimate and the coefficient of correlation. Therefore, we must use the definition to calculate these statistics. The calculations involved are contained in the following table:

y_i	y_i'	$(y_i - y_i')^2$	$(y_i - \bar{y})^2$	$(y_i' - \bar{y})^2$
76	79.22	10.37	3,220.56	2,865.46
92	90.58	2.02	1,660.56	1,778.31
106	103.57	5.90	715.56	851.47
123	118.42	20.98	95.06	205.35
132	135.40	11.56	0.56	7.02
151	154.81	14.52	330.06	486.64
179	177.01	3.96	2,139.06	1,958.95
203	202.40	0.36	4,935.06	4,851.12
1,062		69.66	13,096.48	13,004.32

From the first column, we find that

$$\bar{y} = \frac{1062}{8} = 132.75$$

From the third column, we find that

$$s_{Y,x} = \sqrt{\frac{\sum\limits_{i=1}^{8} (y_i - y_i')^2}{8}} = \sqrt{\frac{69.66}{8}} = 2.95$$

so that the expected y-values differ from the observed y-values by an average of almost three units (three million people). Finally, from columns 4 and 5, we find that

$$r = \sqrt{\frac{\sum\limits_{i=1}^{8} (y_i' - \bar{y})^2}{\sum\limits_{i=1}^{8} (y_i - \bar{y})^2}} = \sqrt{\frac{13,004.32}{13,096.48}} = 0.996$$

indicating an excellent fit from our least-squares exponential curve.

EXERCISES

1. Prove that the observed and the expected y-values have the same mean.
2. Refer to Exercise 3 in the preceding section. Calculate the standard error of the estimate and the coefficient of correlation corresponding to the least-squares regression line. Which does the better job of fitting the observed data, the straight line or the exponential curve?
3. Refer to Exercise 1 in the preceding section. Calculate the standard error of the estimate and the coefficient of correlation for the least-squares exponential curve. Which provides the better fit, the exponential or the least-squares line?
4. Refer to Exercise 5 in the preceding section and calculate the standard error of the estimate and the coefficient of correlation of the least-squares regression line.
5. Derive a formula equivalent to that of Proposition 11.3 in the case that the regression curve is a quadratic.
6. Use the preceding result to calculate the standard error of the estimate for the least-squares quadratic found in Example 3 of the preceding section. Calculate the coefficient of correlation for this curve. Calculate the same two statistics for the least-squares regression line for the same data (see Exercise 2 in the preceding section). Compare the results.

11.4 Normal Regression and Correlation Theory

In Section 11.2, we discussed how we could approximate the true regression line $Y = \alpha + \beta X$ by means of the least-squares line $Y = a + bX$. In Section 11.3, we discussed the coefficient of correlation r, and observed several similarities between r and the true correlation coefficient $\rho(X, Y)$. However, several questions remain to be answered, namely:

1. How closely does the least-squares line approximate the true regression line? In other words, how closely do the sample regression coefficients a and b approximate the true regression coefficients α and β?
2. How closely does the sample coefficient of correlation r approximate the actual correlation coefficient ρ?
3. How closely does a y-value estimated by means of the least-squares line approximate the corresponding y-value estimated by means of the true regression line?

In other words, we have, on the basis of a sample, been able to make estimations concerning the form and degree of the true relationship between the random variables X and Y. Now we would like to know how accurate these estimations might be.

The answers to the above questions are by no means simple. In fact, we must make some very strong assumptions concerning joint and conditional distributions in order to obtain any reasonable answers whatsoever. Even then, the proofs are beyond the scope of this book. Therefore, we will present, in this section, without proofs, a survey of the major results that are known. And we will assume, in all examples, that the "normality" conditions discussed below apply.

In order to find a reasonable answer to question 1 above, we must assume that, given $X = x_0$ (for any fixed x_0), the values that Y may assume are normally distributed with mean $E(Y|x_0)$ and standard deviation σ. That is, for any fixed value x_0, Y has conditional density

$$h_2(y|x_0) = \frac{1}{\sigma\sqrt{2\pi}} \exp\left\{ -\frac{1}{2}\left[\frac{y - E(y|x_0)}{\sigma}\right]^2 \right\}$$

$$= \frac{1}{\sigma\sqrt{2\pi}} \exp\left\{ -\frac{1}{2}\left[\frac{y - (\alpha + \beta x_0)}{\sigma}\right]^2 \right\}$$

Our first result in normal regression theory concerns maximum likelihood estimators for the three unknowns in the conditional density described above: α, β, and σ. To determine what these estimates are, we partially differentiate the natural logarithm of the likelihood function

$$L = \prod_{i=1}^{n} h_2(y_i|x_i) = \left(\frac{1}{\sigma\sqrt{2\pi}}\right)^n \prod_{i=1}^{n} \exp\left\{ -\frac{1}{2}\left[\frac{y_i - (\alpha + \beta x_i)}{\sigma}\right]^2 \right\}$$

with respect to the three unknowns, and obtain the following conclusion:

Proposition 11.10

Under the assumptions of normal regression theory, the maximum likelihood estimates for α, β, and σ are a, b, and $s_{Y,X}$, respectively, where a and b are the coefficients of the least-squares line, and $s_{Y,X}$ is the (sample) standard error of the estimate.

Proof: Exercise

Should we wish to construct confidence interval estimates for either of the parameters α or β, or should we wish to test a hypothesis concerning either, the following two results might prove helpful. We will let $\bar{a}$ and $\bar{b}$ denote the sampling distributions of the two least-squares regression coefficients a and b over all possible bivariate samples of size n.

Proposition 11.11

Under the assumptions of normal regression theory, $\bar{a}$ and $\bar{b}$ are unbiased estimators for α and β, respectively, and:

(a) the statistic $\bar{a}$ has normal distribution with $E(\bar{a}) = \alpha$ and

$$\text{Var}(\bar{a}) = \frac{\sigma^2(s_X^2 + \bar{x}^2)}{ns_X^2}$$

(b) the statistic $\bar{b}$ has normal distribution with $E(\bar{b}) = \beta$ and

$$\text{Var}(\bar{b}) = \frac{\sigma^2}{ns_X^2}$$

Our expressions for both $\text{Var}(\bar{a})$ and $\text{Var}(\bar{b})$ contain the unknown population parameter σ^2, rather than its estimate $s_{Y,X}^2$. The following result, however, will allow us to find a more workable formula from which to test hypotheses or form confidence intervals.

Proposition 11.12

Under the assumptions of normal regression theory, the statistic $ns_{Y,X}^2/\sigma^2$ has a χ^2 distribution with $n - 2$ degrees of freedom. Also, the sampling distribution of this statistic is independent of both $\bar{a}$ and $\bar{b}$.

Combining these two propositions with the definition of the t distribution, we have

Proposition 11.13

Under the assumptions of normal regression theory, the statistics

$$\frac{(\bar{b} - \beta)s_X\sqrt{n-2}}{s_{Y,X}} \quad \text{and} \quad \frac{(\bar{a} - \alpha)s_X\sqrt{n-2}}{s_{Y,X}\sqrt{s_X^2 + s_Y^2}}$$

each have t distribution with $n - 2$ degrees of freedom.

Example 1 Referring back to Example 1 in Section 11.2, suppose that we wish to test, on the basis of the least-squares line obtained from our small sample ($n = 10$), the hypothesis that the regression coefficient $\beta = 0$. According to Proposition 11.13, we must test the statistic

$$t = \frac{(0.07 - 0.00) \cdot 7.159 \cdot \sqrt{8}}{0.445} = 3.185$$

(where the standard deviation $s_X = 7.159$), against the critical values $\pm t_{0.025, 8} = 2.306$. Since our sample statistic does fall in this critical region, we must reject the hypothesis that $\beta = 0$, and conclude that the least squares regression line has nonzero slope (β).

We can construct a $1 - \alpha$ percent confidence interval for α by converting the inequality

$$-t_{\alpha/2, n-2} \leq \frac{(\bar{a} - \alpha)s_X\sqrt{n-2}}{s_{Y,X}\sqrt{s_X^2 + s_Y^2}} \leq t_{\alpha/2, n-2}$$

into the equivalent inequality

$$\bar{a} - t_{\alpha/2, n-2}\frac{s_{Y,X}\sqrt{s_X^2 + s_Y^2}}{s_X\sqrt{n-2}} \leq \alpha \leq \bar{a} + t_{\alpha/2, n-2}\frac{s_{Y,X}\sqrt{s_X^2 + s_Y^2}}{s_X\sqrt{n-2}}$$

Likewise, we can construct a $1 - \alpha$ percent confidence interval for β by converting the inequality

$$-t_{\alpha/2, n-2} \leq \frac{(\bar{b} - \beta)s_X\sqrt{n-2}}{s_{Y,X}} \leq t_{\alpha/2, n-2}$$

into the equivalent inequality

$$\bar{b} - t_{\alpha/2, n-2}\frac{s_{Y,X}}{s_X\sqrt{n-2}} \leq \beta \leq \bar{b} + t_{\alpha/2, n-2}\frac{s_{Y,X}}{s_X\sqrt{n-2}}$$

Example 1 (continued) On the basis of the information contained in our bivariate sample of size 10, we can form the following 95 percent confidence interval for β:

$$0.07 - 2.306 \cdot \frac{0.445}{7.159\sqrt{8}} \leq \beta \leq 0.07 + 2.306 \cdot \frac{0.445}{7.159\sqrt{8}}$$

which is equivalent to

$$0.019 \leq \beta \leq 0.121$$

We turn now to normal correlation theory. In order to give a reasonable answer to question 2 above, we must make the assumption that the joint distribution of the random variables X and Y is bivariate normal. The first thing that we do is find maximum likelihood estimators for the five parameters μ_1, μ_2, σ_1, σ_2, and ρ which define this distribution. This can be accomplished by considering the five partial derivatives of the logarithm of the likelihood function for our sample

$$L = \prod_{i=1}^{n} f(x_i, y_i)$$

where

$$f(x_i, y_i) = \frac{\exp\left\{\frac{-1}{2(1-\rho^2)}\left[\left(\frac{x_i - \mu_1}{\sigma_1}\right)^2 - 2\rho\left(\frac{x_i - \mu_1}{\sigma_1}\right)\left(\frac{y_i - \mu_2}{\sigma_2}\right) + \left(\frac{y_i - \mu_2}{\sigma_2}\right)^2\right]\right\}}{2\pi\sigma_1\sigma_2\sqrt{1-\rho^2}}$$

Upon doing this, we obtain the following results, which are similar to those we obtained in Example 5 of Section 9.2:

Proposition 11.14

Under the assumptions of normal correlation theory, the maximum likelihood estimators for μ_1, μ_2, σ_1, σ_2, and ρ are $\bar{x}$, $\bar{y}$, s_X, s_Y, and r, respectively.

As might be expected, the sample coefficient of correlation r turns out to be the maximum likelihood estimator of the true correlation coefficient ρ. But can r be used to test hypotheses concerning ρ, or to form a confidence interval estimate for ρ? The answer is yes, but we must distinguish two cases. If $\rho = 0$, and there is absolutely no linear correlation between the two variables, then we must use a t statistic. On the other hand, if $|\rho| > 0$, and there is some linear correlation between the two variables, then we must proceed in an entirely different manner, using a normal statistic.

First, let us consider the case $\rho = 0$. We have the following result:

Proposition 11.15

Under the assumptions of normal correlation theory with $\rho = 0$, the statistic

$$\frac{r}{\sqrt{(1 - r^2)/(n - 2)}}$$

has t distribution with $n - 2$ degrees of freedom.

This result is especially useful when testing the null hypothesis

$$H_0: \rho = 0$$

against the alternative

$$H_1: |\rho| > 0$$

at level of significance α. We simply evaluate the sample statistic

$$t = \frac{r}{\sqrt{(1 - r^2)/(n - 2)}}$$

and determine if it falls in the critical region

$$t > t_{\alpha, n-2}$$

Example 2 A bivariate sample of size $n = 20$ yielded a coefficient of correlation $r = 0.25$. Is this solid evidence, at a five percent level of significance, that $|\rho| > 0$; that is, that there is some linear correlation between the two variables in question? To answer this question, we test the hypothesis

$$H_0 : \rho = 0$$

against the alternative

$$H_1 : |\rho| > 0$$

at a five percent level of significance, by considering the statistic

$$t = \frac{0.25}{\sqrt{(1 - 0.0625)/18}} = 1.09$$

Since our sample statistic does not fall in the critical region $t > t_{0.05, \, 18} = 1.734$ we cannot reject the null hypothesis, and therefore cannot conclude that $|\rho| > 0$. How large a sample showing a coefficient of correlation $r = 0.25$ would be sufficient to reject this null hypothesis? We would have to choose n large enough so that

$$t = \frac{0.25}{\sqrt{(1.0 - 0.0625)/(n - 2)}} > t_{0.05, \, n-2} < 1.734$$

That is, we must choose n large enough so that

$$\sqrt{n - 2} > \frac{1.734\sqrt{1.0 - 0.0625}}{0.25}$$

or

$$n - 2 > (1.734)^2 \cdot \frac{1.0 - 0.0625}{0.25^2} = 45.10$$

Therefore, a sample of size $n = 46$ will suffice.

On the other hand, should the true value of ρ be nonzero, the sampling distribution of r becomes quite complicated. However, upon applying what is called Fisher's Z transformation, we obtain a statistic whose sampling distribution is well known, as the following result indicates:

Proposition 11.16

Under the assumptions of normal correlation theory with $\rho \neq 0$, the statistic

$$Z = \tfrac{1}{2} \log_e \left(\frac{1 + r}{1 - r} \right)$$

has a sampling distribution which is approximately normal, with mean

$$E(Z) = \tfrac{1}{2} \log_e\left(\frac{1 + \rho}{1 - \rho}\right)$$

and standard deviation

$$\sigma_Z = \sqrt{\frac{1}{n - 3}}$$

Fisher's Z statistic can be used to test hypotheses concerning the true value of ρ. Should we wish to test the hypothesis

$$H_0: \rho = \rho_0$$

against a one-sided or a two-sided alternative at level of significance α, we would evaluate the (approximately) standard normal statistic

$$z = \frac{\tfrac{1}{2} \log_e\left(\dfrac{1 + r}{1 - r}\right) - \tfrac{1}{2} \log_e\left(\dfrac{1 + \rho}{1 - \rho}\right)}{\sqrt{1/(n - 3)}}$$

and determine if it falls in the corresponding one-sided or two-sided critical region.

Example 3 We wish to determine whether a coefficient of correlation $r = 0.46$ from a bivariate sample of size $n = 28$ supports the contention that $\rho \geq 0.50$. Hence, we test the null hypothesis

$$H_0: \rho = 0.50$$

against the one-sided alternative

$$H_1: \rho < 0.50$$

at a five percent level of significance. Our sample statistic is

$$z = \frac{\tfrac{1}{2} \log_e\left(\dfrac{1 + 0.46}{1 - 0.46}\right) - \tfrac{1}{2} \log_e\left(\dfrac{1 + 0.50}{1 - 0.50}\right)}{\sqrt{1/(28 - 3)}}$$

$$= -0.26$$

which clearly does not fall in the critical region $z < -1.64$. Therefore, we do not have solid evidence that the coefficient of correlation is below 0.5.

Proceeding as we did in Chapter 9, we can derive, from the equality

$$P[|Z - E(Z)| \leq 1.96\sigma_Z] = 0.95$$

the following 95 percent confidence interval for $\tfrac{1}{2} \log_e[(1 + \rho)/(1 - \rho)]$:

$$\tfrac{1}{2} \log_e\left(\frac{1 + r}{1 - r}\right) - \frac{1.96}{\sqrt{n - 3}} \leq \tfrac{1}{2} \log_e\left(\frac{1 + \rho}{1 - \rho}\right) \leq \tfrac{1}{2} \log_e\left(\frac{1 + r}{1 - r}\right) + \frac{1.96}{\sqrt{n - 3}}$$

which is equivalent to

$$\log_e\left(\frac{1+r}{1-r}\right) - \frac{3.92}{\sqrt{n-3}} \leq \log_e\left(\frac{1+\rho}{1-\rho}\right) \leq \log_e\left(\frac{1+r}{1-r}\right) + \frac{3.92}{\sqrt{n-3}}$$

If we set

$$U = \log_e\left(\frac{1+r}{1-r}\right) + \frac{3.92}{\sqrt{n-3}}$$

and

$$L = \log_e\left(\frac{1+r}{1-r}\right) - \frac{3.92}{\sqrt{n-3}}$$

we find that

$$L \leq \log_e\left(\frac{1+\rho}{1-\rho}\right) \leq U$$

Since e^x is an increasing function, this is equivalent to

$$e^L \leq \frac{1+\rho}{1-\rho} \leq e^U$$

Solving this inequality for ρ, we obtain the following 95 percent confidence interval:

$$\frac{e^L - 1}{e^L + 1} \leq \rho \leq \frac{e^U - 1}{e^U + 1}$$

Example 4 We wish to form a 95 percent confidence interval for ρ based on the fact that a bivariate sample of size $n = 28$ showed a coefficient of correlation $r = 0.46$. Since

$$\log_e\left(\frac{1 + 0.46}{1 - 0.46}\right) = 0.9946$$

and $3.92/\sqrt{25} = 0.784$, we have

$$U = 0.9946 + 0.784 = 1.7786$$

and

$$L = 0.9946 - 0.784 = 0.2106$$

Therefore,

$$e^U = 5.924 \quad \text{and} \quad e^L = 1.234$$

so that

$$\frac{0.234}{2.234} \leq \rho \leq \frac{4.924}{6.924}$$

or $0.105 \leq \rho \leq 0.711$.

In order to answer the third question mentioned at the beginning of this section, we must again work under the assumptions of normal regression theory. Our goal is to be able to predict, by means of a confidence interval, the expected (or average) value of Y corresponding to any given value of X:

$$E(Y \mid X = x_0) = \alpha + \beta x_0$$

To do so, we must use the value of Y obtained from the least-squares line $Y = a + bX$. Therefore, it is important to determine the sampling distribution $\bar{Y}$ of this statistic.

Proposition 11.17

Under the assumptions of normal regression theory, the statistic $\bar{Y}$ has t distribution with $n - 2$ degrees of freedom, with mean $E(\bar{Y}) = E(Y \mid x_0)$, and variance

$$\text{Var}(\bar{Y}) = s_{Y,X}^2 \left[\frac{1}{n} + \frac{(x_0 - \bar{x})^2}{n s_X^2} \right]$$

where the unbiased estimator

$$s_{Y,X}^2 = \frac{\sum\limits_{i=1}^{n} (y_i - y_i')^2}{n - 2}$$

is used instead of $s_{Y,X}^2$ as previously defined.

Note that for values of $n > 30$, the sampling distribution of $\bar{Y}$ can be closely approximated by a normal distribution.

Before proceeding any farther, let us take a closer look at the variance of $\bar{Y}$ and the individual components which account for this variance. As might be expected, this variance is directly effected by the size of the standard error of the estimate and inversely effected by the sample size. Also occurring in the numerator of our expression, and therefore having a direct effect on the size of the variance, is the term $x_0 - \bar{x}$, the distance from the point of prediction x_0 to the mean $\bar{x}$ of the sample x-values. Recall that any least-squares regression line must pass through the point $(\bar{x}, \bar{y})$, where $\bar{x}$ and $\bar{y}$ are, respectively, the means of the x-values and y-values in the sample from which the least-squares line was constructed. Likewise, the true regression line must pass through the point $[\mu_X, E(Y \mid X = \mu_X)]$. If we assume that our sample was representative of the population from which it was taken, then we may also assume that the point $(\bar{x}, \bar{y})$ is reasonably close to the point $[\mu_X, E(Y \mid X = \mu_X)]$, as indicated in Figure 11.4. As this diagram indicates, if the point x_0 were close to $\bar{x}$ (or μ_X), the corresponding points $(x_0, \alpha + \beta x_0)$ and $(x_0, a + b x_0)$ on the two lines would also be fairly close. However, due to the difference in the slopes of these lines, the distance between the points $(x_0, \alpha + \beta x_0)$ and $(x_0, a + b x_0)$ grows as the distance between x_0 and $\bar{x}$ increases. Consequently, the reliability of our estimated value $y = a + b x_0$

decreases in direct proportion to the distance between x_0 and $\bar{x}$. This points out the dangers of *extrapolation*, predicting y-values for x-values that fall outside the range of our sample.

One final term occurs, in the denominator, in our expression for the variance of Y, and that is the deviation s_X among the sample x-values. A large value of s_X indicates that there was a wide range of x-values in the sample from which the least-squares line was constructed, and consequently, one would expect to be able to predict y-values more accurately over a larger interval of x-values.

The information contained in Proposition 11.17 indicates that a $1 - \alpha$ percent confidence interval for $E(Y\,|\,X = x_0)$ has the following confidence limits:

$$Y_{x_0} \pm t_{\alpha/2,\, n-2} \cdot s_{Y,\,X} \cdot \sqrt{\frac{1}{n} + \frac{(x_0 - \bar{x})^2}{n s_X^{\,2}}}$$

Example 1 (continued)   We have already seen that we can predict a student's final college GPA based on his IQ by the linear regression formula

GPA $= 0.07 \cdot$ IQ $- 5.147$

Thus, if a student has an IQ of 120, we can predict a

GPA $= 0.07 \cdot 120 - 5.147 = 3.253$

But our least-squares line is only an approximation to the true regression formula expressing IQ in terms of GPA. If we knew this formula, we could simply plug in the value IQ $= 120$ and calculate the expected (or average) value of GPA. The value 3.253 that we have calculated by means of the least

Figure 11.4

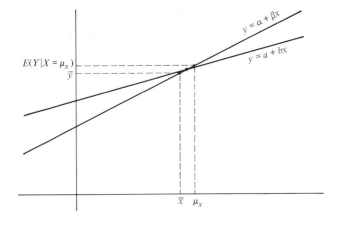

squares line is only an approximation to this "true" expected value of GPA for IQ = 120. However, using the confidence limits derived above, we can use our approximation to form a confidence interval estimation for the "true" GPA. Using the following values

$$x_0 = 120$$
$$\bar{x} = 118.5$$
$$n = 10$$
$$Y_{x_0} = 3.253$$
$$s_X = 7.159$$
$$s_{Y,X} = 0.497 \qquad \text{(the unbiased } s_{Y,X}\text{)}$$
$$t_{0.025,\,8} = 2.306$$

we find the confidence limits to be

$$3.253 \pm 2.306 \cdot 0.497 \sqrt{\frac{1}{10} + \frac{(120 - 118.5)^2}{10 \cdot (7.159)^2}}$$

which equals 3.253 ± 0.370, giving the confidence interval

$$2.883 \le E(Y \,|\, X = 120) \le 3.623$$

If we wish to predict the "true" value of $E(Y \,|\, X = 100)$, based on the approximation $y = 0.07 \cdot 100 - 5.147 = 1.853$ obtained from the least-squares line, we would find our confidence limits to be

$$1.853 \pm 2.306 \cdot 0.497 \sqrt{\frac{1}{10} + \frac{(100 - 118.5)^2}{10 \cdot (7.159)^2}}$$

which equals

$$1.853 \pm 1.004$$

giving the confidence interval

$$0.849 \le E(Y \,|\, X = 100) \le 2.857$$

Notice the lengths of these two confidence intervals. For $x_0 = 120$, which is very close to the average 118.5 of the sample x-values, we can predict to within 0.370 of the true value $E(Y \,|\, X = 120)$. On the other hand, $x_0 = 100$ is not very close to 118.5, and consequently, we can only predict to within 1.004 of the true value $E(Y \,|\, X = 100)$. The confidence interval in the second case is almost three times as wide as the confidence interval in the first case and the width of the confidence interval will increase as we move farther and farther away from the mean 118.5.

To further illustrate the techniques we have developed in this chapter, we will work through the following example in its entirety.

Example 5   The following data indicates the sales (in millions of dollars) of a large department store over the last 11 years:

1964	1965	1966	1967	1968	1969	1970	1971	1972	1973	1974
2.30	2.52	2.77	2.99	3.24	3.46	3.70	3.98	4.26	4.44	4.70

The store would like to use this data to predict its sales in 1975. Since the trend of the sales increase over the last 11 years appears to be linear (make a scatter diagram), it would seem to make sense to find the least-squares line $y = a + bx$ which best approximates this data. To simplify our calculations, we once again "code" the years, this time by means of the transformation

$$x' = \frac{\text{year} - 1969}{1}$$

Our calculations are contained in the following table:

x_i	y_i	x_i^2	$x_i y_i$	y_i^2
-5	2.30	25	-11.50	5.29
-4	2.52	16	-10.08	6.35
-3	2.77	9	-8.31	7.67
-2	2.99	4	-5.98	8.94
-1	3.24	1	-3.24	10.50
0	3.46	0	0.00	11.97
1	3.70	1	3.70	13.69
2	3.98	4	7.96	15.84
3	4.26	9	12.78	18.15
4	4.44	16	17.76	19.71
5	4.70	25	23.50	22.09
0	38.36	110	26.59	140.20

Consequently, the two normal equations are

$$38.36 = 11a + 0b$$
$$26.59 = 0a + 110b$$

Therefore, $a = 3.487$ and $b = 0.242$, and we have the least-squares line

$$y = 3.487 + 0.242x$$

How good a fit does this line provide to the sample data? The standard error of the estimate is

$$s_{Y,x} = \sqrt{\frac{140.20 - 3.487 \cdot 38.36 - 0.242 \cdot 26.59}{11}} = 0.019$$

(the unbiased version equals 0.021), indicating an excellent fit, and the sample coefficient of correlation

$$r = \frac{11 \cdot 26.59 - 0 \cdot 38.36}{\sqrt{11 \cdot 110 - 0^2}\sqrt{11 \cdot 140.2 - (38.36)^2}} = 0.99995$$

is so close to 1 to suggest perfect correlation. If we set

$$U = \log_e\left(\frac{1 + r}{1 - r}\right) + \frac{3.92}{\sqrt{11 - 3}} = 11.983$$

and

$$L = \log_e\left(\frac{1 + r}{1 - r}\right) - \frac{3.92}{\sqrt{11 - 3}} = 9.211$$

then we have the following 95 percent confidence interval for the true correlation:

$$0.9998 = \frac{e^L - 1}{e^L + 1} \le \rho \le \frac{e^U - 1}{e^U + 1} = 0.9999$$

indicating that there is, indeed, almost perfect correlation between these two
variables. How good a job has our least-squares line done in approximating
the slope β of the true regression line? Very well, as the following confidence
interval for β indicates:

$$0.242 - t_{0.025,\, 9} \cdot \frac{0.019}{\sqrt{10}\sqrt{9}} \le \beta \le 0.242 + t_{0.025,\, 9} \cdot \frac{0.019}{\sqrt{10}\sqrt{9}}$$

which equals $0.237 \le \beta \le 0.247$, a confidence interval of length 0.01. Finally,
how reliable is the prediction

$$y_{1975} = 3.487 + 0.242 \cdot 6 = 4.939$$

for 1975? According to the results above, we may expect an error of

$$t_{\alpha/2,\, n-2} \cdot s_{Y,\, X} \sqrt{\frac{1}{n} + \frac{(x_0 - \bar{x})^2}{n s_X{}^2}} = 2.262 \cdot 0.021 \sqrt{\frac{1}{11} + \frac{(6)^2}{11 \cdot 10}} = 0.031$$

so that our 95 percent confidence interval for $E(Y\,|\,X = 1975)$ is

$$4.908 \le E(Y\,|\,X = 1975) \le 4.970$$

EXERCISES

1. Prove Proposition 11.10.
2. Refer to Exercise 4 in Section 11.2. Find confidence interval estimates for β and ρ.
   Obtain a confidence interval estimate for the weight of a person whose height is 6 ft.
3. Refer to Exercise 5 in Section 11.2. Find confidence interval estimates for β and ρ.
   Find a confidence interval estimate for the amount of sales (in units of $1,000) that
   would result if $75,000 were spent on advertising.
4. Refer to Exercise 6 in Section 11.2. Find confidence interval estimates for β and ρ.
   Find a confidence interval estimate for the final grade of a student whose midterm
   grade is 60; for a student whose midterm grade is 90.

ADDITIONAL PROBLEMS

1. The following data has been taken to study the relationship between the speed of a machine, measured in revolutions per minute, and the number of defective items produced by the machine during the 30 minutes the machine was observed:

rpm	Defectives
10.2	8
11.6	7
14.2	13
9.8	6
10.8	9
13.2	10
8.9	5
12.0	12
11.5	10
15.0	2

Find the least-squares regression line that gives the best fit to this data. Calculate the standard error of the estimate and the coefficient of correlation. Form confidence interval estimates for α, β, and ρ. Form a confidence interval estimate for the expected number of defectives produced by a machine which makes 10 rpm; 15 rpm. Determine the effect of the last data point (15, 2) by eliminating it from the sample and repeating the entire process.

2. According to the National Center for Health statistics, the divorce rate per 1000 people in the United States increased during the years 1920–1970 as follows:

1920	1930	1940	1950	1960	1970
1.6	1.6	2.0	2.6	2.2	3.5

Find the least-squares regression line that best fits this data. Calculate the standard error of the estimate and the coefficient of correlation. Construct confidence interval estimates for α, β, and ρ. Also construct a confidence interval estimates for the divorce rate in 1975 and in 1980. Due to the sharp increase in the divorce rate in 1970, perhaps an exponential curve would give a better fit to this data. Find the least-squares exponential curve, and calculate the standard error of the estimate and the coefficient of correlation for this curve. Compare results.

3. The following table gives the batting average of ten leading "sluggers," together with the number of home runs hit by each:

HR's: 36, 29, 38, 47, 34, 28, 33, 36, 31, 27
BA's: 0.267, 0.274, 0.234, 0.287, 0.325, 0.241, 0.278, 0.265, 0.233, 0.298

Find the least squares-line expressing batting average as a function of home runs for these "sluggers." Does this line give a good fit to the observed data? Find a confidence interval estimate for the batting average of a slugger hitting 30, 40, 50, and 60 home runs during any given season.

Appendixes

Appendix A Binomial Distribution

N	X	0.0500	0.1000	0.1500	0.2000	0.2500	0.3000	0.3500	0.4000	0.4500	0.5000
1	0	.9500	.9000	.8500	.8000	.7500	.7000	.6500	.6000	.5500	.5000
	1	.0500	.1000	.1500	.2000	.2500	.3000	.3500	.4000	.4500	.5000
2	0	.9025	.8100	.7225	.6400	.5625	.4900	.4225	.3600	.3025	.2500
	1	.0950	.1800	.2550	.3200	.3750	.4200	.4550	.4800	.4950	.5000
	2	.0025	.0100	.0225	.0400	.0625	.0900	.1225	.1600	.2025	.2500
3	0	.8574	.7290	.6141	.5120	.4219	.3430	.2746	.2160	.1664	.1250
	1	.1354	.2430	.3251	.3840	.4219	.4410	.4436	.4320	.4084	.3750
	2	.0071	.0270	.0574	.0960	.1406	.1890	.2389	.2880	.3341	.3750
	3	.0001	.0010	.0034	.0080	.0156	.0270	.0429	.0640	.0911	.1250
4	0	.8145	.6561	.5220	.4096	.3164	.2401	.1785	.1296	.0915	.0625
	1	.1715	.2916	.3685	.4096	.4219	.4116	.3845	.3456	.2995	.2500
	2	.0135	.0486	.0975	.1536	.2109	.2646	.3105	.3456	.3675	.3750
	3	.0005	.0036	.0115	.0256	.0469	.0756	.1115	.1536	.2005	.2500
	4	.0000	.0001	.0005	.0016	.0039	.0081	.0150	.0256	.0410	.0625
5	0	.7738	.5905	.4437	.3277	.2373	.1681	.1160	.0778	.0503	.0313
	1	.2036	.3281	.3915	.4096	.3955	.3602	.3124	.2592	.2059	.1563
	2	.0214	.0729	.1382	.2048	.2637	.3087	.3364	.3456	.3369	.3125
	3	.0011	.0081	.0244	.0512	.0879	.1323	.1811	.2304	.2757	.3125
	4	.0000	.0005	.0022	.0064	.0146	.0284	.0488	.0768	.1128	.1563
	5	.0000	.0000	.0001	.0003	.0010	.0024	.0053	.0102	.0185	.0313
6	0	.7351	.5314	.3771	.2621	.1780	.1176	.0754	.0467	.0277	.0156
	1	.2321	.3543	.3993	.3932	.3560	.3025	.2437	.1866	.1359	.0938
	2	.0305	.0984	.1762	.2458	.2966	.3241	.3280	.3110	.2780	.2344
	3	.0021	.0146	.0415	.0819	.1318	.1852	.2355	.2765	.3032	.3125
	4	.0001	.0012	.0055	.0154	.0330	.0595	.0951	.1382	.1861	.2344
	5	.0000	.0001	.0004	.0015	.0044	.0102	.0205	.0369	.0609	.0938
	6	.0000	.0000	.0000	.0001	.0002	.0007	.0018	.0041	.0083	.0156
7	0	.6983	.4783	.3206	.2097	.1335	.0824	.0490	.0280	.0152	.0078
	1	.2573	.3720	.3960	.3670	.3115	.2471	.1848	.1306	.0872	.0547
	2	.0406	.1240	.2097	.2753	.3115	.3177	.2985	.2613	.2140	.1641
	3	.0036	.0230	.0617	.1147	.1730	.2269	.2679	.2903	.2918	.2734
	4	.0002	.0026	.0109	.0287	.0577	.0972	.1442	.1935	.2388	.2734
	5	.0000	.0002	.0012	.0043	.0115	.0250	.0466	.0774	.1172	.1641
	6	.0000	.0000	.0001	.0004	.0013	.0036	.0084	.0172	.0320	.0547
	7	.0000	.0000	.0000	.0000	.0001	.0002	.0006	.0016	.0037	.0078
8	0	.6634	.4305	.2725	.1678	.1001	.0576	.0319	.0168	.0084	.0039
	1	.2793	.3826	.3847	.3355	.2670	.1977	.1373	.0896	.0548	.0313
	2	.0515	.1488	.2376	.2936	.3115	.2965	.2587	.2090	.1569	.1094
	3	.0054	.0331	.0839	.1468	.2076	.2541	.2786	.2787	.2568	.2188
	4	.0004	.0046	.0185	.0459	.0865	.1361	.1875	.2322	.2627	.2734
	5	.0000	.0004	.0026	.0092	.0231	.0467	.0808	.1239	.1719	.2188
	6	.0000	.0000	.0002	.0011	.0038	.0100	.0217	.0413	.0703	.1094
	7	.0000	.0000	.0000	.0001	.0004	.0012	.0033	.0079	.0164	.0313
	8	.0000	.0000	.0000	.0000	.0000	.0001	.0002	.0007	.0017	.0039

Column headings (p): 0.5000 0.4500 0.4000 0.3500 0.3000 0.2500 0.2000 0.1500 0.1000 0.0500

N = 9
N = 10
N = 11
N = 12

Appendix A *(continued)*

N = 13

N = 14

N = 15

$p = 0.5000$

$p = 0.4500$

$p = 0.4000$

$p = 0.3500$

$p = 0.3000$

$p = 0.2500$

$p = 0.2000$

$p = 0.1500$

$p = 0.1000$

$p = 0.0500$

$N = 16$

$N = 17$

$N = 18$

Appendix A *(continued)*

N = 19

	0.0500	0.1000	0.1500	0.2000	0.2500	0.3000	0.3500	0.4000	0.4500	0.5000

N = 20

	0.0500	0.1000	0.1500	0.2000	0.2500	0.3000	0.3500	0.4000	0.4500	0.5000

0.5000   0.4500   0.4000   0.3500   0.3000   0.2500   0.2000   0.1500   0.1000   0.0500

N = 21

N = 22

Appendix A (continued)

N = 23

N = 24

0.5000

0.4500

0.4000

0.3500

0.3000

0.2500

0.2000

0.1500

0.1000

0.0500

0 1 2 3 4 5 6 7 8 9 0 1 2 3 4 5 6 7 8 9 0 1 2 3 4 5

X X X X X X X X X X X X X X X X X X X X X X X X X X

N = 25

Appendix B Poisson Distribution

Appendix B (continued)

Appendix D Chi-Square Distribution

df	$\alpha = 0.99$	$\alpha = 0.95$	$\alpha = 0.05$	$\alpha = 0.01$
1	0.0002	0.004	3.841	6.635
2	0.020	0.103	5.991	9.210
3	0.115	0.352	7.815	11.345
4	0.297	0.711	9.488	13.277
5	0.554	1.145	11.070	15.086
6	0.872	1.635	12.592	16.812
7	1.239	2.167	14.067	18.475
8	1.646	2.733	15.507	20.090
9	2.088	3.325	16.919	21.666
10	2.558	3.940	18.307	23.209
11	3.053	4.575	19.675	24.725
12	3.571	5.226	21.026	26.217
13	4.107	5.892	22.362	27.688
14	4.660	6.571	23.685	29.141
15	5.229	7.261	24.996	30.578
16	5.812	7.962	26.296	32.000
17	6.408	8.672	27.587	33.409
18	7.015	9.390	28.869	34.805
19	7.633	10.117	30.144	36.191
20	8.260	10.851	31.410	37.566
21	8.897	11.591	32.671	38.932
22	9.542	12.338	33.924	40.289
23	10.196	13.091	35.172	41.638
24	10.856	13.848	36.415	42.980
25	11.524	14.611	37.652	44.314
26	12.198	15.379	38.885	45.642
27	12.879	16.151	40.113	46.963
28	13.565	16.928	41.337	48.278
29	14.256	17.708	42.557	49.588
30	14.953	18.493	43.773	50.892

Appendix E F Distribution ($\alpha = 0.05$)

| | | | | $df\ num.$ | | | | |
df den.	1	2	3	4	5	6	7	8
1	161.4	199.5	215.7	224.6	230.2	234.0	236.8	238.9
2	18.51	19.00	19.16	19.25	19.30	19.33	19.35	19.37
3	10.13	9.55	9.28	9.12	9.01	8.94	8.89	8.85
4	7.71	6.94	6.59	6.39	6.26	6.16	6.09	6.04
5	6.61	5.79	5.41	5.19	5.05	4.95	4.88	4.82
6	5.99	5.14	4.76	4.53	4.39	4.28	4.21	4.15
7	5.59	4.74	4.35	4.12	3.97	3.87	3.79	3.73
8	5.32	4.46	4.07	3.84	3.69	3.58	3.50	3.44
9	5.12	4.26	3.86	3.63	3.48	3.37	3.29	3.23
10	4.96	4.10	3.71	3.48	3.33	3.22	3.14	3.07
11	4.84	3.98	3.59	3.36	3.20	3.09	3.01	2.95
12	4.75	3.89	3.49	3.26	3.11	3.00	2.91	2.85
13	4.67	3.81	3.41	3.18	3.03	2.92	2.83	2.77
14	4.60	3.74	3.34	3.11	2.96	2.85	2.76	2.70
15	4.54	3.68	3.29	3.06	2.90	2.79	2.71	2.64
16	4.49	3.63	3.24	3.01	2.85	2.74	2.66	2.59
17	4.45	3.59	3.20	2.96	2.81	2.70	2.61	2.55
18	4.41	3.55	3.16	2.93	2.77	2.66	2.58	2.51
19	4.38	3.52	3.13	2.90	2.74	2.63	2.54	2.48
20	4.35	3.49	3.10	2.87	2.71	2.60	2.51	2.45
21	4.32	3.47	3.07	2.84	2.68	2.57	2.49	2.42
22	4.30	3.44	3.05	2.82	2.66	2.55	2.46	2.40
23	4.28	3.42	3.03	2.80	2.64	2.53	2.44	2.37
24	4.26	3.40	3.01	2.78	2.62	2.51	2.42	2.36
25	4.24	3.39	2.99	2.76	2.60	2.49	2.40	2.34
30	4.17	3.32	2.92	2.69	2.53	2.42	2.33	2.27
60	4.00	3.15	2.76	2.53	2.37	2.25	2.17	2.10
120	3.92	3.07	2.68	2.45	2.29	2.17	2.09	2.02
∞	3.84	3.00	2.60	2.37	2.21	2.10	2.01	1.94

Appendix E (*continued*)

| | | | | | *df num.* | | | | |
df den.	9	10	12	15	20	30	60	120	∞
1	240.5	241.9	243.9	245.9	248.0	250.1	252.2	253.3	254.3
2	19.38	19.40	19.41	19.43	19.45	19.46	19.48	19.49	19.50
3	8.81	8.79	8.74	8.70	8.66	8.62	8.57	8.55	8.53
4	6.00	5.96	5.91	5.86	5.80	5.75	5.69	5.66	5.63
5	4.77	4.74	4.68	4.62	4.56	4.50	4.43	4.40	4.36
6	4.10	4.06	4.00	3.94	3.87	3.81	3.74	3.70	3.67
7	3.68	3.64	3.57	3.51	3.44	3.38	3.30	3.27	3.23
8	3.39	3.35	3.28	3.22	3.15	3.08	3.01	2.97	2.93
9	3.18	3.14	3.07	3.01	2.94	2.86	2.79	2.75	2.71
10	3.02	2.98	2.91	2.85	2.77	2.70	2.62	2.58	2.54
11	2.90	2.85	2.79	2.72	2.65	2.57	2.49	2.45	2.40
12	2.80	2.75	2.69	2.62	2.54	2.47	2.38	2.34	2.30
13	2.71	2.67	2.60	2.53	2.46	2.38	2.30	2.25	2.21
14	2.65	2.60	2.53	2.46	2.39	2.31	2.22	2.18	2.13
15	2.59	2.54	2.48	2.40	2.33	2.25	2.16	2.11	2.07
16	2.54	2.49	2.42	2.35	2.28	2.19	2.11	2.06	2.01
17	2.49	2.45	2.38	2.31	2.23	2.15	2.06	2.01	1.96
18	2.46	2.41	2.34	2.27	2.19	2.11	2.02	1.97	1.92
19	2.42	2.38	2.31	2.23	2.16	2.07	1.98	1.93	1.88
20	2.39	2.35	2.28	2.20	2.12	2.04	1.95	1.90	1.84
21	2.37	2.32	2.25	2.18	2.10	2.01	1.92	1.87	1.81
22	2.34	2.30	2.23	2.15	2.07	1.98	1.89	1.84	1.78
23	2.32	2.27	2.20	2.13	2.05	1.96	1.86	1.81	1.76
24	2.30	2.25	2.18	2.11	2.03	1.94	1.84	1.79	1.73
25	2.28	2.24	2.16	2.09	2.01	1.92	1.82	1.77	1.71
30	2.21	2.16	2.09	2.01	1.93	1.84	1.74	1.68	1.62
60	2.04	1.99	1.92	1.84	1.75	1.65	1.53	1.47	1.39
120	1.96	1.91	1.83	1.75	1.66	1.55	1.43	1.35	1.25
∞	1.88	1.83	1.75	1.67	1.57	1.46	1.32	1.22	1.00

Appendix F Student *t* Distribution

df	$\alpha = 0.05$	$\alpha = 0.025$	$\alpha = 0.01$	$\alpha = 0.005$
1	6.314	12.706	31.821	63.657
2	2.920	4.303	6.965	9.925
3	2.353	3.182	4.541	5.841
4	2.132	2.776	3.747	4.604
5	2.015	2.571	3.365	4.032
6	1.943	2.447	3.143	3.707
7	1.895	2.365	2.998	3.499
8	1.860	2.306	2.896	3.355
9	1.833	2.262	2.821	3.250
10	1.812	2.228	2.764	3.169
11	1.796	2.201	2.718	3.106
12	1.782	2.179	2.681	3.055
13	1.771	2.160	2.650	3.012
14	1.761	2.145	2.624	2.977
15	1.753	2.131	2.602	2.947
16	1.746	2.120	2.583	2.921
17	1.740	2.110	2.567	2.898
18	1.734	2.101	2.552	2.878
19	1.729	2.093	2.539	2.861
20	1.725	2.086	2.528	2.845
21	1.721	2.080	2.518	2.831
22	1.717	2.074	2.508	2.819
23	1.714	2.069	2.500	2.807
24	1.711	2.064	2.492	2.797
25	1.708	2.060	2.485	2.787
26	1.706	2.056	2.479	2.779
27	1.703	2.052	2.473	2.771
28	1.701	2.048	2.467	2.763
29	1.699	2.045	2.462	2.756
30	1.697	2.042	2.457	2.750
60	1.671	2.000	2.390	2.660
120	1.658	1.980	2.358	2.617
SNV	1.645	1.960	2.326	2.576

Solutions
to Odd-Numbered
Exercises

Section 1.2

1. $A \cup B$ = horses that are thoroughbreds or are gray
   $A \cap B$ = gray thoroughbreds
   $C(A)$ = nonthoroughbred horses
   $C(B)$ = nongray horses
   $C(A \cup B)$ = nongray, nonthoroughbred horses
   $C(A \cap B)$ = horses that are not thoroughbreds or not gray
   $C(A) \cap B$ = gray nonthoroughbreds
   $A \cap C(B)$ = nongray thoroughbreds
3. 120
5. 40

Section 1.3

3. $26^4 - 26 \cdot 25 \cdot 24 \cdot 23 = 98,176$
5. $2^{10} = 1,024$
7. 288

Section 1.4

1. 210, 6,720, 1,663,200, 670,442,572,800
3. $P_4^8 = 1,680$
5. $P_4^{30} = 657,720$
7. $P_5^{20} = 1,860,480$

Section 1.5

3. 624, 9,216, 1,098,240
7. $C_{10}^{20} = 184,756$
9. $C_5^{50} = 2,118,760$

Section 1.6

3. 4,620
5. $C_7^{14} = 3,432$
7. $C_{100}^{103} = 176,851, 100!/(25!)^4$

Section 1.7

1. $C_5^{20} = 15,504, P_5^{20} = 1,860,480$
3. 771,400
5. $C_3^{18} C_4^{24} C_5^{12} = 6,867,286,272$

Additional problems 1

1. 25
3. $C_{10}{}^{980}, C_{10}{}^{1000} - C_{10}{}^{980}$
5. $P_7{}^{10} = 604,800$
7. $C_3{}^{10} = 120, C_3{}^5 C_1{}^2 = 20$

Section 2.1

1. $S = \{(i, H), (i, T) : i = 1, 2, 3, 4, 5, 6\}$
3. $S =$ the set of the 52 cards in the deck
5. $S = \{00, 01, 02, \ldots, 99\}$
7. $S = \{PN, PD, PQ, ND, NQ, DQ\}$

Section 2.3

1. $\frac{1}{36}, \frac{2}{36}, \frac{3}{36}, \frac{4}{36}, \frac{5}{36}, \frac{6}{36}, \frac{5}{36}, \frac{4}{36}, \frac{3}{36}, \frac{2}{36}, \frac{1}{36}$
3. 0.00144, 0.00198, 0.00355, 0.04754, 0.02113
5. $\frac{1}{3}$
7. $\frac{2}{3}$
9. $\frac{1}{11}, \frac{6}{11}$

Section 2.4

1. $\frac{2}{3}$
3. $\frac{13}{25}$
5. 0.234

Section 2.5

1. $\pi/4$
3. $\frac{1}{8}$
5. $\frac{3}{4}$

Additional problems 2

1. yes
3. 0.8
5. $\frac{5}{12}$, no
7. 0.038
9. $\frac{21}{45}$
11. 0.00065
13. $2/(n-1)$

Section 3.1

5. $\frac{5}{8}$ and $\frac{3}{8}$
7. $(97 \cdot 95)^{-1} = 0.0001$

Section 3.2

3. $\frac{9}{20}$
5. $\frac{11}{108}$
7. all three have probability $\frac{3}{5}$
9. $\frac{3}{5}$

Section 3.3

1. $\frac{20}{41}$

3. $\frac{2}{5}$

5. $\frac{24}{29}$

7. $\frac{3}{11}$

9. $\frac{8}{11}$

Section 3.4

3. $\frac{1}{2}$

5. yes

7. $\frac{3}{5}$

Section 3.5

1. $\frac{1}{9}$

3. $\frac{6}{6400}$

5. 0.1296

Additional problems 3

1. 4; 75 percent to A and 25 percent to B

3. $\frac{3}{4}$

5. 29

7. $\frac{5}{9}$

9. $\frac{2}{3}$

Section 4.2

3. $f(0) = \frac{25}{36}, f(1) = \frac{10}{36}, f(2) = \frac{1}{36}$

5. $f(k) = C_k^{13}C_{13-k}^{39}/C_{13}^{52}$ for $k = 0, 1, 2, \ldots, 13$

7. If n is even and $k = n/2$

$$f(0) = \frac{C_k^n}{2^n} \quad \text{and} \quad f(2i) = \frac{2C_{k-i}^n}{2^n}$$

if n is odd and $k = [n/2] + 1$

$$f(2i - 1) = \frac{2C_{k-i}^n}{2^n}$$

9. $f(1) = \frac{74}{120}, f(5) = \frac{7}{30}, f(10) = \frac{3}{20}; \frac{3}{20}$

Section 4.3

1. (a) $P(A)$

   (b) 2

   (c) $\frac{1}{3}$

   (d) 4.5

   (h) 1.25

   (i) 3.28

   (j) 0.385, 0.385

3. $1.75, $1.42, $2.25

5. 0

Section 4.4

3. 2504, 16, 4

Section 4.5

3. 6, 36
5. 695

Section 4.6

5. $M_X(t) = (25 + 10e^t + e^{2t})/36; \frac{1}{3}, \frac{7}{18}, \frac{1}{2}, \frac{13}{18}, \frac{7}{6}$

Additional problems 4

3. $f(k) = (\frac{3}{5})^{k-1}(\frac{2}{5}), E(X) = \frac{5}{2}; f(1) = \frac{2}{5}, f(2) = \frac{3}{10}, f(3) = \frac{1}{5}, f(4) = \frac{1}{10}$, and $E(X) = 2$
5. $f(k) = 2C_{5-k}^{9-k}(\frac{1}{2})^{10-k}, E(X) = 2.46$

9. $E(X) = (1/n^2) \sum_{k=1}^{n} (2k^2 - k)$

Section 5.1

						Y					
1. X	2	3	4	5	6	7	8	9	10	11	12
0	$\frac{1}{36}$	$\frac{2}{36}$	$\frac{3}{36}$	$\frac{4}{36}$	$\frac{5}{36}$	$\frac{4}{36}$	$\frac{3}{36}$	$\frac{2}{36}$	$\frac{1}{36}$	0	0
1	0	0	0	0	0	$\frac{2}{36}$	$\frac{2}{36}$	$\frac{2}{36}$	$\frac{2}{36}$	$\frac{2}{36}$	0
2	0	0	0	0	0	0	0	0	0	0	$\frac{1}{36}$

			Z			
X	1	2	3	4	5	6
0	$\frac{1}{36}$	$\frac{3}{36}$	$\frac{5}{36}$	$\frac{7}{36}$	$\frac{9}{36}$	0
1	0	0	0	0	0	$\frac{10}{36}$
2	0	0	0	0	0	$\frac{1}{36}$

						Y					
Z	2	3	4	5	6	7	8	9	10	11	12
1	$\frac{1}{36}$	0	0	0	0	0	0	0	0	0	0
2	0	$\frac{2}{36}$	$\frac{1}{36}$	0	0	0	0	0	0	0	0
3	0	0	$\frac{2}{36}$	$\frac{2}{36}$	$\frac{1}{36}$	0	0	0	0	0	0
4	0	0	0	$\frac{2}{36}$	$\frac{2}{36}$	$\frac{2}{36}$	$\frac{1}{36}$	0	0	0	0
5	0	0	0	0	$\frac{2}{36}$	$\frac{2}{36}$	$\frac{2}{36}$	$\frac{2}{36}$	$\frac{1}{36}$	0	0
6	0	0	0	0	0	$\frac{2}{36}$	$\frac{2}{36}$	$\frac{2}{36}$	$\frac{2}{36}$	$\frac{2}{36}$	$\frac{1}{36}$

No pair is independent

3.

X	Y		
	1	*2*	*3*
2	0.1	0	0
3	0	0.4	0
4	0	0.1	0.2
5	0	0	0.2

; no

5. $h(a, b) = C_a^4 C_b^4 C_{5-(a+b)}^{44} / C_5^{52}$ for $0 \le a \le 4$, $0 \le b \le 4$, and $0 \le a + b \le 5$; no

7. $h(a, b) = C_a^8 C_b^6 C_{2-(a+b)}^{10} / C_2^{24}$ for $0 \le a \le 2$, $0 \le b \le 2$, and $0 \le a + b \le 2$; no

Section 5.2

3. $X + Y$:

	2	3	4	5
	$\frac{6}{30}$	$\frac{12}{30}$	$\frac{8}{30}$	$\frac{4}{30}$

$E(X + Y) = \frac{10}{3}$, $\text{Var}(X + Y) = \frac{8}{9}$

XY:

	1	2	3	4	5	6
	$\frac{6}{30}$	$\frac{12}{30}$	$\frac{6}{30}$	$\frac{2}{30}$	0	$\frac{4}{30}$

$E(XY) = \frac{8}{3}$, $\text{Var}(XY) = \frac{2}{356}$

5. $6[1 - (\frac{5}{6})^5] = 3.589$

7. 1

9. $\frac{5}{4}$

Section 5.3

5. (a) 0.52 and 0.886

(b) 0.833 and 0.657; 0.509 and 0.685; 2.917 and 0.859

(c) $-\frac{1}{3}$ and $-\frac{1}{2}$; 0 and 0; 0 and 0

(d) $-\frac{1}{9}$ and $-\frac{1}{5}$

Section 5.4

3. $\bar{X}$:

	1.0	1.5	2.0	2.5	3.0	3.5	4.0
	$\frac{1}{25}$	$\frac{4}{25}$	$\frac{6}{25}$	$\frac{6}{25}$	$\frac{5}{25}$	$\frac{2}{25}$	$\frac{1}{25}$

$E(\bar{X}) = \frac{12}{5}$, $\text{Var}(\bar{X}) = 0.52$

5. at least $1 - (1/n)$

Additional problems 5

1. $\frac{13}{4}$

3. 0.18

5. $\frac{11}{12}$

7. 16

Section 6.1

3. 0.2903, 0.3020, 0.0031, 0.1285

5. 0.3087

7. 0.2241

9. 0.6172

11. 0.0102

13. 0.0010

21. 0.0251, 0.2464

Section 6.2

5. $\frac{5}{13}$

7. 0.0236

Section 6.3

3. 5

5. $2, \frac{1}{32}$

9. 0.1003

Section 6.4

3. 0.5488

5. 0.4232

7. 0.1321

9. 0.0183

11. 0.0183, 0.3679

Additional problems 6

1. 6

3. $\frac{1}{2}$

5. $\sum_{j=0}^{k} (C_j{}^n C_{k-j}^m / C_k{}^{n+m})[(n-j)/(m+n-k)]$

7. $[1 - (5/6)^8]^6 = 0.204$

9. $\frac{4}{7}, \frac{2}{7}, \frac{1}{7}$

11. $\sum_{k=0}^{n} (C_k{}^n)^2 (\frac{1}{2})^{2n}$

13. $\frac{3}{2}$

Section 7.1

1. $\begin{pmatrix} 0 & \frac{2}{3} & 0 & 0 & \frac{1}{3} \\ \frac{1}{3} & 0 & \frac{2}{3} & 0 & 0 \\ 0 & \frac{1}{3} & 0 & \frac{2}{3} & 0 \\ 0 & 0 & \frac{1}{3} & 0 & \frac{2}{3} \\ \frac{2}{3} & 0 & 0 & \frac{1}{3} & 0 \end{pmatrix}, \frac{4}{9}$

3. 0.665

5. 4.508 percent

7. $\begin{matrix} D \\ R \\ A \end{matrix} \begin{pmatrix} \frac{3}{4} & 0 & \frac{1}{4} \\ 0 & \frac{17}{20} & \frac{3}{20} \\ \frac{1}{2} & \frac{1}{2} & 0 \end{pmatrix}, p_{11}{}^{(5)} = 0.505, p_{12}{}^{(5)} = 0.317$

Section 7.2

1. 0.996
3. 0.947

Section 7.3

1. $(\frac{12}{17}, \frac{5}{17}), \frac{12}{17}$
3. $\frac{2}{3}$
5. yes, (0.327, 0.245, 0.134, 0.206, 0.087), 0.087

Additional problems 7

1. 0.062, 0.623, 0.314

Section 8.1

5. infinity
7. $A = 7{,}031{,}250$, 0.3164, $E(X) = 1875$

Section 8.2

1. 6
3. $F(a, b) = (1 - e^{-a}) + (e^{-ab} - 1)/b$, $f_X(x) = e^{-x}$

5. $h(u) = \begin{cases} 0 & \text{elsewhere} \\ u^2 & \text{if } 0 \le u \le 1 \\ 2u - u^2 & \text{if } 1 \le u \le 2 \end{cases}$

Section 8.3

3. $h(u) = \begin{cases} u & \text{if } 0 \le u \le 1 \\ 2 - u & \text{if } 1 \le u \le 2 \\ 0 & \text{otherwise} \end{cases}$

7. (a) $g(x) = \begin{cases} 1/c & \text{if } 0 \le x \le c \\ 0 & \text{otherwise} \end{cases}$.

   (b) $g(x) = \begin{cases} 1/x^2 & \text{if } 0 \le x \le 1 \\ 0 & \text{otherwise} \end{cases}$

   (c) $g(x) = \begin{cases} 2x & \text{if } 0 \le x \le 1 \\ 0 & \text{otherwise} \end{cases}$

   (d) $g(x) = \begin{cases} (1 - x)^{-2} & \text{if } 0 \le x \le \frac{1}{2} \\ 0 & \text{otherwise} \end{cases}$

   (e) $g(x) = \begin{cases} e^{-x} & \text{if } 0 < x < +\infty \\ 0 & \text{otherwise} \end{cases}$

9. 0.2639
13. $\frac{5}{12}$
15. $\beta(\alpha - 1)$, $(\alpha - 1)/(\alpha + \beta - 2)$

Section 8.4

1. 0.9066, 0.2822, 0.0594, 0.5346, 0.2905
5. 38.7 years

7. 13.36 percent

9. 0.0668

Section 8.5

1. 0.7985

3. 0.5704

5. 0.9772

7. 0.0116

9. 0.0241

Additional problems 8

1. 0.465

3. $P(X \le a) = \begin{cases} 2a/3 & \text{if } 0 \le a \le 1 \\ (a+1)/3 & \text{if } 1 \le a \le 2 \end{cases}$

5. $1.25

7. 0.527

Section 9.2

3. yes, yes, yes

5. $\alpha = \dfrac{(\mu_1')^2}{\mu_2' - (\mu_1')^2}, \qquad \beta = \dfrac{\mu_2' - (\mu_1')^2}{\mu_1'}$

Section 9.3

1. 950 ± 19.6, 96.04

3. 8 ± 0.0375 in.

5. 0.184 ± 0.0286, 13.573

7. 0.425 ± 0.0902

Section 9.4

1. $z = 1.98$ is critical at $\alpha = 0.05$, but not at $\alpha = 0.01$.

3. $z = 2.37$ is critical at $\alpha = 0.01$.

5. $z = -2.36$ is critical.

7. $z = -1.14$ is not critical.

9. $z = -1.60$ is not critical.

Section 9.5

1. (a) $\sum\limits_{i=1}^{n} x_i^2 \ge \sigma_0^2 \chi_{\alpha,\,n}^2$

 (b) $\sum\limits_{i=1}^{n} x_i \ge K$

3. $\bar{x} < c_1$ or $\bar{x} > c_2$, where both c_1 and c_2 are functions of α and β_0.

Additional problems 9

1. 90

3. 96.04

5. 545

7. 0.8869, 0.9999, 161

Section 10.1

1. Yes: $z = -3.40$ is critical at $\alpha = 0.01$.
3. Yes: $z = -4.50$ is critical at $\alpha = 0.01$.
5. Yes: $z = 3.60$ is critical at $\alpha = 0.01$.
7. No: $z = 1.12$ is not critical at $\alpha = 0.05$.
9. Yes: $z = 5$ is critical at $\alpha = 0.01$.

Section 10.3

1. No: $\chi^2 = 15.07$ is not critical at $\alpha = 0.05$.
3. $2.460 < \sigma^2 < 7.622$

Section 10.4

1. They don't: $\chi^2 = 86.30$ is critical.
3. $\chi^2 = 4.122$ is critical at $\alpha = 0.05$, but not at $\alpha = 0.01$.
5. $\chi^2 = 3.044$ is not critical.
7. $\chi^2 = 1.718$ is not critical.
9. $\chi^2 = 9.549$ is critical.

Section 10.5

3. $\chi^2 = 1.56$ suggests the data is normal.
5. No: $\chi^2 = 19.32$ is critical at $\alpha = 0.01$.
7. Yes: $\chi^2 = 29.87$ is critical at $\alpha = 0.01$.

Section 10.6

3. $F = 1.125$ is not critical.
5. $F = 5.84$ is critical at $\alpha = 0.01$.

Section 10.7

1. Yes: $t = 2.024$ is critical at $\alpha = 0.05$.
3. Yes: $t = 1.996$ is critical at $\alpha = 0.05$.
7. Yes: $t = 1.667$ is not critical.

Section 10.8

3. The median test gives a $\chi^2 = 4$ which is critical at $\alpha = 0.05$, but the rank-sum test gives a standard normal $U = 1.58$ which is not critical.
5. The sign test is critical at $\alpha = 0.05$.
7. Neither the sign test (probability is 0.183) nor the median test ($\chi^2 = 0.286$) is critical.

Section 10.9

1. $F = 0.16$ is not critical ($SS_{among} = 0.25$, $SS_{within} = 63.50$).
3. $F = 0.309$ is not critical ($SS_{among} = 0.22$, $SS_{within} = 6.04$).

Additional problems 10

1. Yes, at $\alpha = 0.05$ (probability is 0.0287).
3. Yes: $\chi^2 = 23.44$ is critical.
5. $t = 2.19$ is critical at $\alpha = 0.05$.

Section 11.2

1. $y = 0.189e^{0.0237x}$
3. $y = 71.583 + 17.476x$

5. $y = 57.76 + 1.56x$

7. The least-squares line $y = 7.286x - 5.666$ produces a $\chi^2 = 6.11$, while the least-squares exponential $y = 3.206e^{0.447x}$ produces a $\chi^2 = 1.53$. Hence, it would seem that the exponential provides the better fit, and would consequently provide the better predictions:

7	73.25	10	280.04
8	114.54	11	437.87
9	179.10	12	684.67

Unfortunately, extrapolation leads to unreasonable results. Using the least-squares line instead, we get

7	45.34	10	67.19
8	52.62	11	74.48
9	59.91	12	81.77

9. $y = 0.3259x - 2.837$ produces a $\chi^2 = 6.279$, while $y = 0.47e^{0.076x}$ produces a $\chi^2 = 3.481$, so the exponential provides the better fit.

Section 11.3

3. $s_{Y, X} = 0.207, r = 0.998$

Section 11.4

3. $1.02 \le \beta \le 2.10; 0.779 \le \rho \le 0.987; 163.93 \le y_{75} \le 185.59$

Additional problems 11

1. (a) $y = 5.49 + 0.23x; s_{Y, X} = 3.15; r = 0.135;$
$\alpha = 5.49 \pm 5.10; \beta = 0.23 \pm 1.4; 0 \le \rho \le 0.705;$
$y_{10} = 7.79 \pm 0.35; y_{15} = 8.94 \pm 0.52$

(b) $y = 1.38x - 6.77; s_{Y, X} = 1.18; r = 0.859;$
$\alpha = -6.77 \pm 1.98; \beta = 1.38 \pm 0.67; 0.454 \le \rho \le 0.970;$
$y_{10} = 7.03 \pm 0.14; y_{15} = 13.93 \pm 0.26$

3. $y = 0.257 + 0.00039x; s_{Y, X} = 0.028; r = 0.079;$
$y_{30} = 0.269 \pm 0.044; y_{40} = 0.273 \pm 0.063;$
$y_{50} = 0.277 \pm 0.156; y_{60} = 0.280 \pm 0.251$

Index